Student Solutions

for

Applied Calculus

Fourth Edition

Stefan Waner
Hofstra University

Steven R. Costenoble
Hofstra University

THOMSON
BROOKS/COLE

Australia • Brazil • Canada • Mexico • Singapore • Spain • United Kingdom • United States

Printed in the United States of America

1 2 3 4 5 6 7 10 09 08 07 06

Printer: Thomson/West

0-495-01699-3

Cover image: Rings of Douglas Fir Tree © Corbis

Thomson Higher Education
10 Davis Drive
Belmont, CA 94002-3098
USA

For more information about our products, contact us at:
Thomson Learning Academic Resource Center
1-800-423-0563

For permission to use material from this text or product, submit a request online at **http://www.thomsonrights.com.**
Any additional questions about permissions can be submitted by email to **thomsonrights@thomson.com.**

Table of Contents

Chapter 0	1
Chapter 1	11
Chapter 2	48
Chapter 3	72
Chapter 4	113
Chapter 5	137
Chapter 6	179
Chapter 7	205
Chapter 8	240
Chapter 9	274

Chapter 0

0.1

1. $2(4 + (-1))(2 \cdot -4)$
$= 2(3)(-8) = (6)(-8) = -48$

3. `20/(3*4)-1`
$= \frac{20}{12} - 1 = \frac{5}{3} - 1 = \frac{2}{3}$

5. $\frac{3 + ([3 + (-5)])}{3 - 2\times 2} = \frac{3 + (-2)}{3 - 4}$
$= \frac{1}{-1} = -1$

7. `(2-5*(-1))/1-2*(-1)`
$= \frac{2 - 5\cdot(-1)}{1} - 2\cdot(-1)$
$= \frac{2+5}{1} + 2 = 7 + 2 = 9$

9. $2\cdot(-1)^2 / 2 = \frac{2\times(-1)^2}{2} = \frac{2\times 1}{2} = \frac{2}{2} = 1$

11. $2\cdot 4^2 + 1 = 2\times 16 + 1 = 32 + 1 = 33$

13. `3^2+2^2+1`
$= 3^2 + 2^2 + 1 = 9 + 4 + 1 = 14$

15. $\frac{3 - 2(-3)^2}{-6(4 - 1)^2} = \frac{3 - 2\times 9}{-6(3)^2} = \frac{3 - 18}{6\times 9}$
$= \frac{-15}{54} = \frac{5}{18}$

17. `10*(1+1/10)^3`
$= 10\left(1 + \frac{1}{10}\right)^3 = 10(1.1)^3$
$= 10\times 1.331 = 13.31$

19. $3\left[\frac{-2\cdot 3^2}{-(4 - 1)^2}\right] = 3\left[\frac{-2\times 9}{-3^2}\right] = 3\left[\frac{-18}{-9}\right]$
$= 3\times 2 = 6$

21. $3\left[1 - \left(-\frac{1}{2}\right)^2\right]^2 + 1 = 3\left[1 - \frac{1}{4}\right]^2 + 1$
$= 3\left[\frac{3}{4}\right]^2 = 3\left[\frac{9}{16}\right] + 1 = \frac{27}{16} + 1 = \frac{43}{16}$

23. `(1/2)^2-1/2^2`
$= \left[\frac{1}{2}\right]^2 - \frac{1}{2^2} = \frac{1}{4} - \frac{1}{4} = 0$

25. $3\times(2-5) =$ `3*(2-5)`

27. $\frac{3}{2-5} =$ `3/(2-5)`
Note `3/2-5` is wrong, since it corresponds to $\frac{3}{2} - 5$.

29. $\frac{3-1}{8+6} =$ `(3-1)/(8+6)`
Note `3-1/8-6` is wrong, since it corresponds to $3 - \frac{1}{8} - 6$.

31. $3 - \frac{4+7}{8} =$ `3-(4+7)/8`

33. $\frac{2}{3+x} - xy^2 =$ `2/(3+x)-x*y^2`

35. $3.1x^3 - 4x^{-2} - \frac{60}{x^2-1}$
= `3.1x^3-4x^(-2)-60/(x^2-1)`

37. $\frac{\left[\frac{2}{3}\right]}{5} =$ `(2/3)/5`
Note that we use only (round) parentheses in technology formulas, and not brackets.

39. $3^{4-5}\times 6 =$ 3^(4–5)*6
Note that the entire exponent is in parentheses.

41. $3\left[1 + \frac{4}{100}\right]^{-3} =$ `3*(1+4/100)^(-3)`
Note that we use only (round) parentheses in technology formulas, and not brackets.

43. $3^{2x-1} + 4^x - 1$ = `3^(2*x-1)+4^x-1`
Note that the entire exponent of 3 is in parentheses.

45. 2^{2x^2-x+1} = `2^(2x^2-x+1)`
Note that the entire exponent is in parentheses.

47. $\dfrac{4e^{-2x}}{2-3e^{-2x}}$
= `4*e^(-2*x)/(2-3e^(-2*x))`
or `4(*e^(-2*x))/(2-3e^(-2*x))`
or `(4*e^(-2*x))/(2-3e^(-2*x))`

49. $3\left[1 - \left(-\frac{1}{2}\right)^2\right]^2 + 1$
= `3(1-(-1/2)^2)^2+1`
Note that we use only (round) parentheses in technology formulas, and not brackets.

0.2

1. $3^3 = 27$

3. $-(2 \cdot 3)^2 = -(2^2 \cdot 3^2) = -(4 \cdot 9) = -36$ or $-(2 \cdot 3)^2 = -(6^2) = -36$

5. $\left(\frac{-2}{3}\right)^2 = \frac{(-2)^2}{3^2} = \frac{4}{9}$

7. $(-2)^{-3} = \frac{1}{(-2)^3} = \frac{1}{-8} = -\frac{1}{8}$

9. $\left(\frac{1}{4}\right)^{-2} = \frac{1}{(1/4)^2} = \frac{1}{1/4^2} = \frac{1}{1/16} = 16$

11. $2 \cdot 3^0 = 2 \cdot 1 = 2$

13. $2^3\,2^2 = 2^{3+2} = 2^5 = 32$ or $2^3\,2^2 = 8 \cdot 4 = 32$

15. $2^2\,2^{-1}\,2^4\,2^{-4} = 2^{2-1+4-4} = 2^1 = 2$

17. $x^3x^2 = x^{3+2} = x^5$

19. $-x^2x^{-3}y = -x^{2-3}y = -x^{-1}y = -\frac{y}{x}$

21. $\frac{x^3}{x^4} = x^{3-4} = x^{-1} = \frac{1}{x}$

23. $\frac{x^2y^2}{x^{-1}y} = x^{2-(-1)}y^{2-1} = x^3y$

25. $\frac{(xy^{-1}z^3)^2}{x^2yz^2} = \frac{x^2(y^{-1})^2(z^3)^2}{x^2yz^2} = x^{2-2}y^{-2-1}z^{6-2} =$ $y^{-3}z^4 = \frac{z^4}{y^3}$

27. $\left(\frac{xy^{-2}z}{x^{-1}z}\right)^3 = \frac{(xy^{-2}z)^3}{(x^{-1}z)^3} = \frac{x^3y^{-6}z^3}{x^{-3}z^3} = x^{3-(-3)}y^{-6}z^{3-3}$ $= x^6y^{-6} = \frac{x^6}{y^6}$

29. $\left(\frac{x^{-1}y^{-2}z^2}{xy}\right)^{-2} = (x^{-1-1}y^{-2-1}z^2)^{-2} = (x^{-2}y^{-3}z^2)^{-2}$ $= x^4y^6z^{-4} = \frac{x^4y^6}{z^4}$

31. $3x^{-4} = \frac{3}{x^4}$

33. $\frac{3}{4}x^{-2/3} = \frac{3}{4x^{2/3}}$

35. $1 - \frac{0.3}{x^{-2}} - \frac{6}{5}x^{-1} = 1 - 0.3x^2 - \frac{6}{5x}$

37. $\sqrt{4} = 2$

39. $\sqrt{\frac{1}{4}} = \frac{\sqrt{1}}{\sqrt{4}} = \frac{1}{2}$

41. $\sqrt{\frac{16}{9}} = \frac{\sqrt{16}}{\sqrt{9}} = \frac{4}{3}$

43. $\frac{\sqrt{4}}{5} = \frac{2}{5}$

45. $\sqrt{9} + \sqrt{16} = 3 + 4 = 7$

47. $\sqrt{9 + 16} = \sqrt{25} = 5$

49. $\sqrt[3]{8 - 27} = \sqrt[3]{-19} \approx -2.668$

51. $\sqrt[3]{\frac{27}{8}} = \frac{\sqrt[3]{27}}{\sqrt[3]{8}} = \frac{3}{2}$

53. $\sqrt{(-2)^2} = \sqrt{4} = 2$

55. $\sqrt{\frac{1}{4}(1+15)} = \sqrt{\frac{16}{4}} = \frac{\sqrt{16}}{\sqrt{4}} = \frac{4}{2} = 2$

57. $\sqrt{a^2b^2} = \sqrt{a}\,\sqrt{b} = ab$

59. $\sqrt{(x+9)^2} = x + 9$ ($x + 9 > 0$ because x is positive)

61. $\sqrt[3]{x^3(a^3+b^3)} = \sqrt[3]{x^3}\,\sqrt[3]{a^3+b^3} = x\sqrt[3]{a^3+b^3}$
(Notice: *Not* $x(a+b)$.)

63. $\sqrt{\frac{4xy^3}{x^2y}} = \sqrt{\frac{4y^2}{x}} = \frac{\sqrt{4}\,\sqrt{y^2}}{\sqrt{x}} = \frac{2y}{\sqrt{x}}$

65. $\sqrt{3} = 3^{1/2}$

67. $\sqrt{x^3} = x^{3/2}$

69. $\sqrt[3]{xy^2} = (xy^2)^{1/3}$

71. $\frac{x^2}{\sqrt{x}} = \frac{x^2}{x^{1/2}} = x^{2-1/2} = x^{3/2}$

73. $\frac{3}{5x^2} = \frac{3}{5}x^{-2}$

75. $\frac{3x^{-1.2}}{2} - \frac{1}{3x^{2.1}} = \frac{3}{2}x^{-1.2} - \frac{1}{3}x^{-2.1}$

77. $\frac{2x}{3} - \frac{x^{0.1}}{2} + \frac{4}{3x^{1.1}} = \frac{2}{3}x - \frac{1}{2}x^{0.1} + \frac{4}{3}x^{-1.1}$

79. $\frac{1}{(x^2+1)^3} - \frac{3}{4\sqrt[3]{(x^2+1)}} =$
$\frac{1}{(x^2+1)^3} - \frac{3}{4(x^2+1)^{1/3}} =$
$(x^2+1)^{-3} - \frac{3}{4}(x^2+1)^{-1/3}$

81. $2^{2/3} = \sqrt[3]{2^2}$

83. $x^{4/3} = \sqrt[3]{x^4}$

85. $(x^{1/2}y^{1/3})^{1/5} = \sqrt[5]{\sqrt{x}\,\sqrt[3]{y}}$

87. $-\frac{3}{2}x^{-1/4} = -\frac{3}{2x^{1/4}} = -\frac{3}{2\sqrt[4]{x}}$

89. $0.2x^{-2/3} + \frac{3}{7x^{-1/2}} = \frac{0.2}{x^{2/3}} + \frac{3x^{1/2}}{7} = \frac{0.2}{\sqrt[3]{x^2}} + \frac{3\sqrt{x}}{7}$

91. $\frac{3}{4(1-x)^{5/2}} = \frac{3}{4\sqrt{(1-x)^5}}$

93. $4^{-1/2}4^{7/2} = 4^{-1/2+7/2} = 4^3 = 64$

95. $3^{2/3}3^{-1/6} = 3^{2/3-1/6} = 3^{1/2} = \sqrt{3}$

97. $\frac{x^{3/2}}{x^{5/2}} = x^{3/2-5/2} = x^{-1} = \frac{1}{x}$

99. $\frac{x^{1/2}y^2}{x^{-1/2}y} = x^{1/2+1/2}y^{2-1} = xy$

101. $\left(\frac{x}{y}\right)^{1/3}\left(\frac{y}{x}\right)^{2/3} = \left(\frac{y}{x}\right)^{-1/3}\left(\frac{y}{x}\right)^{2/3} = \left(\frac{y}{x}\right)^{1/3}$

103. $x^2 - 16 = 0$, $x^2 = 16$, $x = \pm\sqrt{16} = \pm 4$

105. $x^2 - \frac{4}{9} = 0$, $x^2 = \frac{4}{9}$, $x = \pm\sqrt{\frac{4}{9}} = \pm\frac{2}{3}$

107. $x^2 - (1 + 2x)^2 = 0$, $x^2 = (1 + 2x)^2$, $x = \pm(1 + 2x)$; if $x = 1 + 2x$ then $-x = 1$, $x = -1$; if $x = -(1 + 2x)$ then $3x = -1$, $x = -1/3$. So, $x = -1$ or $-1/3$.

109. $x^5 + 32 = 0$, $x^5 = -32$, $x = \sqrt[5]{-32} = -2$

111. $x^{1/2} - 4 = 0$, $x^{1/2} = 4$, $x = 4^2 = 16$

113. $1 - \frac{1}{x^2} = 0$, $1 = \frac{1}{x^2}$, $x^2 = 1$, $x = \pm\sqrt{1} = \pm 1$

115. $(x - 4)^{-1/3} = 2$, $x - 4 = 2^{-3} = \frac{1}{8}$,

$x = 4 + \frac{1}{8} = \frac{33}{8}$

0.3

1. $x(4x + 6) = 4x^2 + 6x$

3. $(2x - y)y = 2xy - y^2$

5. $(x + 1)(x - 3) = x^2 + x - 3x - 3 = x^2 - 2x - 3$

7. $(2y + 3)(y + 5) = 2y^2 + 3y + 10y + 15 = 2y^2 + 13y + 15$

9. $(2x - 3)^2 = 4x^2 - 12x + 9$

11. $\left(x + \frac{1}{x}\right)^2 = x^2 + 2 + \frac{1}{x^2}$

13. $(2x - 3)(2x + 3) = (2x)^2 - 3^2 = 4x^2 - 9$

15. $\left(y - \frac{1}{y}\right)\left(y + \frac{1}{y}\right) = y^2 - \left(\frac{1}{y}\right)^2 = y^2 - \frac{1}{y^2}$

17. $(x^2 + x - 1)(2x + 4) = (x^2 + x - 1)2x + (x^2 + x - 1)4 = 2x^3 + 2x^2 - 2x + 4x^2 + 4x - 4 = 2x^3 + 6x^2 + 2x - 4$

19. $(x^2 - 2x + 1)^2 = (x^2 - 2x + 1)(x^2 - 2x + 1) = x^2(x^2 - 2x + 1) - 2x(x^2 - 2x + 1) + (x^2 - 2x + 1) = x^4 - 2x^3 + x^2 - 2x^3 + 4x^2 - 2x + x^2 - 2x + 1 = x^4 - 4x^3 + 6x^2 - 4x + 1$

21. $(y^3 + 2y^2 + y)(y^2 + 2y - 1) = y^3(y^2 + 2y - 1) + 2y^2(y^2 + 2y - 1) + y(y^2 + 2y - 1) = y^5 + 2y^4 - y^3 + 2y^4 + 4y^3 - 2y^2 + y^3 + 2y^2 - y = y^5 + 4y^4 + 4y^3 - y$

23. $(x + 1)(x + 2) + (x + 1)(x + 3) = (x + 1)(x + 2 + x + 3) = (x + 1)(2x + 5)$

25. $(x^2 + 1)^5(x + 3)^4 + (x^2 + 1)^6(x + 3)^3 = (x^2 + 1)^5(x + 3)^3(x + 3 + x^2 + 1) = (x^2 + 1)^5(x + 3)^3(x^2 + x + 4)$

27. $(x^3 + 1)\sqrt{x+1} - (x^3 + 1)^2\sqrt{x+1} = (x^3 + 1)\sqrt{x+1}\,[1 - (x^3 + 1)] = -x^3(x^3 + 1)\sqrt{x+1}$

29. $\sqrt{(x+1)^3} + \sqrt{(x+1)^5} = \sqrt{(x+1)^3}\cdot[1 + \sqrt{(x+1)^2}\,] = \sqrt{(x+1)^3}\,(1 + x + 1) = (x + 2)\sqrt{(x+1)^3}$

31. (a) $2x + 3x^2 = x(2 + 3x)$ **(b)** $x(2 + 3x) = 0$; $x = 0$ or $2 + 3x = 0$; $x = 0$ or $-2/3$

33. (a) $6x^3 - 2x^2 = 2x^2(3x - 1)$
(b) $2x^2(3x - 1) = 0$; $x = 0$ or $3x - 1 = 0$; $x = 0$ or $1/3$

35. (a) $x^2 - 8x + 7 = (x - 1)(x - 7)$
(b) $(x - 1)(x - 7) = 0$; $x - 1 = 0$ or $x - 7 = 0$; $x = 1$ or 7

37. (a) $x^2 + x - 12 = (x - 3)(x + 4)$
(b) $(x - 3)(x + 4) = 0$; $x - 3 = 0$ or $x + 4 = 0$; $x = 3$ or -4

39. (a) $2x^2 - 3x - 2 = (2x + 1)(x - 2)$
(b) $(2x + 1)(x - 2) = 0$; $2x + 1 = 0$ or $x - 2 = 0$; $x = -1/2$ or 2

41. (a) $6x^2 + 13x + 6 = (2x + 3)(3x + 2)$
(b) $(2x + 3)(3x + 2) = 0$; $2x + 3 = 0$ or $3x + 2 = 0$; $x = -3/2$ or $-2/3$

43. (a) $12x^2 + x - 6 = (3x - 2)(4x + 3)$
(b) $(3x - 2)(4x + 3) = 0$; $3x - 2 = 0$ or $4x + 3 = 0$; $x = 2/3$ or $-3/4$

45. (a) $x^2 + 4xy + 4y^2 = (x + 2y)^2$
(b) $(x + 2y)^2 = 0$; $x + 2y = 0$; $x = -2y$

47. **(a)** $x^4 - 5x^2 + 4 = (x^2 - 1)(x^2 - 4) =$
$(x - 1)(x + 1)(x - 2)(x + 2)$
(b) $(x - 1)(x + 1)(x - 2)(x + 2) = 0$; $x - 1 = 0$ or $x + 1 = 0$ or $x - 2 = 0$ or $x + 2 = 0$; $x = \pm 1$ or ± 2

0.4

1. $\dfrac{x-4}{x+1}\cdot\dfrac{2x+1}{x-1} = \dfrac{(x-4)(2x+1)}{(x+1)(x-1)} =$
$\dfrac{2x^2-7x-4}{x^2-1}$

3. $\dfrac{x-4}{x+1}+\dfrac{2x+1}{x-1} =$
$\dfrac{(x-4)(x-1)+(x+1)(2x+1)}{(x+1)(x-1)} =$
$\dfrac{3x^2-2x+5}{x^2-1}$

5. $\dfrac{x^2}{x+1}-\dfrac{x-1}{x+1} = \dfrac{x^2-(x-1)}{x+1} = \dfrac{x^2-x+1}{x+1}$

7. $\dfrac{1}{\left(\dfrac{x}{x-1}\right)}+x-1 = \dfrac{x-1}{x}+x-1 =$
$\dfrac{x-1+x(x-1)}{x} = \dfrac{x^2-1}{x}$

9. $\dfrac{1}{x}\left(\dfrac{x-3}{xy}+\dfrac{1}{y}\right) = \dfrac{1}{x}\left(\dfrac{x-3+x}{xy}\right) = \dfrac{2x-3}{x^2y}$

11. $\dfrac{(x+1)^2(x+2)^3-(x+1)^3(x+2)^2}{(x+2)^6} =$
$\dfrac{(x+1)^2(x+2)^2[(x+2)-(x+1)]}{(x+2)^6} = \dfrac{(x+1)^2}{(x+2)^4}$

13. $\dfrac{(x^2-1)\sqrt{x^2+1}-\dfrac{x^4}{\sqrt{x^2+1}}}{x^2+1} =$
$\dfrac{(x^2-1)(x^2+1)-x^4}{(x^2+1)\sqrt{x^2+1}} = \dfrac{-1}{\sqrt{(x^2+1)^3}}$

15. $\dfrac{\dfrac{1}{(x+y)^2}-\dfrac{1}{x^2}}{y} = \dfrac{x^2-(x+y)^2}{yx^2(x+y)^2} =$
$\dfrac{x^2-x^2-2xy-y^2}{yx^2(x+y)^2} = \dfrac{-y(2x+y)}{yx^2(x+y)^2} = \dfrac{-(2x+y)}{x^2(x+y)^2}$

0.5

1. $x + 1 = 0$, $x = -1$

3. $-x + 5 = 0$, $x = 5$

5. $4x - 5 = 8$, $4x = 13$, $x = 13/4$

7. $7x + 55 = 98$, $7x = 43$, $x = 43/7$

9. $x + 1 = 2x + 2$, $-x = 1$, $x = -1$

11. $ax + b = c$, $ax = c - b$, $x = (c - b)/a$

13. $2x^2 + 7x - 4 = 0$, $(2x - 1)(x + 4) = 0$, $x = -4, \frac{1}{2}$

15. $x^2 - x + 1 = 0$, $\Delta = -3 < 0$, so this equation has no real solutions

17. $2x^2 - 5 = 0$, $x^2 = \frac{5}{2}$, $x = \pm\sqrt{\frac{5}{2}}$

19. $-x^2 - 2x - 1 = 0$, $-(x + 1)^2 = 0$, $x = -1$

21. $\frac{1}{2}x^2 - x - \frac{3}{2} = 0$, $x^2 - 2x - 3 = 0$, $(x + 1)(x - 3) = 0$, $x = -1, 3$

23. $x^2 - x = 1$, $x^2 - x - 1 = 0$, $x = \frac{1 \pm \sqrt{5}}{2}$ by the quadratic formula

25. $x = 2 - \frac{1}{x}$, $x^2 = 2x - 1$, $x^2 - 2x + 1 = 0$, $(x - 1)^2 = 0$, $x = 1$

27. $x^4 - 10x^2 + 9 = 0$, $(x^2 - 1)(x^2 - 9) = 0$, $x^2 = 1$ or $x^2 = 0$, $x = \pm 1, \pm 3$

29. $x^4 + x^2 - 1 = 0$, $x^2 = \frac{-1 \pm \sqrt{5}}{2}$ by the quadratic formula, $x = \pm\sqrt{\frac{-1 \pm \sqrt{5}}{2}}$

31. $x^3 + 6x^2 + 11x + 6 = 0$, $(x + 1)(x + 2)(x + 3) = 0$, $x = -1, -2, -3$

33. $x^3 + 4x^2 + 4x + 3 = 0$, $(x + 3)(x^2 + x + 1) = 0$, $x = -3$ (For $x^2 + x + 1 = 0$, $\Delta = -3 < 0$, so there are no real solutions to this quadratic equation.)

35. $x^3 - 1 = 0$, $x^3 = 1$, $x = \sqrt[3]{1} = 1$

37. $y^3 + 3y^2 + 3y + 2 = 0$, $(y + 2)(y^2 + y + 1) = 0$, $y = -2$ (For $y^2 + y + 1 = 0$, $\Delta = -3 < 0$, so there are no real solutions to this quadratic equation.)

39. $x^3 - x^2 - 5x + 5 = 0$, $(x - 1)(x^2 - 5) = 0$, $x = 1, \pm\sqrt{5}$

41. $2x^6 - x^4 - 2x^2 + 1 = 0$, $(2x^2 - 1)(x^4 - 1) = 0$, [or $(2x^2 - 1)(x^2 - 1)(x^2 + 1) = 0$; in any case, think of the cubic you get by substituting y for x^2], $x = \pm 1, \pm\frac{1}{\sqrt{2}}$

43. $(x^2 + 3x + 2)(x^2 - 5x + 6) = 0$, $(x + 2)(x + 1)(x - 2)(x - 3) = 0$, $x = -2, -1, 2, 3$

0.6

1. $x^4 - 3x^3 = 0,\ x^3(x - 3) = 0,\ x = 0, 3$

3. $x^4 - 4x^2 = -4,\ x^4 - 4x^2 + 4 = 0,\ (x^2 - 2)^2 = 0,\ x = \pm\sqrt{2}$

5. $(x + 1)(x + 2) + (x + 1)(x + 3) = 0,$ $(x + 1)(x + 2 + x + 3) = 0,\ (x + 1)(2x + 5) = 0,\ x = -1, -5/2$

7. $(x^2 + 1)^5(x + 3)^4 + (x^2 + 1)^6(x + 3)^3 = 0,$ $(x^2 + 1)^5(x + 3)^3(x + 3 + x^2 + 1) = 0,$ $(x^2 + 1)^5(x + 3)^3(x^2 + x + 4) = 0,\ x = -3$ (Neither $x^2 + 1 = 0$ nor $x^2 + x + 4 = 0$ has a real solution.)

9. $(x^3 + 1)\sqrt{x + 1} - (x^3 + 1)^2\sqrt{x + 1} = 0,$ $(x^3 + 1)\sqrt{x + 1}\,[1 - (x^3 + 1)] = 0,$ $-x^3(x^3 + 1)\sqrt{x + 1} = 0,\ x = 0, -1$

11. $\sqrt{(x + 1)^3} + \sqrt{(x + 1)^5} = 0,\ \sqrt{(x + 1)^3}\,(1 + x + 1) = 0,\ (x + 2)\sqrt{(x + 1)^3} = 0,\ x = -1$ ($x = -2$ is not a solution because $\sqrt{(x + 1)^3}$ is not defined for $x = -2$.)

13. $(x + 1)^2(2x + 3) - (x + 1)(2x + 3)^2 = 0,$ $(x + 1)(2x + 3)(x + 1 - 2x - 3) = 0,$ $(x + 1)(2x + 3)(-x - 2) = 0,\ x = -2, -3/2, -1$

15. $\dfrac{(x + 1)^2(x + 2)^3 - (x + 1)^3(x + 2)^2}{(x + 2)^6} = 0,$ $\dfrac{(x + 1)^2(x + 2)^2[(x + 2) - (x + 1)]}{(x + 2)^6} = 0,\ \dfrac{(x + 1)^2}{(x + 2)^4} = 0,\ (x + 1)^2 = 0,\ x = -1$

17. $\dfrac{2(x^2 - 1)\sqrt{x^2 + 1} - \dfrac{x^4}{\sqrt{x^2 + 1}}}{x^2 + 1} = 0,$ $\dfrac{2(x^2 - 1)(x^2 + 1) - x^4}{(x^2 + 1)\sqrt{x^2 + 1}} = 0,\ \dfrac{x^4 - 2}{(x^2 + 1)\sqrt{x^2 + 1}} = 0,$ $x^4 - 2 = 0,\ x = \pm\sqrt[4]{2}$

19. $x - \dfrac{1}{x} = 0,\ x^2 - 1 = 0,\ x = \pm 1$

21. $\dfrac{1}{x} - \dfrac{9}{x^3} = 0,\ x^2 - 9 = 0,\ x = \pm 3$

23. $\dfrac{x - 4}{x + 1} - \dfrac{x}{x - 1} = 0,$ $\dfrac{(x - 4)(x - 1) - x(x + 1)}{(x + 1)(x - 1)} = 0,$ $\dfrac{-6x + 4}{(x + 1)(x - 1)} = 0,\ -6x + 4 = 0,\ x = 2/3$

25. $\dfrac{x + 4}{x + 1} + \dfrac{x + 4}{3x} = 0,$ $\dfrac{3x(x + 4) + (x + 1)(x + 4)}{3x(x + 1)} = 0,$ $\dfrac{(x + 4)(3x + x + 1)}{3x(x + 1)} = 0,\ \dfrac{(x + 4)(4x + 1)}{3x(x + 1)} = 0,$ $(x + 4)(4x + 1) = 0,\ x = -4, -1/4$

Chapter 1

1.1

1. Using the table,

(a) $f(0) = 2$ **(b)** $f(2) = 0.5$.

3. Using the table,

(a) $f(2) - f(-2) = 0.5 - 2 = -1.5$

(b) $f(-1)f(-2) = (4)(2) = 8$

(c) $-2f(-1) = -2(4) = -8$

5. $f(x) = 4x - 3$

(a) $f(-1) = 4(-1) - 3 = -4 - 3 = -7$

(b) $f(0) = 4(0) - 3 = 0 - 3 = -3$

(c) $f(1) = 4(1) - 3 = 4 - 3 = 1$

(d) Substitute y for x to obtain

$$f(y) = 4y - 3$$

(e) Substitute $(a+b)$ for x to obtain

$$f(a+b) = 4(a+b) - 3$$

7. $f(x) = x^2 + 2x + 3$

(a) $f(0) = (0)^2 + 2(0) + 3$
$= 0 + 0 + 3 = 3$

(b) $f(1) = 1^2 + 2(1) + 3$
$= 1 + 2 + 3 = 6$

(c) $f(-1) = (-1)^2 + 2(-1) + 3$
$= 1 - 2 + 3 = 2$

(d) $f(-3) = (-3)^2 + 2(-3) + 3$
$= 9 - 6 + 3 = 6$

(e) Substitute a for x to obtain

$f(a) = a^2 + 2a + 3$

(f) Substitute $(x+h)$ for x to obtain

$f(x+h) = (x+h)^2 + 2(x+h) + 3$

9. $g(s) = s^2 + \frac{1}{s}$

(a) $g(1) = 1^2 + \frac{1}{1} = 1 + 1 = 2$

(b) $g(-1) = (-1)^2 + \frac{1}{(-1)} = 1 - 1 = 0$

(c) $g(4) = 4^2 + \frac{1}{4} = 16 + \frac{1}{4}$

$= 16\frac{1}{4}$ or $\frac{65}{4}$ or 16.25

(d) Substitute x for s to obtain

$$g(x) = x^2 + \frac{1}{x}$$

(e) Substitute $(s+h)$ for s to obtain

$$g(s+h) = (s+h)^2 + \frac{1}{s+h}$$

(f) $g(s+h) - g(s)$

= Answer to part (e) – Original function

$$= \left((s+h)^2 + \frac{1}{s+h}\right) - \left(s^2 + \frac{1}{s}\right)$$

11. $f(t) = \begin{cases} -t & \text{if } t < 0 \\ t^2 & \text{if } 0 \le t < 4 \\ t & \text{if } t \ge 4 \end{cases}$

(a) $f(-1) = -(-1) = 1$ (using the first formula, since $-1 < 0$).

(b) $f(1) = 1^2 = 1$ (using the second formula, since $0 \le 1 < 4$).

(c) $f(4) = 4$ (using the third formula, since $4 \ge 4$).

$f(2) = 2^2 = 4$ (using the second formula, since $0 \le 2 < 4$).

Therefore,

$f(4) - f(2) = 4 - 4 = 0$

(d) $f(3) = 3^2 = 9$ (using the second formula, since $0 \le 3 < 4$).

$f(-3) = -(-3) = 3$ (using the first formula, since $-3 < 0$).

Therefore,

$f(3)f(-3) = (9)(3) = 27$.

13. $f(x) = x - \frac{1}{x^2}$, with domain $(0, +\infty)$

(a) Since 4 is in $(0, +\infty)$, $f(4)$ is defined.

$f(4) = 4 - \frac{1}{4^2} = 4 - \frac{1}{16} = \frac{63}{16}$

(b) Since 0 is not in $(0, +\infty)$, $f(0)$ is not defined.

(c) Since -1 is not in $(0, +\infty)$, $f(-1)$ is not defined.

15. $f(x) = \sqrt{x+10}$, with domain $[-10, 0)$

(a) Since 0 is not in $[-10, 0)$, $f(0)$ is not defined.

(b) Since 9 is not in $[-10, 0)$, $f(9)$ is not defined.

(c) Since -10 is in $[-10, 0)$, $f(-10)$ is defined.

$f(-10) = \sqrt{-10+10} = \sqrt{0} = 0$

17. $f(x) = x^2$

(a) $f(x+h) = (x+h)^2$

Therefore,

$$\begin{aligned} f(x+h) - f(x) &= (x+h)^2 - x^2 \\ &= x^2+2xh+h^2 - x^2 \\ &= 2xh + h^2 \\ &= h(2x + h) \end{aligned}$$

(b) Using the answer to part (a)

$$\begin{aligned} \frac{f(x+h) - f(x)}{h} &= \frac{h(2x + h)}{h} \\ &= 2x + h \end{aligned}$$

19. $f(x) = 2-x^2$

(a) $f(x+h) = 2-(x+h)^2$

Therefore,

$f(x+h) - f(x)$

$= 2-(x+h)^2 - (2-x^2)$

$= 2-x^2-2xh-h^2 - 2+x^2$

$= -2xh-h^2$

$= -h(2x + h)$

(b) Using the answer to part (a)

$$\begin{aligned} \frac{f(x+h) - f(x)}{h} &= -\frac{h(2x + h)}{h} \\ &= -(2x + h) \end{aligned}$$

21. $f(x) = 0.1x^2 - 4x + 5$

Technology formula: `0.1*x^2-4*x+5`

Table of Values:

x	0	1	2	3
$f(x)$	5	1.1	−2.6	−6.1
x	4	5	6	7
$f(x)$	−9.4	−12.5	−15.4	−18.1
x	8	9	10	
$f(x)$	−20.6	−22.9	−25	

23. $h(x) = \dfrac{x^2-1}{x^2+1}$

Technology formula: `(x^2-1)/(x^2+1)`

Table of Values (rounded to four decimal places):

x	0.5	1.5	2.5	3.5
$h(x)$	−0.6000	0.3846	0.7241	0.8491
x	4.5	5.5	6.5	7.5
$h(x)$	0.9059	0.9360	0.9538	0.9651
x	8.5	9.5	10.5	
$h(x)$	0.9727	0.9781	0.9820	

25.

(a) Using the table,

$P(5) = 117 \qquad P(10) = 132$

Since $P(9) = 130$ and $P(10) = 132$, we estimate $P(9.5)$ to be midway between these 130 and 132:

$P(9.5) \approx \dfrac{130+132}{2} = 131.$

Interpretation:

Since $P(t)$ gives the number of million people employed in July 1 of year t, we have:

$P(5) = 117$: Approximately 117 million people were employed in the US on July 1, 1995.

$P(10) = 132$: Approximately 132 million people were employed in the US on July 1, 2000.

$P(9.5) \approx 131$: Approximately 131 million people were employed in the US on January 1, 2000. (Note that January 1, 2000 is midway between July 1, 1999 and July 1, 2000.)

(b) The table gives $P(t)$ for $5 \le t \le 11$. Hence, the domain of P is the set of numbers t with $5 \le t \le 11$, that is, $[5, 11]$.

27. **(a)** The model is valid for the range 1994 ($t = 0$) through 2004 ($t = 10$). Thus, an appropriate domain is [0, 10]. $t \geq 0$ is not an appropriate domain because it would predict U.S. trade with China into the indefinite future with no basis.

(b) $C(t) = 3t^2 - 7t + 50$

$C(10) = 3(10)^2 - 7(10) + 50 = 300 - 70 + 50 =$ \$280 billion.

Since $t = 10$ represents 2004, the answer tells us that U.S. trade with China in 2004 was valued at approximately \$280 billion.

29. The technology formulas are:

(1) `-0.2*t^2+t+16`

(2) `0.2*t^2+t+16`

(3) `t+16`

The following table shows the values predicted by the three models.

t	0	2	4	6	7
S	16	18	22	28	30
(1)	16	17.2	16.8	14.8	13.2
(2)	16	18.8	23.2	29.2	32.8
(3)	16	18	20	22	23

As shown in the table, the values predicted by model (2) are much closer to the observed values S than those predicted by the other models.

(b) Since 1998 corresponds to $t = 8$,

$$S(t) = 0.2\,t^2 + t + 16$$
$$S(8) = 0.2(8)^2 + 8 + 16 = 36.8$$

So the spending on corrections in 1998 was predicted to be \$36.8 billion.

31. $q(p) = 361{,}201 - (p+1)^2$

(a) Note p is expressed in cents, not dollars, so 50¢ is represented by $p = 50$, not 0.50.

$$\begin{aligned} q(50) &= 361{,}201 - (50+1)^2 \\ &= 361{,}201 - 2601 \\ &= 358{,}600 \text{ brownie dishes} \end{aligned}$$

(b) If they give them away, $p = 0$, so

$$\begin{aligned} q(0) &= 361{,}201 - (0+1)^2 \\ &= 361{,}201 - 1 \\ &= 361{,}200 \text{ brownie dishes} \end{aligned}$$

(c) If they sell no dishes, $q = 0$, and so

$0 = 361{,}201 - (p+1)^2$

$(p+1)^2 = 361{,}201$

$p+1 = \sqrt{361{,}201} = 601$

$p = 601 - 1 = 600$¢, or \$6.00.

33. $P(t) = \begin{cases} 75t + 200 & \text{if } 0 \leq t \leq 4 \\ 600t - 1900 & \text{if } 4 < t \leq 9 \end{cases}$

(a) $P(0) = 75(0) + 200 = 200$. We used the first formula, since 0 is in [0, 4].

$P(4) = 75(4) + 200 = 300 + 200 = 500$. We used the first formula, since 4 is in [0, 4].

$P(5) = 600(5) - 1900 = 3000 - 1900 = 1100$. We used the second formula, since 5 is in (4, 9].

Interpretation:

Since $P(t)$ gives the speed of Intel processors at the start of year $1995 + t$, we have:

$P(0) = 200$: At the start of 1995, the processor speed was 200 megahertz.

$P(4) = 500$: At the start of 1999, the processor speed was 500 megahertz.

$P(5) = 1100$: At the start of 2000, the processor speed was 1100 megahertz.

(b) Since 2.0 gigahertz = 2000 megahertz, we want $P(t) = 2000$. The first formula gives speeds no higher than 500 megahertz (see part (a)), so we use the second formula instead:

$$\begin{aligned} P(t) &= 2000 \\ 600t - 1900 &= 2000 \\ 600t &= 3900 \\ t &= 3900/600 = 6.5, \end{aligned}$$

or midway through 2001

(c) A technology formula for P is

```
(75*t+200)*(t<=4)+
        (600*t-1900)*(t>4)
```

(For a graphing calculator, use x instead of t, and ≤ instead of <=)

Table of Values:

t	P(t)
0	200
1	275
2	350
3	425
4	500
5	1100
6	1700
7	2300
8	2900
9	3500

35. $C(t) = \begin{cases} 0.08t + 0.6 & \text{if } 0 \le t < 8 \\ 0.355t - 1.6 & \text{if } 8 \le t \le 11 \end{cases}$

(a) A technology formula for C is

```
(0.08*t+0.6)*(t<8)+(0.355*t-
1.6)*(t>=8)
```

(For a graphing calculator, use x instead of t, and ≥ instead of >=)

Table of Values:

t	0	1	2	3	4	5
$C(t)$	0.6	0.68	0.76	0.84	0.92	1
t	6	7	8	9	10	11
$C(t)$	1.08	1.16	1.24	1.595	1.95	2.305

(b) The costs of a Superbowl ad in 1998, 1999, and 2000 were:

1998: $C(8) = \$1.24$ million

1999: $C(9) = \$1.595$ million

2000: $C(10) = \$1.95$ million

From 1998 to 1999, the cost increased by $1.595 – $1.24 = $0.355 million dollars.

From 1999 to 2000, the cost increased by $1.95 – $1.595 = $0.355 million dollars.

Thus, the cost increased at a rate of $0.355 million dollars (or $355,000) per year between 1998 and 2000.

37. A taxable income of $26,000 falls in the category "Over $7,300 but not over $29,700" and so we use the formula "$730.00 + 15% of the amount over $7,300."

$$T(26{,}000) = \$730 + 0.15(26{,}000 - 7300) = \$3535.00$$

A taxable income of $65,000 falls in the category "Over $29,700 but not over $71,950" and so we use the formula "$4,090.00 + 25% of the amount over $29,700."

$T(65{,}000) = \$4090 + 0.25(65{,}000–29{,}700)$

$= \$12{,}915$

39. $C(q) = 2000 + 100q^2$

(a) $C(10) = 2000+100(10)^2 = 2000+10{,}000$

$= \$12{,}000$

(b) Net Cost = Total Cost – Subsidy

$N(q) = C(q) - S(q)$

$= 2000 + 100q^2 - 500q$

$N(20) = 2000 + 100(20)^2 - 500(20)$

$= 2000 + 40{,}000 - 10{,}000$

$= \$32{,}000$

41. $p(t) = 100\left(1 - \frac{12{,}200}{t^{4.48}}\right)$ $(t \ge 8.5)$

(a) Technology formula:

```
100*(1-12200/t^4.48)
```

(b) Table of values:

t	9	10	11	12	13	14
$p(t)$	35.2	59.6	73.6	82.2	87.5	91.1
t	15	16	17	18	19	20
$p(t)$	93.4	95.1	96.3	97.1	97.7	98.2

(c) From the table, $p(12) = 82.2$, so that 82.2% of children are able to speak in at least single words by the age of 12 months.

(b) We seek the first value of t such that $p(t)$ is at least 90. Since $t = 14$ has this property ($p(14) = 91.1$) we conclude that, at 14 months, 90% or more children are able to speak in at least single words.

43. The dependent variable is a function of the independent variable. Here, the market price of gold m is a function of time t. Thus, the independent variable is t and the dependent variable is m.

45. To obtain the function notation, write the dependent as a function of the independent variable. Thus $y = 4x^2 - 2$ can be written as
$f(x) = 4x^2 - 2$ or $y(x) = 4x^2 - 2$

47. Number of sound files
= Starting number + New files
= 200 + 10×Number of days
So, $N(t) = 200 + 10t$
(N = number of sound files, t = time in days)

49. As the text reminds us: to evaluate f of a quantity (such as $x+h$) replace x everywhere by the *whole quantity* $x+h$:

$$f(x) = x^2 - 1$$
$$f(x+h) = (x+h)^2 - 1.$$

51. Functions with infinitely many points in their domain (such as $f(x) = x^2$) cannot be specified numerically. So, the assertion is false.

1.2

1. From the graph, we find

(a) $f(1) = 20$ **(b)** $f(2) = 30$

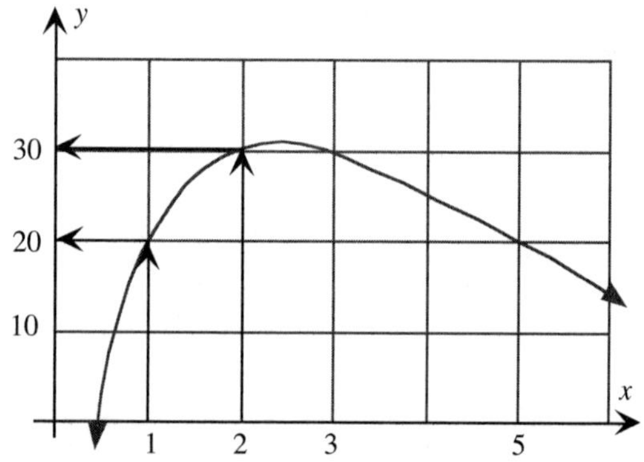

In a similar way, we find

(c) $f(3) = 30$ **(d)** $f(5) = 20$

(e) $f(3) - f(2) = 30 - 30 = 0$

3. From the graph,

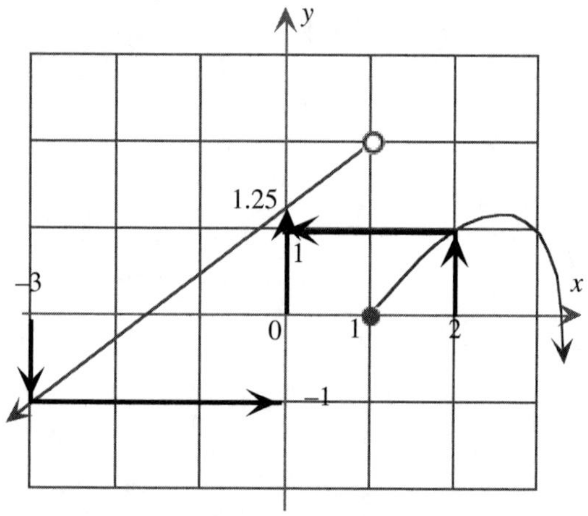

(a) $f(-3) = -1$ **(b)** $f(0) = 1.25$

(c) $f(1) = 0$ since the solid dot is on (1, 0) ($f(1) \neq 2$ since the hollow dot at (1, 2) indicates that (1, 2) is not a point of the graph.)

(d) $f(2) = 1$

(e) Since $f(3) = 1$ and $f(2) - 1$,

$$\frac{f(3) - f(2)}{3 - 2} = \frac{1 - 1}{3 - 2} = 0$$

5.

(a) $f(x) = x \quad (-1 \le x \le 1)$

Since the graph of $f(x) = x$ is a diagonal 45° line through the origin inclining up from left to right, the correct graph is (I)

(b) $f(x) = -x \quad (-1 \le x \le 1)$

Since the graph of $f(x) = -x$ is a diagonal 45° line through the origin inclining down from left to right, the correct graph is (IV)

(c) $f(x) = \sqrt{x} \quad (0 < x < 4)$

Since the graph of $f(x) = \sqrt{x}$ is the top half of a sideways parabola, the correct graph is (V)

(d) $f(x) = x + \frac{1}{x} - 2 \quad (0 < x < 4)$

If we plot a few points like $x = 0.1$, 1, 2, and 3 we find that the correct graph is (VI).

(e) $f(x) = |x| \quad (-1 \le x \le 1)$

Since the graph of $f(x) = |x|$ is a "V"-shape with its vertex at the origin, the correct graph is (III).

(f) $f(x) = x - 1 \ (-1 \le x \le 1)$

Since the graph of $f(x) = x-1$ is a straight line through (0, –1) and (1, 0), the correct graph is (II).

7. $f(x) = -x^3$ (domain $(-\infty, +\infty)$)

Technology formula: `-(x^3)`

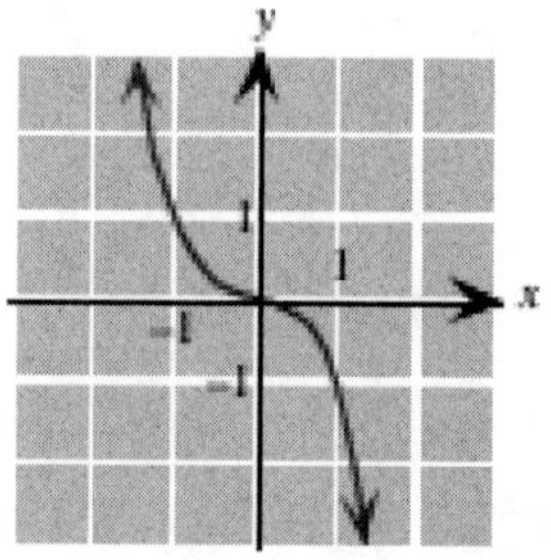

9. $f(x) = x^4$ (domain $(-\infty, +\infty)$)

Technology formula: `x^4`

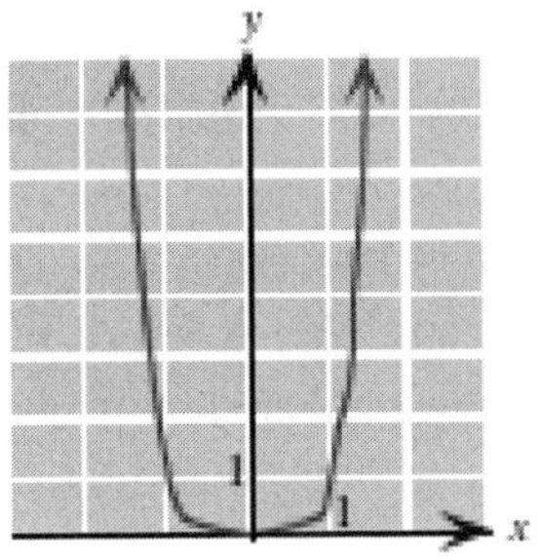

11. $f(x) = \frac{1}{x^2}$ $(x \neq 0)$

Technology formula: `1/x^2`

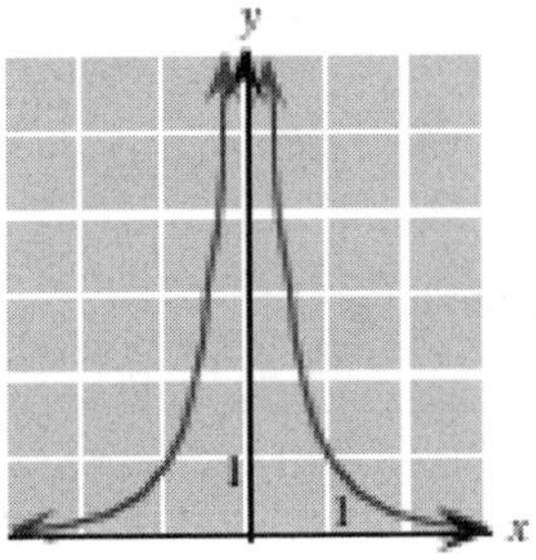

13. $f(x) = \begin{cases} x & \text{if } -4 \le x < 0 \\ 2 & \text{if } 0 \le x \le 4 \end{cases}$

Technology formula: `x*(x<0)+2*(x>=0)`

(For a graphing calculator, use ≥ instead of >=.)

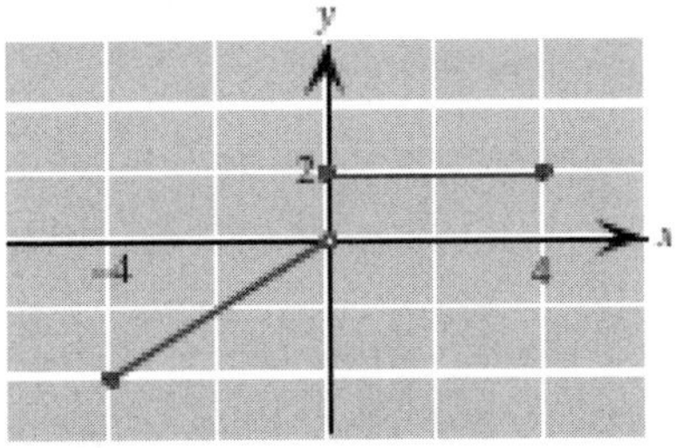

(a) $f(-1) = -1$. We used the first formula, since -1 is in $[-4, 0)$.

(b) $f(0) = 2$. We used the second formula, since 0 is in $[0, 4]$.

(c) $f(1) = 2$. We used the second formula, since 1 is in $[0, 4]$.

15. $f(x) = \begin{cases} x^2 & \text{if } -2 < x \le 0 \\ 1/x & \text{if } 0 < x \le 4 \end{cases}$

Technology formula:

`(x^2)*(x<=0)+(1/x)*(0<x)`

(For a graphing calculator, use ≤ instead of <=.)

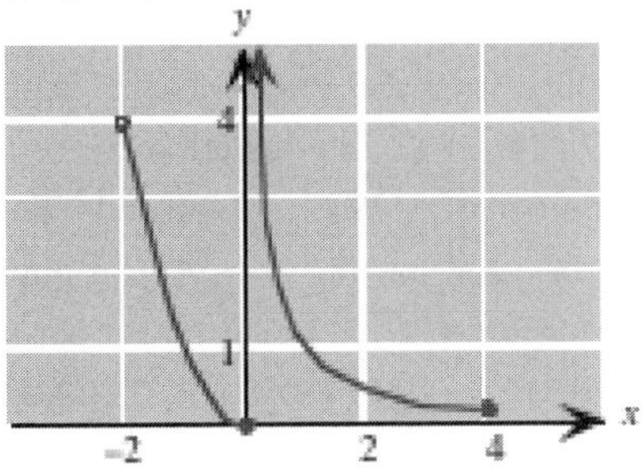

(a) $f(-1) = 1^2 = 1$. We used the first formula, since -1 is in $(-2, 0]$.

(b) $f(0) = 0^2 = 0$. We used the first formula, since 0 is in $(-2, 0]$.

(c) $f(1) = 1/1 = 1$. We used the second formula, since 1 is in $(0, 4]$.

17. $f(x) = \begin{cases} x & \text{if } -1 < x \le 0 \\ x+1 & \text{if } 0 < x \le 2 \\ x & \text{if } 2 < x \le 4 \end{cases}$

Technology formula:

`x*(x<=0)+(x+1)*(0<x)*(x<=2)+x*(2<x)`

(For a graphing calculator, use ≤ instead of <=.)

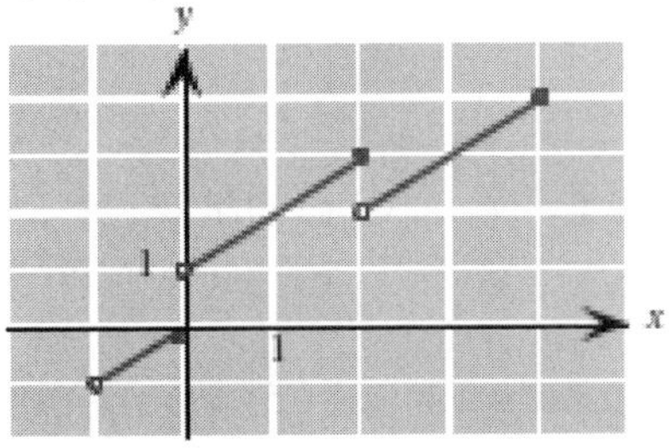

(a) $f(0) = 0$. We used the first formula, since 0 is in $(-1, 0]$.

(b) $f(1) = 1+1 = 2$. We used the second formula, since 1 is in $(0, 2]$.

(c) $f(2) = 2+1 = 3$. We used the second formula, since 2 is in (0, 2].

(d) $f(3) = 3$. We used the third formula, since 3 is in (2, 4].

19. Reading from the graph, $f(6) \approx 2000$, $f(9) \approx 2800$, $f(7.5) \approx 2500$. Interpretation: Since $f(t)$ is the is the number of SUVs (in thousands) sold in the U.S. in the year starting $t+1990$, we interpret the results as follows:

$f(6) \approx 2000$: In 1996 (1990+6), 2,000,000 SUVs were sold.

$f(9) \approx 2800$: In 1999 (1990+9), 2,800,000 were sold.

$f(7.5) \approx 2500$: Since 1990+7.5 = 1997.5, or half way through 1997, we interpret this by saying that, in the year beginning July, 1997, 2,500,000 were sold.

21. From the graph,

$f(6)-f(5) \approx 2000-1750 = 250$

$f(10)-f(9) \approx 2900-2750 = 150$

Therefore, $f(6)-f(5)$ is larger.

Interpretation: In general, the difference $f(b)-f(b)$ measures the change in sales from year a to year b. So, SUV sales increased more from 1995 to 1996 than from 1999 to 2000.

23.

(a) From the graph, we see that $N(t)$ is defined for $-1.5 \le t \le 1.5$. Thus, the domain of N is $[-1.5, 1.5]$.

(b) From the graph, $N(-0.5) \approx 131$, $N(0) \approx 132$, $N(1) \approx 132$. Since $N(t)$ gives the number of million people employed in the U.S. in year Jan. 2000+t, we interpret the results as follows:

$N(-0.5) \approx 131$: In July, 1999 (Jan. 2000 − 0.5), approximately 131 million people were employed.

$N(0) \approx 132$: In January 2000, approximately 132 million people were employed.

$N(1) \approx 132$: In January 2001 Jan. 2000 + 1), approximately 132 million people were employed.

(c) The graph is descending from around $t = 0.5$ to $t = 1.5$. Thus, $N(t)$ is falling on the interval [0.5, 1.5], showing that employment was falling during the period July, 2000–July, 2001.

25.

(a) The technology formulas are:

(A): `0.005*x+2.75`

(B): `0.01*x+20+25/x`

(C): `0.0005*x^2-0.07*x+23.25`

(D): `25.5*1.08^(x-5)`

The following table shows the values predicted by the four models.

x	5	25	40	100	125
A(x)	22.91	21.81	21.25	21.25	22.31
(A)	2.775	2.875	2.95	3.25	3.375
(B)	25.05	21.25	21.025	21.25	21.45
(C)	22.913	21.813	21.25	21.25	22.313
(D)	25.5	118.85	377.03	38177	261451

Model (C) fits the data almost perfectly—more closely that any of the other models.

(b) Graph of model (C):

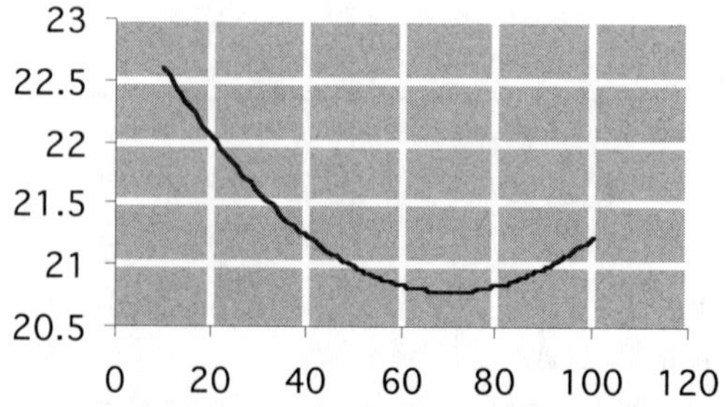

`0.0005*x^2-0.07*x+23.25`

The lowest point on the graph occurs at $x = 70$ with a y-coordinate of 20.8. Thus, the lowest cost per shirt is \$20.80, which the team can obtain by buying 70 shirts.

27. A plot of the given data suggests a curve with a low point somewhere between $t = 0$ and $t = 5$. A linear model would predict perpetually increasing or decreasing value of the euro (depending on whether the slope is positive or negative) and an exponential model $n(t) = Ab^t$ would also be perpetually increasing or decreasing (depending whether b is larger than 1 or less than 1). This leaves a quadratic model as the only possible choice. In fact, a quadratic can always be found that passes through any three points not on the same straight line with different x-coordinates. Therefore, a quadratic model would give an exact fit.

29. $p(t) = 100\left(1 - \frac{12{,}200}{t^{4.48}}\right) \qquad (t \geq 8.5)$

(a) Technology formula:

```
100*(1-12200/t^4.48)
```

(b) Graph:

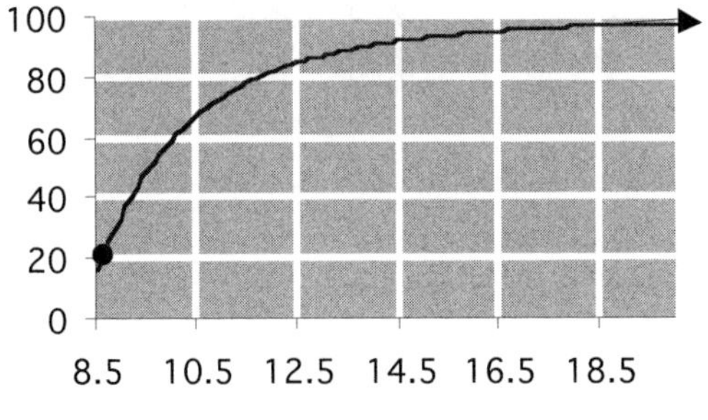

(c) If we zoom in to the graph near $x = 12$ we find that $p(12) \approx 82$. Thus, 82% of children are able to speak in at least single words by the age of 12 months.

(d) We need to find the approximate value of t so that $p(t) = 90$. From the graph above, it occurs somewhere between $t = 12.5$ and 14.5. Here is a close-up of the graph near that point:

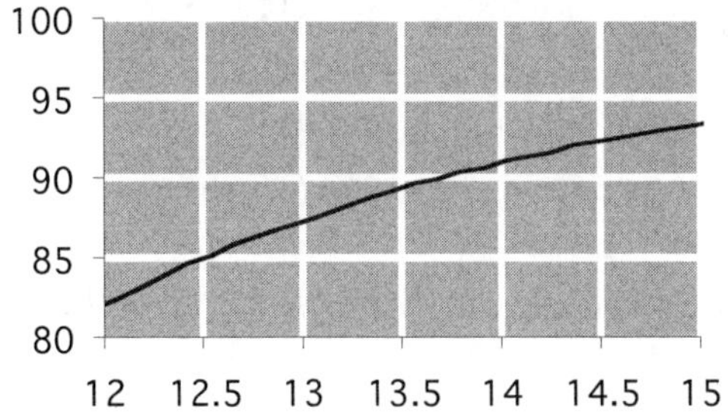

This close-up shows that $f(13.7) \approx 90$. Therefore, 90% of children speaking in at least single words at approximately 13.7 months of age, or 14 months if we round to the nearest month.

31. (a) Technology formula:

```
(75*t+200)*(t<=4)+
        (600*t-1900)*(t>4)
```

(For a graphing calculator, use x instead of t, and ≥ instead of >=)

Graph:

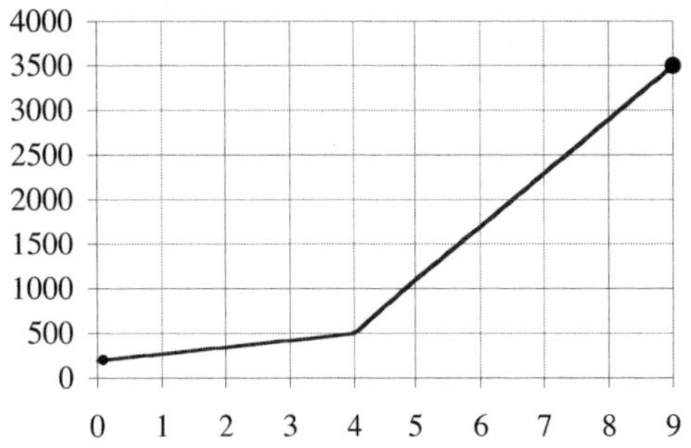

(b) To answer the question, we need to find the value of t such that $P(t) = 2000$ (because 2.0 gigahertz = 2000 megahertz). From the graph, $P(6.5) \approx 2000$, and 6.5 years since the start of 1995 was midway through 2001. Thus, processor speeds first reached 2.0 gigahertz midway through 2001.

33. $C(t) = \begin{cases} 0.08t + 0.6 & \text{if } 0 \le t < 8 \\ 0.355t - 1.6 & \text{if } 8 \le t \le 11 \end{cases}$

(a) Technology formula:

```
(0.08*t+0.6)*(t<8)+(0.355*t-1.6)*(t>=8)
```

(For a graphing calculator, use x instead of t, and ≥ instead of >=)

Graph:

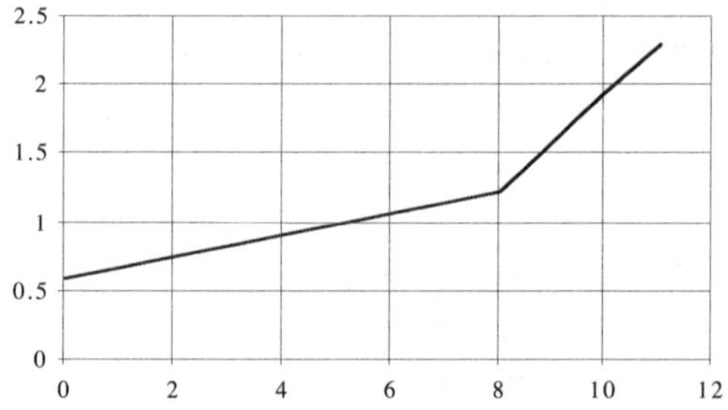

(b) To answer the question, we need to find the first integer value of t such that $C(t)$ exceeds 2. Although $C(10) \approx 2$, it is a little less than 2. On the other hand, $C(11) \ge 2$, and $t = 11$ represents the first year such that $C(t) \ge 2$. Thus, a Superbowl ad first exceeded \$2 million in 1990+11 = 2001.

35. True. A graphically specified function is specified by a graph. Given a graph, we can read off a set of values to construct a table, and hence specify the function numerically. (The more accurate the graph is, the more accurate the numerical values are.)

37. False. A numerically specified function with domain [0, 10] is specified by a table of some values between 0 and 10. Since only certain values of the function are specified, we can obtain only certain points on the graph.

39. If two functions are specified by the same formula $f(x)$ say, their graphs must follow the same curve $y = f(x)$. However, it is the domain of the function that specifies what portion of the curve appears on the graph. Thus, if the functions have different domains, their graphs will be different portions of the curve $y = f(x)$.

41. Suppose we already have the graph of f and want to construct the graph of g. We can plot a point of the graph of g as follows: Choose a value for x ($x = 7$, say) and then "look back" 5 units to read off $f(x-5)$ ($f(2)$ in this instance). This value gives the y-coordinate we want. In other words, points on the graph of g are obtained by "looking back 5 units" to the graph of f and then copying that portion of the curve. Put another way, the graph of g is the same as the graph of f, but shifted 5 units to the right.

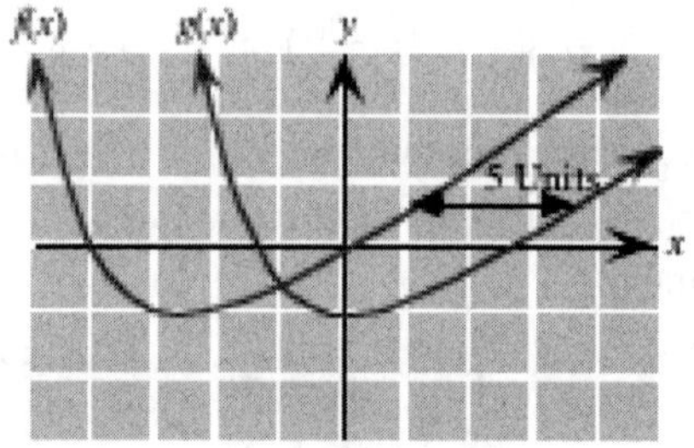

1.3

1.

x	-1	0	1
y	5	8	

We calculate the slope m first. The first two points shown give a changes in x and y of

$\Delta x = 0 - (-1) = 1$

$\Delta y = 8 - 5 = 3$

This gives a slope of

$$m = \frac{\Delta y}{\Delta x} = \frac{3}{1} = 3.$$

Now look at the second and third points: The change in x is again

$\Delta x = 1 - 0 = 1$

and so Δy must be given by the formula

$\Delta y = m\Delta x$

$\Delta y = 3(1) = 3.$

This means that the missing value of y is

$8 + \Delta y = 8 + 3 = 11.$

3.

x	2	3	5
y	-1	-2	

We calculate the slope m first. The first two points shown give a changes in x and y of

$\Delta x = 3 - 2 = 1$

$\Delta y = -2 - (-1) = -1$

This gives a slope of

$$m = \frac{\Delta y}{\Delta x} = \frac{-1}{1} = -1.$$

Now look at the second and third points: The change in x is

$\Delta x = 5 - 3 = 2$

and so Δy must be given by the formula

$\Delta y = m\Delta x$

$\Delta y = (-1)(2) = -2.$

This means that the missing value of y is

$-2 + \Delta y = -2 + (-2) = -4.$

5.

x	-2	0	2
y	4		10

We calculate the slope m first. The first and third points shown give a changes in x and y of

$\Delta x = 2 - (-2) = 4$

$\Delta y = 10 - 4 = 6$

This gives a slope of

$$m = \frac{\Delta y}{\Delta x} = \frac{6}{4} = \frac{3}{2}.$$

Now look at the first and second points: The change in x is

$\Delta x = 0 - (-2) = 2$

and so Δy must be given by the formula

$\Delta y = m\Delta x$

$\Delta y = (\frac{3}{2})(2) = 3.$

This means that the missing value of y is

$4 + \Delta y = 4 + 3 = 7.$

7. From the table,

$b = f(0) = -2.$

The slope (using the first two points) is

$$m = \frac{y_2 - y_2}{x_2 - x_1} = \frac{-2 - (-1)}{0 - (-2)} = \frac{-1}{2} = -\frac{1}{2}.$$

Thus, the linear equation is

$$f(x) = mx + b = -\frac{1}{2}x - 2,$$

or $f(x) = -\frac{x}{2} - 2.$

9. The slope (using the first two points) is

$$m = \frac{y_2 - y_2}{x_2 - x_1} = \frac{-2 - (-1)}{-3 - (-4)} = \frac{-1}{1} = -1.$$

To obtain $f(0) = b$, notice that, since the slope is -1, y decreases by 1 for every one-unit increase in x. Thus,

$f(0) = f(-1) + m = -4 - 1 = -5.$

This gives

$f(x) = mx + b = -x - 5.$

11. In the table, x increases in steps of 1 and f increases in steps of 4, showing that f is linear with slope

$$m = \frac{\Delta y}{\Delta x} = \frac{4}{1} = 4$$

and intercept

$$b = f(0) = 6$$

giving

$$f(x) = mx + b = 4x + 6.$$

The function g does not increase in equal steps, so g is not linear.

13. In the first three points listed in the table, x increases in steps of 3, but f does not increase in equal steps, whereas g increases in steps of 6. Thus, based on the first three points, only g could possibly be linear, with slope

$$m = \frac{\Delta y}{\Delta x} = \frac{6}{3} = 2$$

and intercept

$$b = g(0) = -1$$

giving

$$g(x) = mx + b = 2x - 1.$$

We can now check that the remaining points in the table fit the formula $g(x) = 2x-1$, showing that g is indeed linear.

15. Slope = coefficient of $x = -\frac{3}{2}$

17. Write the equation as

$$y = \frac{x}{6} + \frac{1}{6}$$

Slope = coefficient of $x = \frac{1}{6}$

19. If we solve for x we find that the given equation represents the vertical line $x = -1/3$, and so its slope is infinite (undefined).

21. $3y + 1 = 0$

Solving for y:

$$3y = -1$$

$$y = -\frac{1}{3}$$

Slope = coefficient of $x = 0$

23. $4x + 3y = 7$

Solve for y:

$$3y = -4x + 7$$

$$y = -\frac{4}{3}x + \frac{7}{3}$$

Slope = coefficient of $x = -\frac{4}{3}$

25. $y = 2x - 1$

y-intercept $= -1$, slope $= 2$

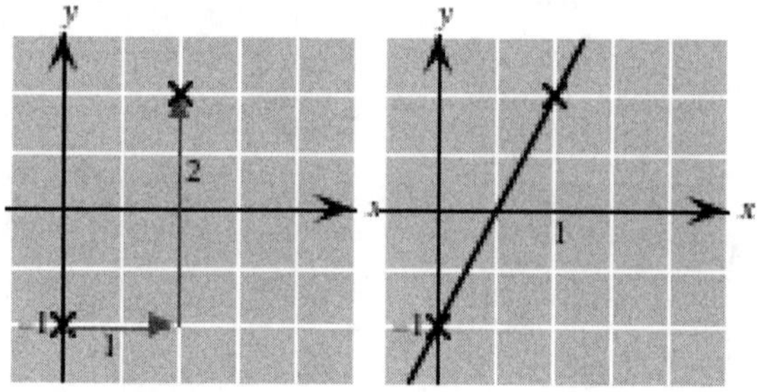

27. $y = -\frac{2}{3}x + 2$

y-intercept $= 2$, slope $= -\frac{2}{3}$

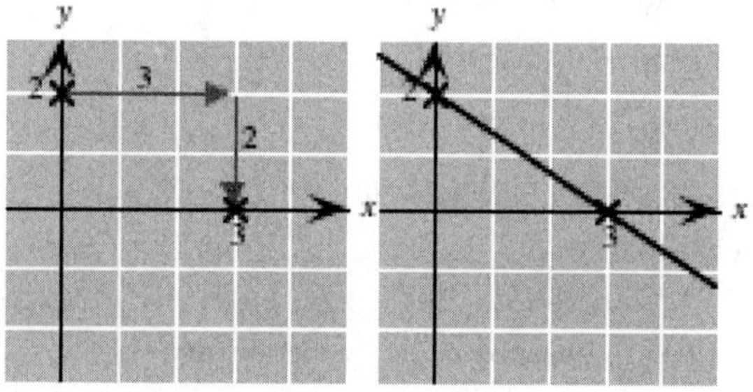

29. $y + \frac{1}{4}x = -4$

Solve for y to obtain $y = -\frac{1}{4}x - 4$

y−intercept = −4, slope = $-\frac{1}{4}$

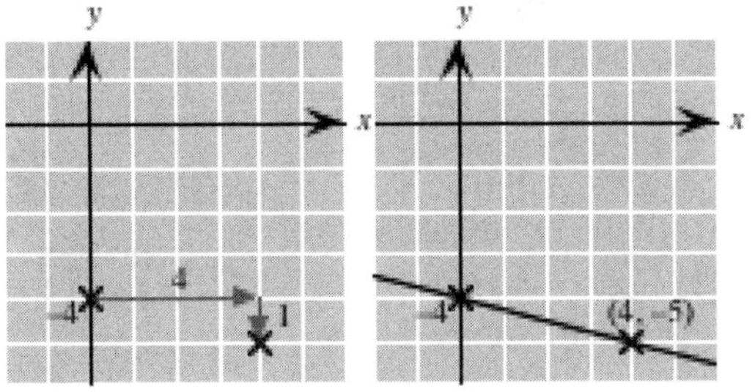

31. $7x - 2y = 7$

Solve for y:

$$-2y = -7x + 7$$

$$y = \frac{7}{2}x - \frac{7}{2}$$

y−intercept = $-\frac{7}{2}$ = −3.5, slope = $\frac{7}{2}$ = 3.5

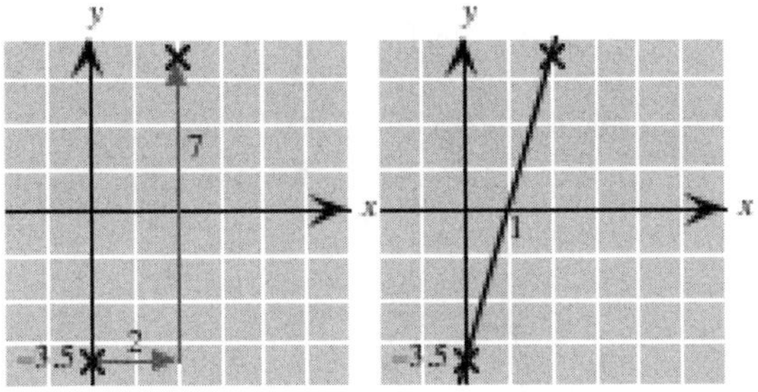

33. $3x = 8$

Solve for x to obtain $x = \frac{8}{3}$.

The graph is a vertical line:

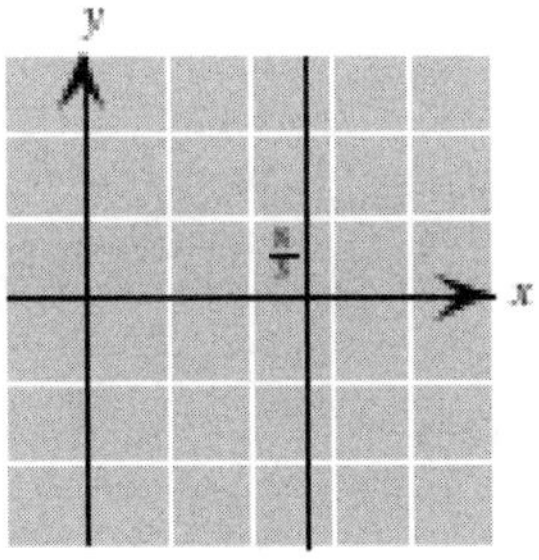

35. $6y = 9$

Solve for y to obtain $y = \frac{9}{6} = \frac{3}{2} = 1.5$

y−intercept = $\frac{3}{2}$ = 1.5, slope = 0

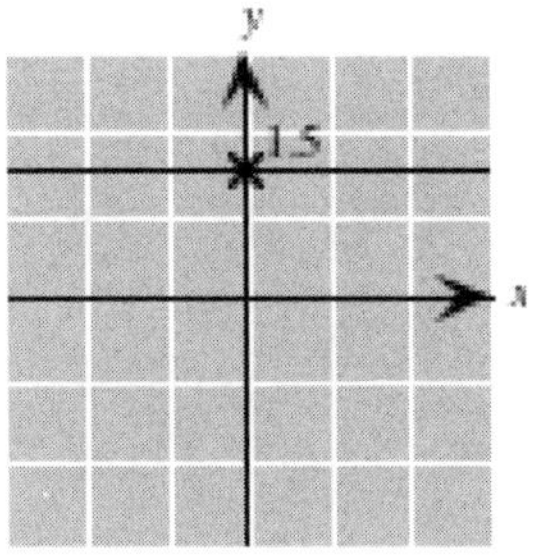

37. $2x = 3y$

Solve for y to obtain $y = \frac{2}{3}x$

y−intercept = 0, slope = $\frac{2}{3}$

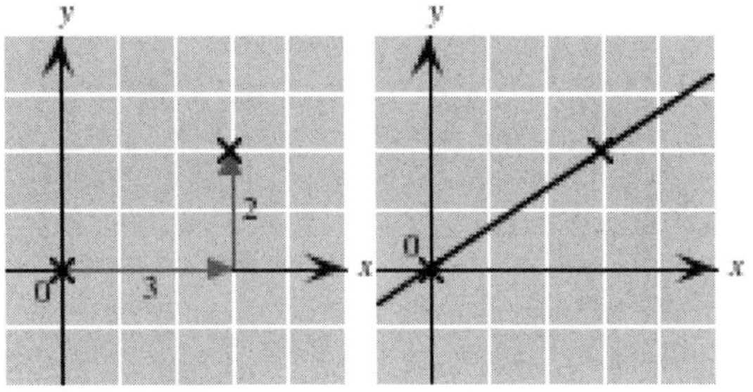

39. (0, 0) and (1, 2)

$$m = \frac{y_2 - y_1}{x_2 - x_1} = \frac{2 - 0}{1 - 0} = 2$$

41. (−1, −2) and (0, 0)

$$m = \frac{y_2 - y_1}{x_2 - x_1} = \frac{0 - (-2)}{0 - (-1)} = \frac{2}{1} = 2$$

43. (4, 3) and (5, 1)

$$m = \frac{y_2 - y_1}{x_2 - x_1} = \frac{1 - 3}{5 - 4} = \frac{-2}{1} = -2$$

45. $(1, -1)$ and $(1, -2)$

$$m = \frac{y_2 - y_1}{x_2 - x_1} = \frac{-2 - (-1)}{1 - 1} \quad \text{Undefined}$$

47. $(2, 3.5)$ and $(4, 6.5)$

$$m = \frac{y_2 - y_1}{x_2 - x_1} = \frac{6.5 - 3.5}{4 - 2} = \frac{3}{2} = 1.5$$

49. $(300, 20.2)$ and $(400, 11.2)$

$$m = \frac{y_2 - y_1}{x_2 - x_1} = \frac{11.2 - 20.2}{400 - 300}$$
$$= \frac{-9}{100} = -0.09$$

51. $(0, 1)$ and $(-\frac{1}{2}, \frac{3}{4})$

$$m = \frac{y_2 - y_1}{x_2 - x_1} = \frac{\frac{3}{4} - 1}{-\frac{1}{2} - 0}$$
$$= \frac{-\frac{1}{4}}{-\frac{1}{2}} = \frac{1}{4} \cdot 2 = \frac{1}{2}$$

53. (a, b) and (c, d) $(a \neq c)$

$$m = \frac{y_2 - y_1}{x_2 - x_1} = \frac{d - b}{c - a}$$

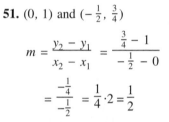

55.

(a)

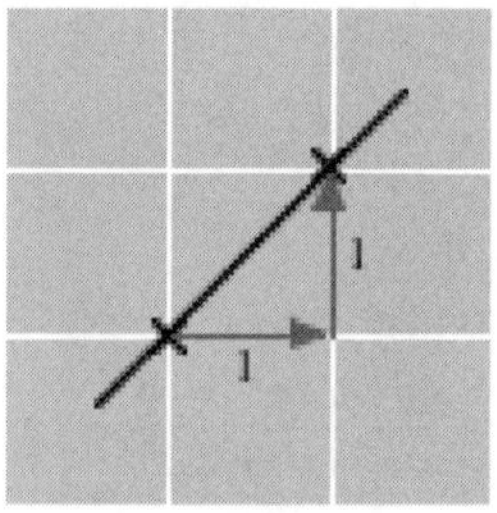

$$m = \frac{\Delta y}{\Delta x} = \frac{1}{1}$$
$$= 1$$

(b)

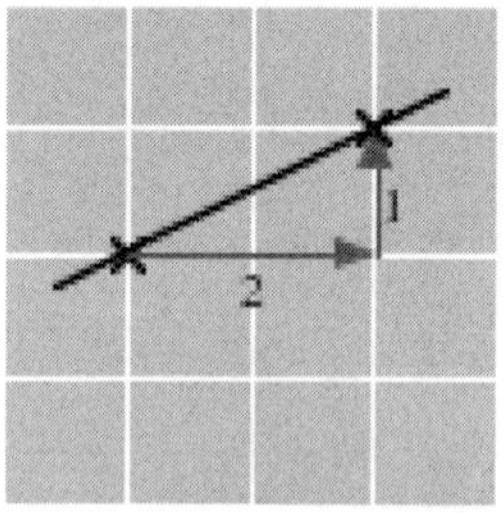

$$m = \frac{\Delta y}{\Delta x} = \frac{1}{2}$$

(c)

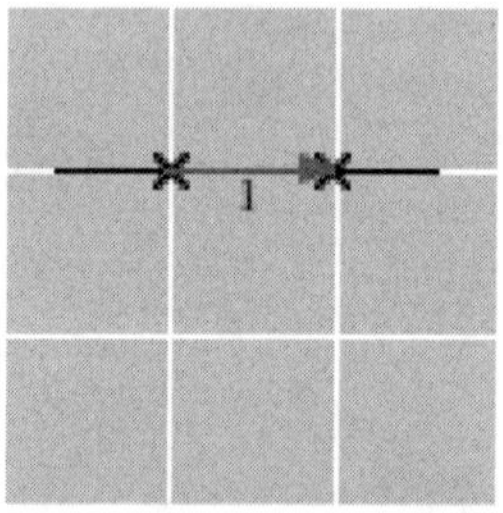

$$m = \frac{\Delta y}{\Delta x} = \frac{0}{1}$$
$$= 0$$

(d)

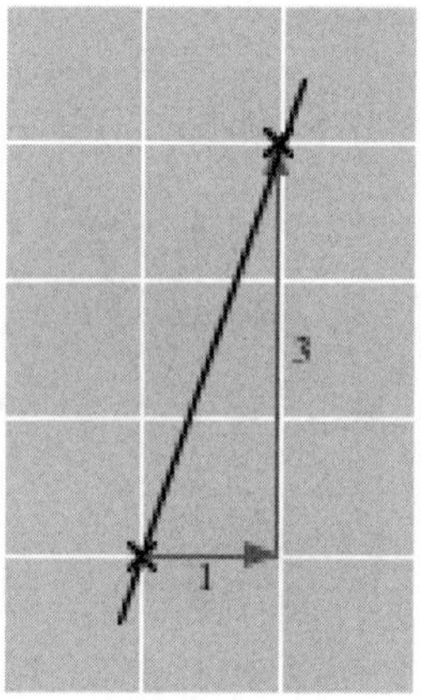

$$m = \frac{\Delta y}{\Delta x} = \frac{3}{1}$$
$$= 3$$

(e)

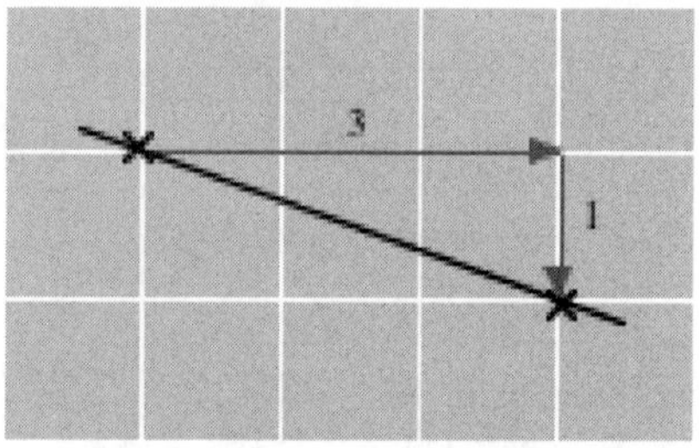

$$m = \frac{\Delta y}{\Delta x} = \frac{-1}{3} = -\frac{1}{3}$$

(f)

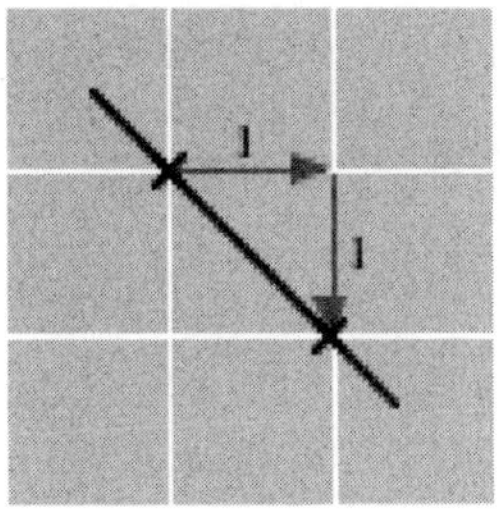

$$m = \frac{\Delta y}{\Delta x} = \frac{-1}{1}$$
$$= -1$$

(g)

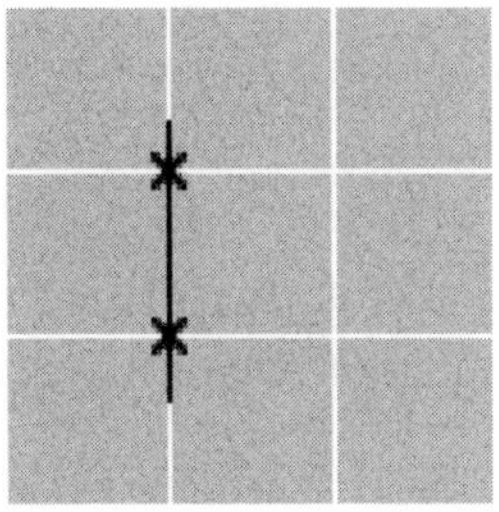

Vertical line; undefined slope

(h)

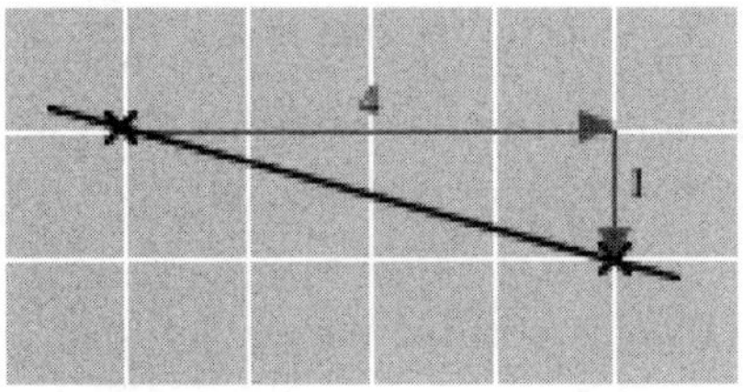

$$m = \frac{\Delta y}{\Delta x} = \frac{-1}{4} = -\frac{1}{4}$$

(i)

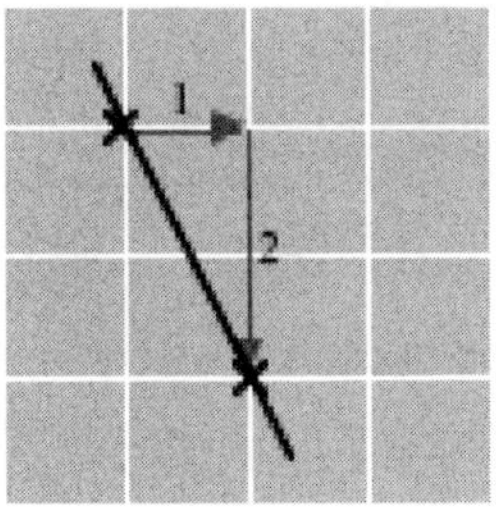

$$m = \frac{\Delta y}{\Delta x} = \frac{-2}{1}$$
$$= -2$$

57. Through (1, 3) with slope 3

Point: (1, 3) **Slope:** $m = 3$

$$b = y_1 - mx_1$$
$$= 3 - 3\cdot 1 = 0$$

Thus, the equation is

$$y = mx + b$$
$$y = 3x + 0$$
$$y = 3x$$

59. Through $(1, -\frac{3}{4})$ with slope $\frac{1}{4}$

Point: $(1, -\frac{3}{4})$ **Slope:** $m = \frac{1}{4}$

$$b = y_1 - mx_1$$
$$= -\tfrac{3}{4} - \tfrac{1}{4}\cdot 1 = -1$$

Thus, the equation is

$$y = mx + b$$
$$y = \tfrac{1}{4}x - 1$$

61. Through (20, −3.5) and increasing at a rate of 10 units of y per unit of x

Point: (20, −3.5)

Slope: $m = \dfrac{\Delta y}{\Delta x} = \dfrac{10}{1} = 10$

$$b = y_1 - mx_1$$
$$= -3.5 - (10)(20)$$
$$= -3.5 - 200 = -203.5$$

Thus, the equation is

$$y = mx + b$$
$$y = 10x - 203.5$$

63. Through (2, −4) and (1, 1)

Point: (2, −4)

Slope: $m = \dfrac{y_2 - y_1}{x_2 - x_1} = \dfrac{1 - (-4)}{1 - 2} = \dfrac{5}{-1} = -5$

$$b = y_1 - mx_1$$
$$= -4 - (-5)(2) = 6$$

Thus, the equation is

$$y = mx + b$$
$$y = -5x + 6$$

65. Through (1,−0.75) and (0.5, 0.75)

Point: (1,−0.75)

Slope: $m = \dfrac{y_2 - y_1}{x_2 - x_1} = \dfrac{0.75 - (-0.75)}{0.5 - 1}$

$$= \frac{1.5}{-0.5} = -3$$

$b = y_1 - mx_1$

$= -0.75 - (-3)(1) = -0.75+3 = 2.25$

Thus, the equation is

$y = mx + b$

$y = -3x + 2.25$

67. Through (6, 6) and parallel to the line $x + y = 4$

Point: (6, 6)

Slope: Same as slope of $x + y = 4$. To find the slope, solve for y, getting

$y = -x + 4$

Thus, $m = -1$.

$b = y_1 - mx_1$

$= 6 - (-1)(6) = 6+6 = 12$

Thus, the equation is

$y = mx + b$

$y = -x + 12$

69. Through (0.5, 5) and parallel to the line $4x - 2y = 11$

Point: (0.5, 5)

Slope: Same as slope of $4x - 2y = 11$. To find the slope, solve for y, getting

$2y = 4x - 11$

$$y = 2x - \frac{11}{2}$$

Thus, $m = 2$.

$b = y_1 - mx_1$

$= 5 - (2)(0.5) = 5 - 1 = 4$

Thus, the equation is

$y = mx + b$

$y = 2x + 4$

71. A table of values of x and y comes from a linear equation precisely when the successive ratios $\Delta y/\Delta x$ are all the same. Thus, to test such a table of values, compute the corresponding successive changes Δx in x and Δy in y, and compute the ratios $\Delta y/\Delta x$. If the answer is always the same number, then the values in the table come from a linear function.

73. To find the linear function, solve the equation $ax + by = c$ for y:

$by = -ax + c$

$$y = -\frac{a}{b}x + \frac{c}{b}$$

Thus, the desired function is $f(x) = -\frac{a}{b}x + \frac{c}{b}$.

If $b = 0$, then $\frac{a}{b}$ and $\frac{c}{b}$ are undefined, and y cannot be specified as a function of x. (The graph of the resulting equation would be a vertical line.)

75. The slope of the line is $m = \frac{\Delta y}{\Delta x} = \frac{3}{1} = 3$.

Therefore, if, in a straight line, y is increasing three times as fast as x, then its slope is 3.

77. If m is positive then y will increase as x increases; if m is negative then y will decrease as x increases; if m is zero then y will not change as x changes.

79.

	A	B	C	D
1	x	y	m	b
2	1	2	=(B3-B2)/(A3-A2)	=B2-C2*A2
3	3	-1	Slope	Intercept

The slope computed in cell C2 is given by

$$m = \frac{y_2 - y_1}{x_2 - x_1} = \frac{-1 - 2}{3 - 1} = -1.5$$

If we increase the y-coordinate in cell B3, this increases y_2, and thus increases the numerator $\Delta y = y_2 - y_1$ without effecting the denominator Δx. Thus the slope will increase.

1.4

1. For a linear cost function,

$C(x) = mx + b$

m = marginal cost = \$1500 per piano

b = fixed cost = \$1200

Thus, the daily cost function is

$C(x) = 1500x + 1200.$

(a) The cost of manufacturing 3 pianos is

$C(3) = 1500(3) + 1200$

$= 4500 + 1200 = \$5700$

(b) The cost of manufacturing each additional piano (such as the third one or the 11th one) is the marginal cost, $m = \$1500$.

(c) Same answer as (b).

3. We are given two points on the graph of the linear cost function: (100, 10,500) and (120, 11,0000) (x is the number of items, and y is the cost C).

Marginal cost:

$$m = \frac{C_2 - C_1}{x_2 - x_1} = \frac{11{,}000 - 10{,}500}{120 - 100}$$

$$= \frac{500}{20} = \$25 \text{ per bicycle.}$$

Fixed cost:

$$b = C_1 - mx_1 = 10{,}500 - (25)(100)$$

$$= 10{,}500 - 2500 = \$8000$$

5.

(a) For a linear cost function,

$C(x) = mx + b.$

m = marginal cost = \$0.40 per copy

b = fixed cost = \$70

Thus, the cost function is

$C(x) = 0.4x + 70.$

The revenue function is

$R(x) = 0.50x$ (x copies @ 50¢ per copy)

The profit function is

$$P(x) = R(x) - C(x)$$

$$= 0.5x - (0.4x + 70)$$

$$= 0.5x - 0.4x - 70$$

$$= 0.1x - 70$$

(b) $P(500) = 0.1(500) - 70$

$= 50-70 = -20$

Since P is negative, this represents a loss of \$20.

(c) For break-even,

$P(x) = 0$

$0.1x - 70 = 0$

$0.1x = 70$

$$x = \frac{70}{0.1} = 700 \text{ copies}$$

7. A linear demand function has the form

$q = mp + b.$

(x is the price p, and y is the demand q). We are given two points on its graph: (1, 1960) and (5, 1800).

Slope:

$$m = \frac{q_2 - q_1}{p_2 - p_1} = \frac{1800-1960}{5 - 1} = \frac{-160}{4} = -40$$

Intercept:

$$b = q_1 - mp_1 = 1960 - (-40)(1)$$

$$= 1960 + 40 = 2000$$

Thus, the demand equation is

$q = mp + b$

$q = -40p + 2000$

9. (a) A linear demand function has the form

$q = mp + b.$

(x is the price p, and y is the demand q). We are given two points on its graph:

2004 second quarter data: (111, 45.4)

2004 fourth quarter data: (105, 51.4)

Slope:

$$m = \frac{q_2 - q_1}{p_2 - p_1} = \frac{51.4 - 45.4}{105 - 111} = \frac{6}{-6} = -1$$

Intercept:

$$b = q_1 - mp_1 = 45.4 - (-1)(111) = 156.4$$

Thus, the demand equation is

$q = mp + b$

$q = -p + 156.4$

If $p = \$103$, then

$q = -p + 156.4 = -103 + 156.4$

$= 53.4$ million phones

(b) Since the slope is −1 million phones per unit increase in price, we interpret of the slope as follows: For every __$1__ increase in price, sales of cellphones decrease by __1 million__ units.

11.

(a) Demand Function: The given points are

$(p, q) = (1, 90)$ and $(2, 30)$

Slope:

$$m = \frac{q_2 - q_1}{p_2 - p_1} = \frac{30 - 90}{2 - 1} = -60$$

Intercept:

$b = q_1 - mp_1 = 90 - (-60(1) = 150$

Thus, the demand equation is

$q = mp + b$

$q = -60p + 150$

Supply Function: The given points are

$(p, q) = (1, 20)$ and $(2, 100)$

Slope:

$$m = \frac{q_2 - q_1}{p_2 - p_1} = \frac{100 - 20}{2 - 1} = 80$$

Intercept:

$b = q_1 - mp_1 = 20 - (80)1 = -60$

Thus, the supply equation is

$q = mp + b$

$q = 80p - 60$

(b) For equilibrium,

Supply = Demand

$80p - 60 = -60p + 150$

$140p = 210$

$$p = \frac{210}{140} = 1.5$$

Thus, the chias should be marked at \$1.50 each.

13.

(a) Here is the given graph with successive points joined by line segments.

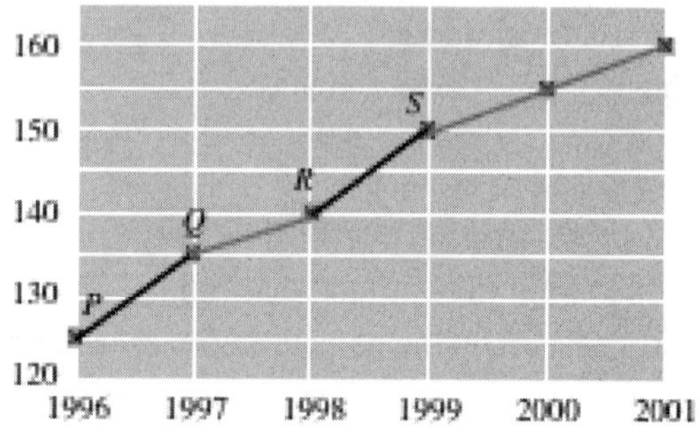

The segments PQ and RS are the steepest (they have the largest Δy ($\Delta y = 10$) for the fixed $\Delta x = 1$), and hence have the largest slope ($m = \Delta y/\Delta x = 10$). (You can check that all line segments joining non-successive pairs of points—for example P and R—have smaller slopes.) Thus, we have two possible answers:

PQ: Points $P(1996, 125)$ and $Q(1997, 135)$

RS: Points $R(1998, 140)$ and $(1999, 150)$.

(b) Consulting the textbook, we find that the slope m measures the rate of change of the number of in-ground swimming pools, and is measured in thousands of pools per year (units of y per unit of x). Thus, the number of new in-ground pools increased most rapidly during the periods 1996–1997 and 1998–1999, when it rose by 10,000 new pools in a year.

15. We are asked for a linear model of N as a function of t. That is, $N = mt + b$. (N is playing the role of y and t is playing the role of x.) The two points we are given are

$(t, N) = (-1, 350)$ $t = -1$ represents 1999

and

$(t, N) = (1, 450)$ $t = 1$ represents 2001

(Notice that we are also thinking of N in millions of transactions.)

Slope:

$$m = \frac{N_2 - N_1}{t_2 - t_1} = \frac{450 - 350}{1 - (-1)} = \frac{100}{2} = 50$$

Intercept:

$$b = N_1 - mt_1 = 350 - (50)(-1) = 400$$

Thus, the linear model is

$N = mt + b$

$N = 50t + 400$ million transactions

Consulting the textbook, we find that the slope m measures the rate of change of the number of online shopping transactions, and is measured in millions of transactions per year (units of N per unit of t).

17.

(a) We are asked for a linear model of s as a function of t. That is, $s = mt + b$. The two points we are given are

$(t, s) = (0, 240)$ $\qquad t = 0$ represents 2000

and

$(t, s) = (25, 600)$ $\qquad t = 1$ represents 2001

(Notice that we are also thinking of s in billions of dollars.)

Slope:

$$m = \frac{s_2 - s_1}{t_2 - t_1} = \frac{600 - 240}{25 - 0} = \frac{360}{25} = 14.4$$

Intercept:

$b = 240$ (The point $(0, 240)$ gives us the y-intercept.)

Thus, the linear model is

$s = mt + b$

$s = 14.4t + 240$ billion dollars.

The quantity s increases by $m = 14.4$ for every one-unit increase in t. Thus, Medicare spending is predicted to rise at a rate of \$14.4 billion per year.

(b) In 2040, $t = 40$, so

$$s(40) = 14.4(40) + 240 = 576+240 = 816.$$

Thus, Medicare spending in 2040 will be \$816 billion.

19. $s(t) = 2.5t + 10$

(a) Velocity = slope = 2.5 feet/sec.

(b) After 4 seconds, $t = 4$, so

$$s(4) = 2.5(4) + 10 = 10+10 = 20$$

Thus the model train has moved 20 feet along the track.

(c) The train will be 25 feet along the track when $s = 25$. Substituting gives

$$25 = 2.5t + 10$$

Solving for time t gives

$$2.5t = 25-10 = 15$$

$$t = \frac{15}{2.5} = 6 \text{ seconds}$$

21.

(a) Take s to be displacement from Jones Beach, and t to be time in hours. We are given two points

$(t, s) = (10, 0)$ $\qquad s = 0$ for Jones Beach.

$(t, s) = (10.1, 13)$ $\qquad$ 6 minutes = 0.1 hours

We are asked for the speed, which equals the magnitude of the slope.

$$m = \frac{s_2 - s_1}{t_2 - t_1} = \frac{13 - 0}{10.1 - 10} = \frac{13}{0.1} = 130$$

Units of slope = units of s per unit of t

= miles per hour

Thus, the police car was traveling at 130 mph.

(b) For the displacement from Jones Beach at time t, we want to express s as a linear function of t; namely, $s = mt + b$. We already know $m = 130$ from part (a). For the intercept, use

$$b = s_1 - mt_1 = 0 - 130(10) = -1300$$

Therefore, the displacement at time t is

$s = mt + b$

$s = 130t - 1300$

23. F = Fahrenheit temperature,

C = Celsius temperature,

and we want F as a linear function of C. That is,

$F = mC + b$

(F plays the role of y and C plays the role of x.)

We are given two points:

$(C, F) = (0. 32)$ Freezing point

$(C, F) = (100. 212)$ Boiling point

Slope:

$$m = \frac{F_2 - F_1}{C_2 - C_1} = \frac{212 - 32}{100 - 0} = \frac{180}{100} = 1.8$$

Intercept:

$b = F_1 - mC_1 = 32 - 1.8(0) = 32.$

Thus, the linear relation is

$F = mC + b$

$F = 1.8C + 32$

When $C = 30°$

$F = 1.8(30) + 32 = 54 + 32 = 86°$

When $C = 22°$

$F = 1.8(22) + 32 = 39.6 + 32 = 71.6°$

Rounding to the nearest degree gives 72°F.

When $C = -10°$

$F = 1.8(-10) + 32 = -18 + 32 = 14°$

When $C = -14°$

$F = 1.8(-14) + 32 = -25.2 + 32 = 6.8°$

Rounding to the nearest degree gives 7°F.

25. Income = royalties + screen rights

I = 5% of net profits + 50,000

$I = 0.05N + 50,000$ Equation notation

$I(N) = 0.05N + 50,000$ Function notation

For an income of $100,000,

$100,000 = 0.05N + 50,000$

$0.05N = 50,000$

$$N = \frac{50,000}{0.05} = \$1,000,000$$

Her marginal income is her increase in income per $1 increase in net profit. This is the slope, m = 0.05 dollars of income per dollar of net profit, or 5¢ per dollar of net profit.

27. We want w as a linear function of n:

$w = mn + b$

Thus, w plays the role of y and n plays the role of x. Here is the (milk) data listed in the customary way ($x = n$ first, and $y = w$ second).

n	57	59
w	56	60

Slope:

$$m = \frac{w_2 - w_1}{n_2 - n_1} = \frac{60-56}{59-57} = \frac{4}{2} = 2$$

Intercept:

$b = w_1 - mn_1 = 56-(2)(57)$

$= 56-114 = -58$

Thus, the linear function is

$w = 2n - 58.$

For the second part of the question, we are told that $n = 50$. Thus,

$w = 2(50) - 58 = 100-58$

= 42 billion pounds of milk

29. We want c (cheese production) as a linear function of m (milk production) in the western states.

$c = km + b$ We are using k for the slope

Thus, c plays the role of y and m plays the role of x. Here is the (western states) data listed in the customary way ($x = m$ first, and $y = c$ second).

m	56	60
c	2.7	3.0

Slope:

$$k = \frac{c_2 - c_1}{m_2 - m_1} = \frac{3.0 - 2.7}{60 - 56} = \frac{0.3}{4} = 0.075$$

$b = c_1 - km_1 = 2.7-(0.075)(56)$

$= 2.7-4.2 = -1.5$

Thus, the linear equation is

$c = km + b$

$c = 0.075m - 1.5$

For the second part of the question, we want the number of pounds of cheese produced for every

10 pounds of milk. The slope $k = 0.075$ gives us the number of pounds of cheese produced per (additional) *one* pound of milk. Multiplying this by 10 gives 0.75 pounds of cheese per 10 pounds of milk.

31. We want the temperature T as a linear function of the rate r of chirping. That is,

$$T(r) = mr + b.$$

Thus, T plays the role of y and r plays the role of x. We are given two points:

$$(r, T) = (140, 80) \text{ and } (120, 75)$$

Slope:

$$m = \frac{T_2 - T_1}{r_2 - r_1} = \frac{75-80}{120-140} = \frac{-5}{-20} = \frac{1}{4}$$

Intercept:

$$b = T_1 - mr_1 = 80 - \tfrac{1}{4}(140) = 80-35 = 45$$

Thus, the linear function is

$$T(r) = mr + b$$
$$T(r) = \tfrac{1}{4}r + 45$$

When the chirping rate is 100 chirps per minute, $r = 100$, and so the temperature is

$$T(100) = \tfrac{1}{4}(100) + 45 = 25+45 = 70°\text{F}$$

33. The hourly profit function is given by

Profit = Revenue – Cost

$$P(x) = R(x) - C(x)$$

(Hourly) cost function: This is a fixed cost of \$5132 only:

$$C(x) = 5132$$

(Hourly) revenue function: This is a variable of \$100 per passenger cost only:

$$R(x) = 100x$$

Thus, the profit function is

$$P(x) = R(x) - C(x)$$
$$P(x) = 100x - 5132$$

For the domain of $P(x)$, the number of passengers x cannot exceed the capacity: 405. Also, x cannot be negative. Thus, the domain is given by $0 \le x \le 405$, or [0. 405].

For break-even, $P(x) = 0$

$$100x - 5132 = 0$$
$$100x = 5132, \text{ or } x = \frac{5132}{100} = 51.32$$

If x is larger than this, then the profit function is positive, and so there should be at least 52 passengers; $x \ge 52$, for a profit.

35. To compute the break-even point, we use the profit function:

Profit = Revenue – Cost

$$P(x) = R(x) - C(x)$$

$R(x) = 2x$ \$2 per unit

$$\begin{aligned} C(x) &= \text{Variable Cost} + \text{Fixed Cost} \\ &= 40\% \text{ of Revenue} + 6000 \\ &= 0.4(2x) + 6000 \\ &= 0.8x + 6000 \end{aligned}$$

Thus,

$$P(x) = R(x) - C(x)$$
$$P(x) = 2x - (0.8x + 6000)$$
$$P(x) = 1.2x - 6000$$

For break-even, $P(x) = 0$

$$1.2x - 6000 = 0$$
$$1.2x = 6000$$
$$x = \frac{6000}{1.2} = 5000$$

Therefore, 5000 units should be made to break even.

37. To compute the break-even point, we use the revenue and cost functions:

$$\begin{aligned} R(x) &= \text{Selling price} \times \text{Number of units} \\ &= SPx \end{aligned}$$

$$\begin{aligned} C(x) &= \text{Variable Cost} + \text{Fixed Cost} \\ &= VCx + FC \end{aligned}$$

(Note that “variable cost per unit” is marginal cost.) For break-even

$R(x) = C(x)$

$SPx = VCx + FC$

$SPx - VCx = FC$

$x(SP - VC) = FC$

$x = \frac{FC}{SP - VC}$

39. Take x to be the number of grams of perfume he buys and sells. The profit function is given by

Profit = Revenue – Cost

$P(x) = R(x) - C(x)$

Cost function $C(x)$:

Fixed costs:	20,000
Cheap perfume @ $20 per g:	$20x$
Transportation @ $30 per 100 g:	$0.3x$

Thus the cost function is

$C(x) = 20x + 0.3x + 20{,}000$

$C(x) = 20.3x + 20{,}000$

Revenue function $R(x)$

$R(x) = 600x$ $600 per gram

Thus, the profit function is

$P(x) = R(x) - C(x)$

$P(x) = 600x - (20.3x + 20{,}000)$

$P(x) = 579.7x - 20{,}000,$

with domain $x \geq 0$.

For break-even, $P(x) = 0$

$579.7x - 20{,}000 = 0$

$579.7x = 20{,}000$

$x = \frac{20{,}000}{579.7} \approx 34.50$

Thus, he should buy and sell 34.50 grams of perfume per day to break even.

41. $C(t) = \begin{cases} 0.08t + 0.6 & \text{if } 0 \leq t < 8 \\ 0.355t - 1.6 & \text{if } 8 \leq t \leq 11 \end{cases}$

million dollars.

Since 1999 is represented by $t = 9$, we use the second formula,

$C(t) = 0.355t - 1.6.$

Since $m = 0.355$ million dollars per year, or $355,000 per year, the cost of an ad was increasing by $355,000 per year.

43. The data is

t	0	5	9
y	200	50	250

(a) 1995–2000 (first two data points):

Slope: $m = \frac{y_2 - y_1}{t_2 - t_1} = \frac{50-200}{5-0} = -30$

Intercept: $b = 200$ Specified in first data point

Thus, the linear model is

$y = mt + b$

$y = -30t + 200$

(b) 2000–2004 (second and third data points):

Slope: $m = \frac{y_2 - y_1}{t_2 - t_1} = \frac{250-50}{9-5} = 50$

Intercept: $b = y_1 - mt_1 = 50 - 50(5) = -200$

Thus, the linear model is

$y = mt + b$

$y = 50t - 200$

(c) Since the first model is valid for $0 \leq t \leq 5$ and the second one for $5 \leq t \leq 9$, we put them together as

$y = \begin{cases} -30t + 200 & \text{if } 0 \leq t \leq 5 \\ 50t - 200 & \text{if } 5 < t \leq 9 \end{cases}$

Notice that, since both formulas agree at $t = 5$, we can also say

$y = \begin{cases} -30t + 200 & \text{if } 0 \leq t < 5 \\ 50t - 200 & \text{if } 5 \leq t \leq 9 \end{cases}$.

(d) Since 2002 is represented by $t = 7$, we use the second formula to obtain

$y = 50(7) - 200 = 150$

45.

1989–1994 ($0 \leq t \leq 5$)

Points: $(t, C) = (0, 30{,}000)$ 1989 data

$(t, C) = (5, 23{,}000)$ 1994 data

Slope: $m = \dfrac{C_2 - C_1}{t_2 - t_1} = \dfrac{23{,}000 - 30{,}000}{5 - 0}$

$= \dfrac{-7000}{5} = -1400$

Intercept: $b = 30{,}000$

Thus, the linear model is

$C = mt + b$

$C = -1400t + 30{,}000$

1994–1999 ($5 \le t \le 10$)

Points: $(t, C) = (5, 23{,}000)$ 1994 data

$(t, C) = (10, 60{,}000)$ 1999 data

Slope: $m = \dfrac{C_2 - C_1}{t_2 - t_1} = \dfrac{60{,}000 - 23{,}000}{10 - 5}$

$= \dfrac{37{,}000}{5} = 7400$

Intercept: $b = C_1 - mt_1 = 23{,}000 - 7400(5)$

$= 23{,}000 - 37{,}000 = 14{,}000$

Thus, the linear model is

$C = mt + b$

$C = 7400t + 14{,}000$

Putting them together gives

$$C(t) = \begin{cases} -1400t + 30{,}000 & \text{if } 0 \le t \le 5 \\ 7400t - 14{,}000 & \text{if } 5 < t \le 10 \end{cases}$$

Since 1992 corresponds to $t = 3$, we use the first formula to obtain

$C(3) = -1400(3) + 30{,}000$

$= -4200 + 30{,}000 = 25{,}800$ students

47. We want d as a piecewise-linear function of r, using the three points (r, d) given:

$(r, d) = (1.3, 22), (1.6, 35)$, and $(1.1, 30)$

Caution: These points are not given in ascending order of r. We rearrange them in increasing order of r:

$(r, d) = (1.1, 30), (1.3, 22)$, and $(1.6, 35)$

First pair of points:

$(r, d) = (1.1, 30)$ and $(1.3, 22)$

Slope $m = \dfrac{d_2 - d_1}{r_2 - r_1} = \dfrac{22 - 30}{1.3 - 1.1} = \dfrac{-8}{0.2} = -40$

Intercept: $b = d_1 - mr_1 = 30 - (-40)1.1$

$= 30 + 44 = 74$

Thus, when $1.1 \le r \le 1.3$, the linear model is

$d = mr + b$

$d = -40 + 74$

Second pair of points:

$(r, d) = (1.3, 22)$ and $(1.6, 35)$

Slope $m = \dfrac{d_2 - d_1}{r_2 - r_1} = \dfrac{35 - 22}{1.6 - 1.3} = \dfrac{13}{0.3} = \dfrac{130}{3}$

Intercept: $b = d_1 - mr_1 = 22 - \dfrac{130}{3}(1.3)$

$= 22 - \dfrac{169}{3} = -\dfrac{103}{3}$

Thus, when $1.3 \le r \le 1.6$, the linear model is

$d = mr + b$

$d = \dfrac{130}{3}r - \dfrac{103}{3}$

Putting the two linear models together gives

$$d(r) = \begin{cases} -40r + 74 & \text{if } 1.1 \le r \le 1.3 \\ \dfrac{130r}{3} - \dfrac{103}{3} & \text{if } 1.3 < r \le 1.6 \end{cases}$$

When there are the same number of available men as women, the ratio r of available men to women is 1, so we are asked to calculate $d(1)$. Since $1 \le 1.3$, we extrapolate the first formula:

$d(1) = -40(1) + 74 = 34\%$

Thus, the divorce rate is 34%.

49. The units of the slope m are units of y (bootlags) per unit of x (zonars). The intercept b is on the y-axis, and is thus measured in units of y (bootlags). Thus, m is measured in bootlags per zonar and b is measured in bootlags.

51. If a quantity changes linearly with time, it must change by the same amount for every unit change in time. Thus, since it increases by 10 units in the first day, it must increase by 10 units each day, including the third.

53. $v = 0.1t + 20$ m/sec
Since the slope is 0.1, the velocity is increasing at a rate of 0.1 m/sec every second. Since the velocity is increasing, the object is accelerating (choice B).

55. Increasing the number of items from the breakeven results in a profit: Because the slope of the revenue graph is larger than the slope of the cost graph, it is higher than the cost graph to the right of the point of intersection, and hence corresponds to a profit.

1.5

1. (1, 1), (2, 2), (3, 4) ; $y = x-1$

x	y	Predicted $\hat{y} = x-1$	Residual $y - \hat{y}$	Residual^2 $(y - \hat{y})^2$
1	1	0	1	1
2	2	1	1	1
3	4	2	2	4

SSE = Sum of squares of residuals

= 4+1+1 = 6

3. (0,−1), (1,3), (4,6), (5,0); $y = -x+2$

x	y	Predicted $\hat{y} = -x+2$	Residual $y - \hat{y}$	Residual^2 $(y - \hat{y})^2$
0	−1	2	−3	9
1	3	1	2	4
4	6	−2	8	64
5	0	−3	3	9

SSE = Sum of squares of residuals

= 9 + 4 + 64 + 9 = 86

5. (1, 1), (2, 2), (3, 4)

(a) $y = 1.5x-1$

x	y	$\hat{y}$	$y - \hat{y}$	$(y - \hat{y})^2$
1	1	0.5	0.5	0.25
2	2	2	0	0
3	4	3.5	0.5	0.25

SSE = Sum of squares of residuals = 0.5

(b) $y = 2x - 1.5$

x	y	$\hat{y}$	$y - \hat{y}$	$(y - \hat{y})^2$
1	1	0.5	0.5	0.25
2	2	2.5	−0.5	0.25
3	4	4.5	−0.5	0.25

SSE = Sum of squares of residuals = 0.75

The model that gives the better fit is (a) because it gives the smaller value of SSE.

7. (0, −1), (1, 3), (4, 6), (5, 0)

(a) $y = 0.3x + 1.1$

x	y	$\hat{y}$	$y - \hat{y}$	$(y - \hat{y})^2$
0	−1	1.1	−2.1	4.41
1	3	1.4	1.6	2.56
4	6	2.3	3.7	13.69
5	0	2.6	−2.6	6.76

SSE = Sum of squares of residuals = 27.42

(b) $y = 0.4x+0.9$

x	y	$\hat{y}$	$y - \hat{y}$	$(y - \hat{y})^2$
0	−1	0.9	−1.9	3.61
1	3	1.3	1.7	2.89
4	6	2.5	3.5	12.25
5	0	2.9	−2.9	8.41

SSE = Sum of squares of residuals = 27.16

The model that gives the better fit is (b) because it gives the smaller value of SSE.

9. (1,1), (2,2), (3,4)

	x	y	xy	x^2
	1	1	1	1
	2	2	4	4
	3	4	12	9
Σ (Sum)	6	7	17	14

$n = 3$ (number of data points)

Slope: $m = \dfrac{n(\Sigma xy) - (\Sigma x)(\Sigma y)}{n(\Sigma x^2) - (\Sigma x)^2}$

$= \dfrac{3(17) - (6)(7)}{3(14) - 6^2} = \dfrac{9}{6} = 1.5$

Intercept: $b = \dfrac{\Sigma y - m(\Sigma x)}{n}$

$= \dfrac{7 - 1.5(6)}{3} = -\dfrac{2}{3} \approx -0.6667$

Thus, the regression line is

$y = mx + b$

$y = 1.5x - 0.6667$

Graph:

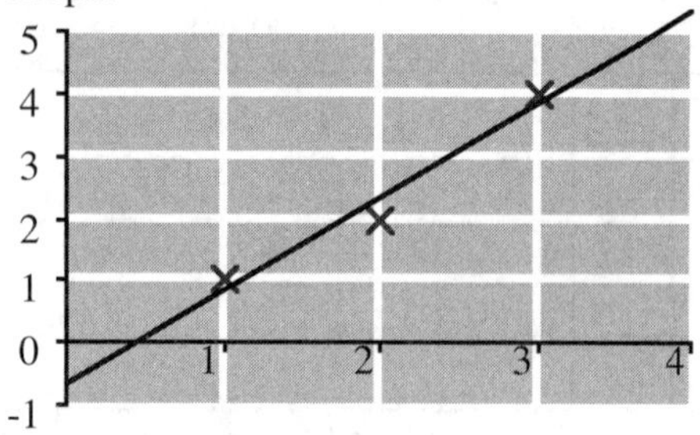

11. (0, −1), (1 ,3), (4, 6), (5, 0)

	x	y	xy	x^2
	0	−1	0	0
	1	3	3	1
	4	6	24	16
	5	0	0	25
Σ (Sum)	10	8	27	42

$n = 4$ (number of data points)

Slope: $m = \dfrac{n(\Sigma xy) - (\Sigma x)(\Sigma y)}{n(\Sigma x^2) - (\Sigma x)^2}$

$= \dfrac{4(27) - (10)(8)}{4(42) - 10^2} = \dfrac{28}{68} \approx 0.4118$

Intercept: $b = \dfrac{\Sigma y - m(\Sigma x)}{n}$

$= \dfrac{8 - \left(\frac{28}{68}\right)(10)}{4} \approx 0.9706$

Thus, the regression line is

$y = mx + b$

$y = 0.4118x + 0.9706$

Graph:

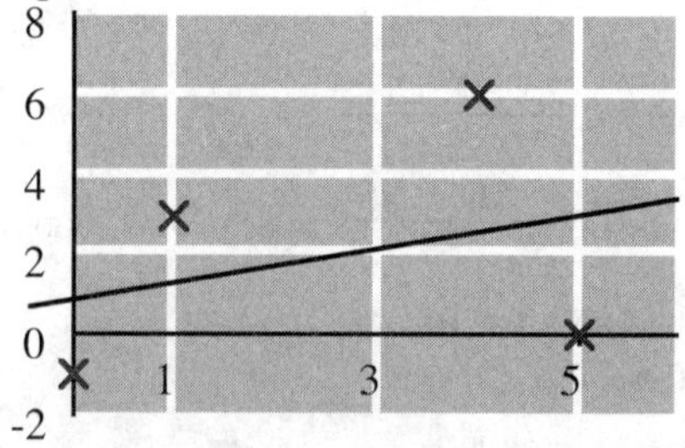

13. (a) {(1, 3), (2, 4), (5, 6)}

	x	y	xy	x^2	y^2
	1	3	3	1	9
	2	4	8	4	16
	5	6	30	25	36
Σ	8	13	41	30	61

$n = 3$ (number of data points)

$r = \dfrac{n(\Sigma xy) - (\Sigma x)(\Sigma y)}{\sqrt{n(\Sigma x^2) - (\Sigma x)^2} \cdot \sqrt{n(\Sigma y^2) - (\Sigma y)^2}}$

$= \dfrac{3(41) - (8)(13)}{\sqrt{3(30) - (8)^2}\sqrt{3(61) - (13)^2}}$

$\approx \dfrac{19}{19.078784} \approx 0.9959$

(b) {(0, −1), (2, 1), (3, 4)}

	x	y	xy	x^2	y^2
	0	−1	0	0	1
	2	1	2	4	1
	3	4	12	9	16
Σ	5	4	14	13	18

$n = 3$ (number of data points)

$r = \dfrac{n(\Sigma xy) - (\Sigma x)(\Sigma y)}{\sqrt{n(\Sigma x^2) - (\Sigma x)^2} \cdot \sqrt{n(\Sigma y^2) - (\Sigma y)^2}}$

$= \dfrac{3(14) - (5)(4)}{\sqrt{3(13) - (5)^2}\sqrt{3(18) - (4)^2}}$

$\approx \dfrac{22}{23.0651252} \approx 0.9538$

(c) {(4, −3), (5, 5), (0, 0)}

	x	y	xy	x^2	y^2
	4	−3	−12	16	9
	5	5	25	25	25
	0	0	0	0	0
Σ	9	2	13	41	34

$n = 3$ (number of data points)

$$r = \frac{n(\Sigma xy) - (\Sigma x)(\Sigma y)}{\sqrt{n(\Sigma x^2) - (\Sigma x)^2} \cdot \sqrt{n(\Sigma y^2) - (\Sigma y)^2}}$$
$$= \frac{3(13) - (9)(2)}{\sqrt{3(41)-(9)^2}\sqrt{3(34)-(2)^2}}$$
$$\approx \frac{21}{64.1560597} \approx 0.3273$$

The value of r in part (a) has the largest absolute value. Therefore, the regression line for the data in part (a) is the best fit.

The value of r in part (c) has the smallest absolute value. Therefore, the regression line for the data in part (c) is the worst fit.

Since r is not ±1 for any of these lines, none of them is a perfect fit.

15. The entries in the xy column are obtained by multiplying the entries in the x column by the corresponding entries in the y column. The entries in the x^2 column are the squares of the entries in the x column.

	x	y	xy	x^2
	3	500	1500	9
	5	600	3000	25
	7	800	5600	49
Σ (Sum)	15	1900	10100	83

$n = 3$ (number of data points)

Slope: $m = \frac{n(\Sigma xy) - (\Sigma x)(\Sigma y)}{n(\Sigma x^2) - (\Sigma x)^2}$
$= \frac{3(10100) - (15)(1900)}{3(83)-(15)^2}$
$= \frac{1800}{24} = 75$

Intercept: $b = \frac{\Sigma y - m(\Sigma x)}{n}$
$= \frac{1900 - (75)(15)}{3}$
$= \frac{775}{3} \approx 258.33$ (to 2 decimal places)

Thus, the regression line is

$y = mx + b$

$y = 75x + 258.33$

To estimate the 2008 sales we put $x = 8$:

$y = 75(8) + 258.33 \approx 858.33$ million

17. Calculation of the regression line:

	x	y	xy	x^2
	0	6	0	0
	2	10	20	4
	4	16	64	16
Σ (Sum)	6	32	84	20

$n = 3$ (number of data points)

Slope: $m = \frac{n(\Sigma xy) - (\Sigma x)(\Sigma y)}{n(\Sigma x^2) - (\Sigma x)^2}$
$= \frac{3(84) - (6)(32)}{3(20) - (6)^2}$
$= \frac{60}{24} = 2.5$

Intercept: $b = \frac{\Sigma y - m(\Sigma x)}{n}$
$= \frac{32 - (2.5)(6)}{3}$
$= \frac{17}{3} \approx 5.67$ (to 2 decimal places)

Thus, the regression line is

$y = mt + b$ Independent variable is called t

$y = 2.5t + 5.67$

$y(3) \approx 2.5(3) + 5.67 = \$13.17$ billion

19. Calculation of the regression line:

	x	y	xy	x^2
	20	3	60	400
	40	6	240	1,600
	80	9	720	6,400
	100	15	1500	10,000
Σ (Sum)	240	33	2520	18,400

$n = 4$ (number of data points)

$$\text{Slope: } m = \frac{n(\Sigma xy) - (\Sigma x)(\Sigma y)}{n(\Sigma x^2) - (\Sigma x)^2}$$
$$= \frac{4(2520) - (240)(33)}{4(18400) - (240)^2}$$
$$= \frac{2160}{16{,}000} = 0.135$$
$$\text{Intercept: } b = \frac{\Sigma y - m(\Sigma x)}{n}$$
$$= \frac{33 - (0.135)(240)}{4}$$
$$= \frac{0.6}{4} = 0.15$$

Thus, the regression line is

$y = mx + b$

$y = 0.135x + 0.15$

$y(50) = 0.135(50) + 0.15$

$= 6.9$ million jobs

21. (a) Since production is a function of cultivated area, we take x as cultivated area, and y as production:

x	25	30	32	40	52
y	15	25	30	40	60

Using the method of Example 3, we obtain the following regression line and plot (coefficients rounded to two decimal places):

$y = 1.62x - 23.87$

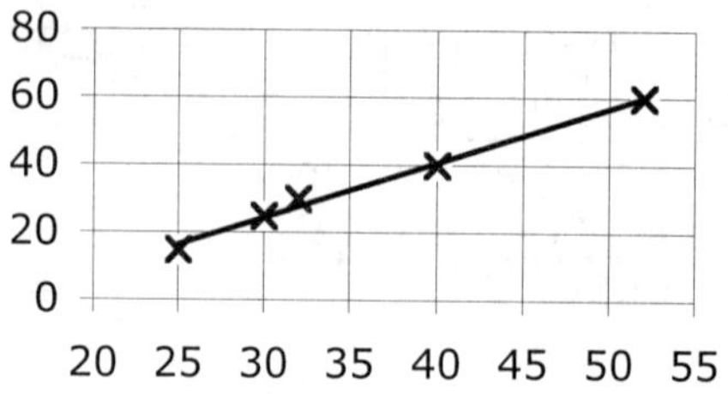

(b) To interpret the slope $m = 1.62$, recall that units of m are units of y per unit of x; That is, millions of tons of production of soybeans per million acres of cultivated land. Thus, production increases by 1.62 million tons of soybeans per million acres of cultivated land. More simply, each acre of cultivated land produces about 1.62 tons of soybeans.

23. (a) Using x = median household income and y = poverty rate gives the following table of values:

x	y
39	13
38	14
39	14
41	13
44	12
42	11
41	12
40	12

Using the method of Example 3, we obtain the following regression line and plot (coefficients rounded to two significant digits):

$y = -0.40x + 29$

Graph:

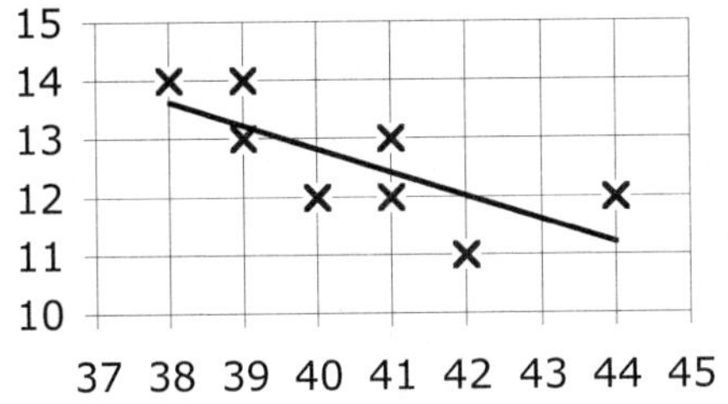

Notice from the graph that y decreases and x increases. Therefore, the graph suggests a relationship between x and y.

(b) To interpret the slope, recall that units of the slope are units of y (percentage points of the poverty rate) per unit of x (median house income in thousands of dollars). Thus,

$m = -0.40$ percentage points per \$1000 dollars,

indicating that the poverty rate declines by 0.40% for each \$1000 increase in the median household income.

(c) Using the method of Example 4, we can use technology to show the value of r^2:

$$r^2 \approx 0.5385$$

so $r = \sqrt{r^2} \approx -\sqrt{0.5385} \approx -0.7338$.

(We used the negative square root because the slope of the regression equation is negative.)

Since r is not close to 1, the correlation between x and y is not a "strong" one.

25. (a) Using the method of Example 3, we obtain the following regression line and plot (coefficients rounded to two significant digits):

$p = 0.13t + 0.22$

Graph:

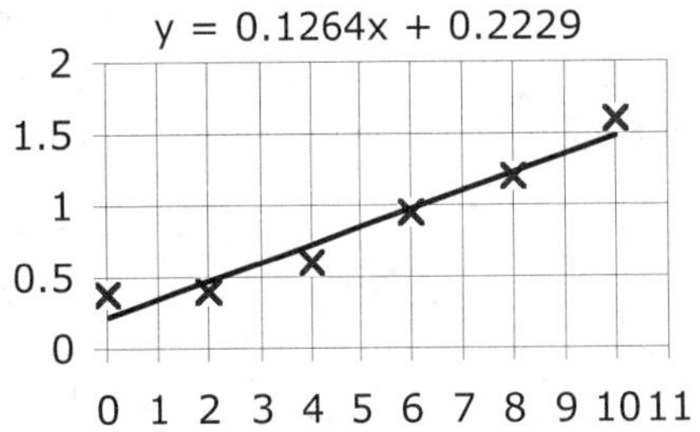

(b) The first and last points lie above the regression line, while the central points lie below it, suggesting a curve.

(c) Here is an Excel worksheet showing the computation of the residuals (based on Example 1 in the text):

	A	B	C	D
1	t	p (Observed)	p (predicted)	Residual
2	0	0.38	=0.13*A2+0.22	=B2-C2
3	2	0.4		
4	4	0.6		
5	6	0.95		
6	8	1.2		
7	10	1.6		

↓

	A	B	C	D
1	t	p (Observed)	p (predicted)	Residual
2	0	0.38	0.22	0.16
3	2	0.4	0.48	-0.08
4	4	0.6	0.74	-0.14
5	6	0.95	1	-0.05
6	8	1.2	1.26	-0.06
7	10	1.6	1.52	0.08

Notice that the residuals are positive at first, become negative, and then become positive, confirming the impression from the graph.

27. The regression line is defined to be the line that gives the lowest sum-of-squares error, SSE. If we are given two points, (a, b) and (c, d) with $a \neq c$ then there is a line that passes through these two points, giving SSE = 0. Since 0 is the smallest value possible, this line must be the regression line.

29. If the points $(x_1, y_1), (x_2, y_2), \ldots, (x_n, y_n)$ lie on a straight line, then the sum-of-squares error, SSE, for this line is zero. Since 0 is the smallest value possible, this line must be the regression line.

31. Calculation of the regression line:

	x	y	xy	x^2
	0	0	0	0
	$-a$	a	$-a^2$	a^2
	a	a	a^2	a^2
Σ	0	$2a$	0	$2a^2$

$n = 3$ (number of data points)

$$\text{Slope: } m = \frac{n(\Sigma xy) - (\Sigma x)(\Sigma y)}{n(\Sigma x^2) - (\Sigma x)^2}$$

$$= \frac{3(0) - (0)(2a)}{3(2a^2) - 0^2} = 0$$

Regression Coefficient:

$$r = \frac{n(\Sigma xy) - (\Sigma x)(\Sigma y)}{\sqrt{n(\Sigma x^2) - (\Sigma x)^2} \cdot \sqrt{n(\Sigma y^2) - (\Sigma y)^2}}$$

has the same numerator as m, and we have just seen that this numerator is zero, Hence, $r = 0$.

33. No. The regression line through $(-1, 1)$, $(0, 0)$, and $(1, 1)$ passes through none of these points.

Chapter 1 Review Exercises

1.

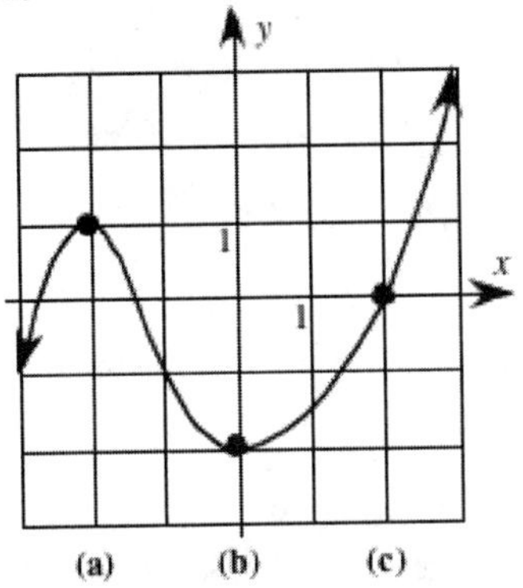

(a) 1 **(b)** −2 **(c)** 0

(d) $f(2) - f(-2) = 0 - 1 = -1$

2.

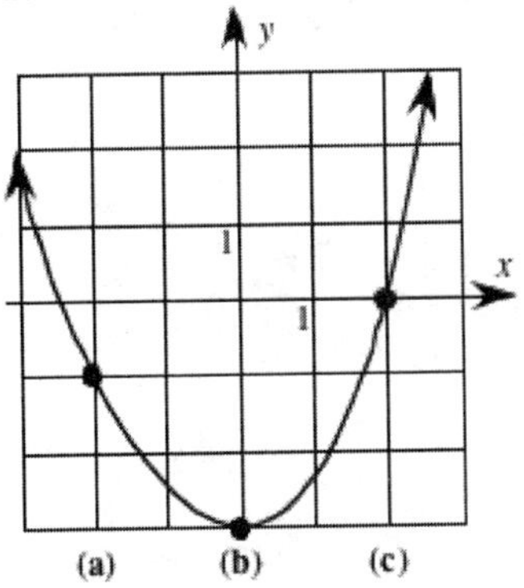

(a) −1 **(b)** −3 **(c)** 0

(d) $f(2) - f(-2) = 0 - (-1) = 1$

3.

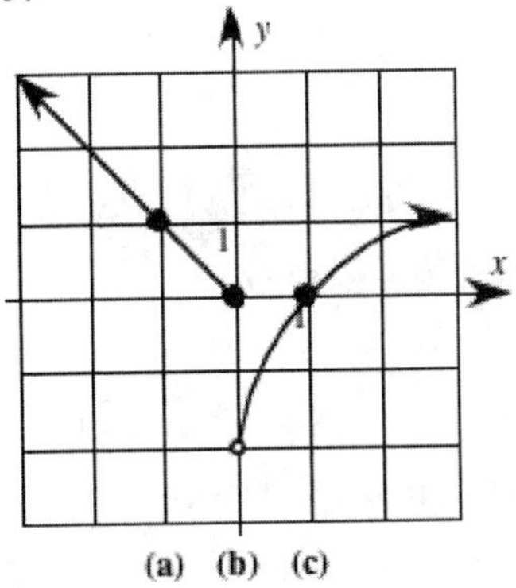

(a) 1 **(b)** 0 **(c)** 0

(d) $f(1) - f(-1) = 0 - 1 = -1$

4.

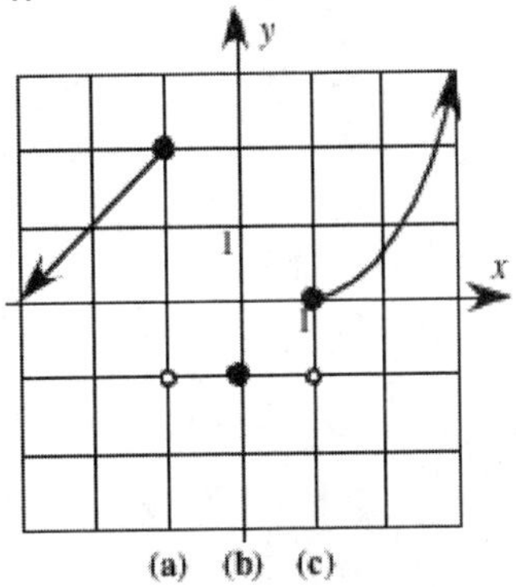

(a) 2 **(b)** −1 **(c)** 0

(d) $f(1) - f(-1) = 0 - 2 = -2$

5. $y = -2x + 5$

y−intercept = 5 slope = −2

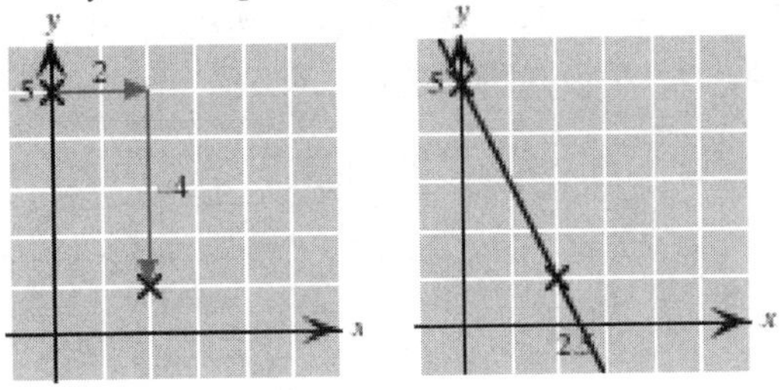

6. $2x - 3y = 12$

Solving for y gives

$$-3y = -2x + 12$$

$$y = \frac{2}{3}x - 4$$

y−intercept = −4 slope = $\frac{2}{3}$

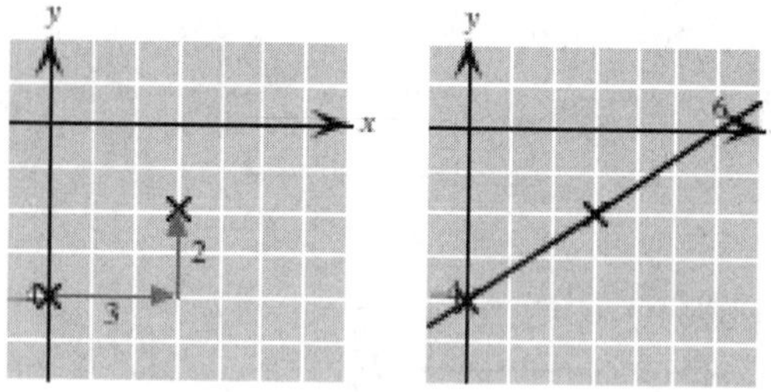

7. $y = \begin{cases} \frac{1}{2}x & \text{if } -1 \le x \le 1 \\ x - 1 & \text{if } 1 < x \le 3 \end{cases}$

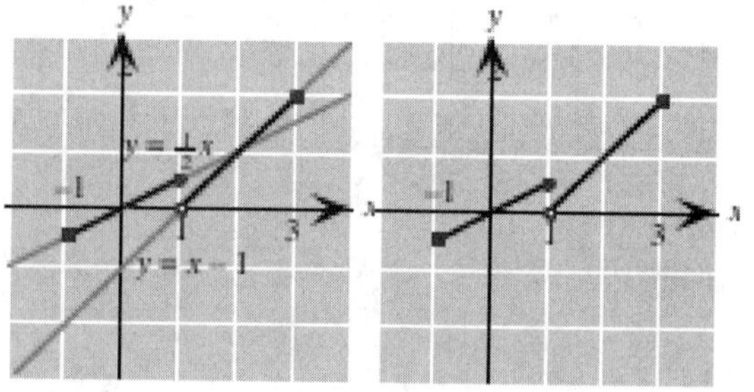

8. $f(x) = 4x - x^2$ with domain [0, 4]

Technology formula: `4*x-x^2`

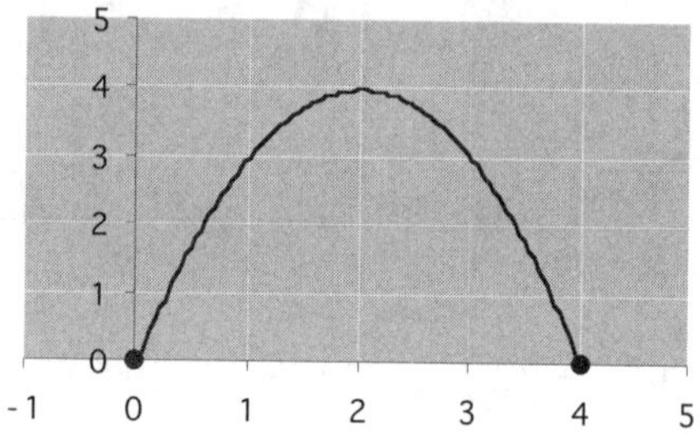

x	−2	0	1	2	4
$f(x)$	4	2	1	0	2
$g(x)$	−5	−3	−2	−1	1
$h(x)$	1.5	1	0.75	0.5	0
$k(x)$	0.25	1	2	4	16
$u(x)$	0	4	3	0	−12

Here are the graphs of the five functions, with the points connected:

9.

$f(x)$

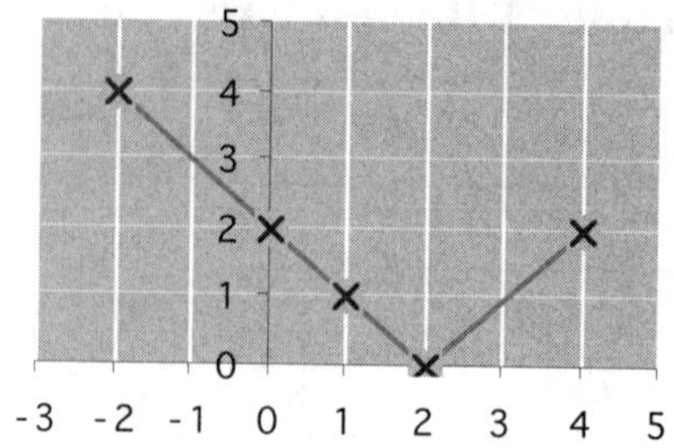

V-shape indicates **absolute value** model.

10.

$g(x)$

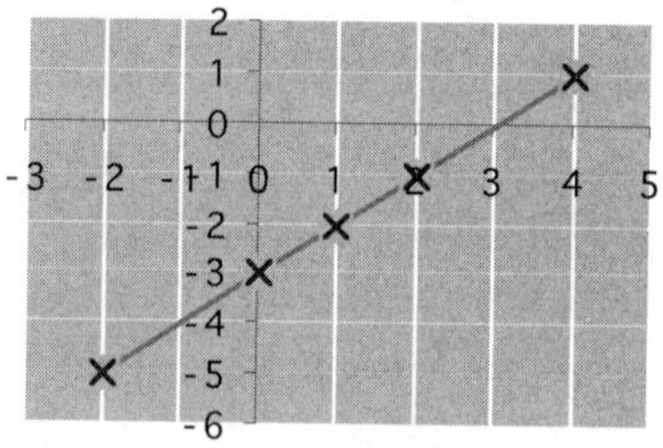

Linear

11.

$h(x)$

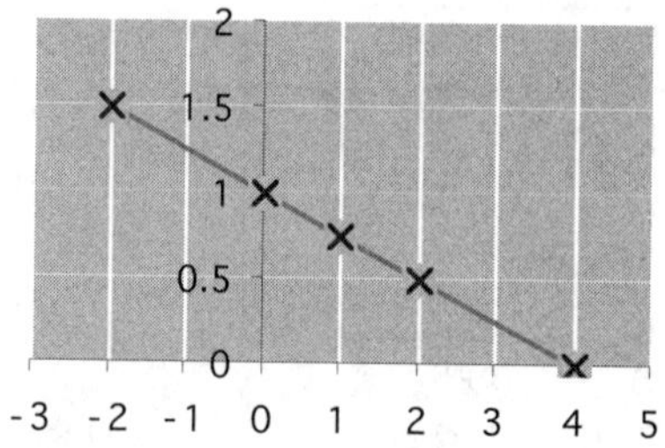

Linear

12.

$k(x)$

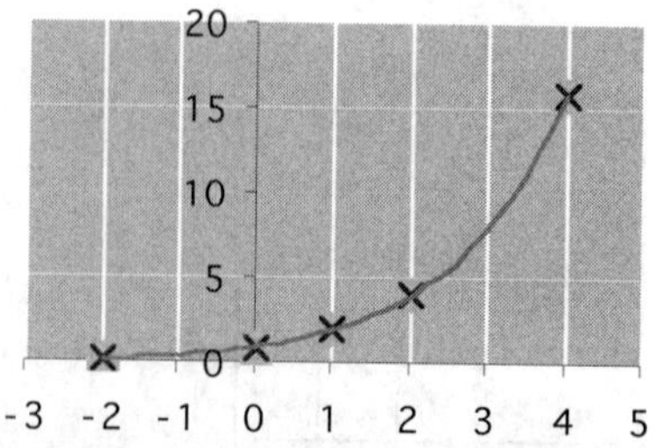

y doubles for each 1-unit increase in x, indicating an **exponential** model.

13.

$u(x)$

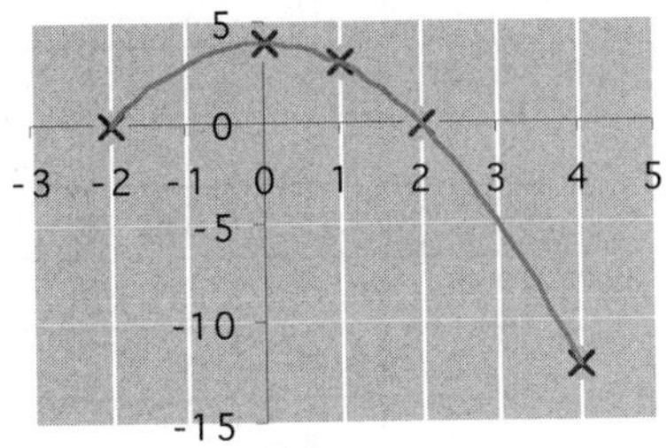

Parabolic shape of graph indicates a **quadratic** model.

14. Line through (3, 2) with slope −3

Point: (3, 2)

Slope: $m = -3$

Intercept: $b = y_1 - mx_1$

$= 2-(-3)(3) = 2+9 = 11$

Thus, the equation is

$y = mx + b$

$y = -3x + 11$

15. Through (−1, 2) and (1, 0)

Point: (1, 0)

Slope: $m = \frac{y_2 - y_1}{x_2 - x_1} = \frac{0 - 2}{1 - (-1)} = \frac{-2}{2} = -1$

Intercept: $b = y_1 - mx_1 = 0-(-1)(1) = 1$

Thus, the equation is

$y = mx + b$

$y = -x + 1$

16. Through (1, 2) parallel to $x - 2y = 2$

Point: (1, 2)

Slope: Parallel to the line $x - 2y = 2$, and so has the same slope. Solve for y to obtain

$-2y = -x - 2$

$y = \frac{1}{2}x + 1$

Thus the slope is $m = \frac{1}{2}$

Intercept: $b = y_1 - mx_1 = 2-\frac{1}{2}(1) = \frac{3}{2}$

Thus, the equation is

$y = mx + b$

$y = \frac{1}{2}x + \frac{3}{2}$

17. With slope 1/2 crossing $3x + y = 6$ at its x-intercept

Point: We need the x-intercept of $3x + y = 6$. This is given by setting $y = 0$ and solving for x:

$3x + 0 = 6$

$x = 2.$

Thus, the point is (2, 0) On the x-axis, so y=0

Slope: $m = \frac{1}{2}$

Intercept: $b = y_1 - mx_1 = 0 - \frac{1}{2}(2) = -1$

Thus, the equation is

$y = mx + b$

$y = \frac{1}{2}x - 1$

18. $y = -0.5x + 1$:

x	observed y	predicted y	residual^2
−1	1	1.5	0.25
1	1	0.5	0.25
2	0	0	0
		SSE:	0.5

$y = -x/4 + 1$:

x	observed y	predicted y	residual^2
−1	1	1.25	0.0625
1	1	0.75	0.0625
2	0	0.5	0.25
		SSE:	0.375

The second line, $y = -x/4 + 1$, is a better fit.

19. $y = x + 1$:

x	observed y	predicted y	residual^2
−2	−1	−1	0
−1	1	0	1
0	1	1	0
1	2	2	0
2	4	3	1
3	3	4	1
		SSE:	3

$y = x/2 + 1$:

x	observed y	predicted y	residual^2
−2	−1	0	1
−1	1	0.5	0.25
0	1	1	0
1	2	1.5	0.25
2	4	2	4
3	3	2.5	0.25
		SSE:	5.75

The first line, $y = x + 1$, is the better fit.

20.

	x	y	xy	x^2	y^2
	−1	1	−1	1	1
	1	2	2	1	4
	2	0	0	4	0
Σ (Sum)	2	3	1	6	5

$n = 3$ (number of data points)

$$\text{Slope: } m = \frac{n(\Sigma xy) - (\Sigma x)(\Sigma y)}{n(\Sigma x^2) - (\Sigma x)^2} = \frac{3(1) - (2)(3)}{3(6) - 2^2} = \frac{-3}{14} \approx -0.214$$

$$\text{Intercept: } b = \frac{\Sigma y - m(\Sigma x)}{n} = \frac{3 - (-0.214)(2)}{3} \approx 1.14$$

Thus, the regression line is

$y = mx + b$

$y = -0.214x + 1.14$

The correlation coefficient is

$$r = \frac{n(\Sigma xy) - (\Sigma x)(\Sigma y)}{\sqrt{n(\Sigma x^2) - (\Sigma x)^2} \cdot \sqrt{n(\Sigma y^2) - (\Sigma y)^2}} = \frac{3(1) - (2)(3)}{\sqrt{3(6)-(2)^2}\sqrt{3(5)-(3)^2}} \approx -0.33$$

21.

	x	y	xy	x^2	y^2
	−2	−1	2	4	1
	−1	1	−1	1	1
	0	1	0	0	1
	1	2	2	1	4
	2	4	8	4	16
	3	3	9	9	9
Σ (Sum)	3	10	20	19	32

$n = 6$ (number of data points)

$$\text{Slope: } m = \frac{n(\Sigma xy) - (\Sigma x)(\Sigma y)}{n(\Sigma x^2) - (\Sigma x)^2} = \frac{6(20) - (3)(10)}{6(19) - 3^2} = \frac{90}{105} \approx 0.857$$

$$\text{Intercept: } b = \frac{\Sigma y - m(\Sigma x)}{n} = \frac{10 - (0.857)(3)}{6} \approx 1.24$$

Thus, the regression line is

$y = mx + b$

$y = 0.857x + 1.24$

The correlation coefficient is

$$r = \frac{n(\Sigma xy) - (\Sigma x)(\Sigma y)}{\sqrt{n(\Sigma x^2) - (\Sigma x)^2} \cdot \sqrt{n(\Sigma y^2) - (\Sigma y)^2}} = \frac{6(20) - (3)(10)}{\sqrt{6(19)-(3)^2}\sqrt{6(32)-(10)^2}} \approx 0.92$$

22. $n(x) = \begin{cases} 0.02x & \text{if } 0 \le x \le 1000 \\ 0.025x - 5 & \text{if } 1000 < x \le 2000 \end{cases}$

(a) $n(500) = 0.02(500) = 10$ books per day

We used the first formula, since 500 is in [0, 1000].

$n(1000) = 0.02(1000) = 20$ books per day

We used the first formula, since 1000 is in [0, 1000].

$n(1500) = 0.025(1500) - 5$

$= 37.6 - 5 = 32.5$ books per day

We used the second formula, since 1500 is in (1000, 2000].

(b) The coefficient 0.025 is the slope of the second formula, indicating that, for web site traffic of more than 1000 hits per day and up to 2000 hits per day ($1000 < x \le 2000$) book sales are increasing by 0.025 books per additional hit.

(c) To sell an average of 30 books per day, we desire $n(x) = 30$. Of we try the first formula, we get

$$0.02x = 30$$

giving $x = 1500$, which is not in the domain of the first formula. So, we try the second formula:

$$0.025x - 5 = 30$$

$$0.025x = 35$$

$$x = \frac{35}{0.025} = 1400 \text{ hits,}$$

which is in the domain of the second formula. Thus, 1400 hits per day will result in average sales of 30 books per day.

23.

t	1	2	3	4	5	6
$S(t)$	12.5	37.5	62.5	72.0	74.5	75.0

(a) Technology formulas:

(A): `300/(4+100*5^(-t))`

(B): `13.3*t+8.0`

(C): `-2.3*t^2+30.0*t-3.3`

(D): `7*3^(0.5*t)`

Here are the values for the three given models (rounded to 1 decimal place):

t	1	2	3	4	5	6
(A)	12.5	37.5	62.5	72.1	74.4	74.9
(B)	21.3	34.6	47.9	61.2	74.5	87.8
(C)	24.4	47.5	66.0	79.9	89.2	93.9
(D)	12.1	21.0	36.4	63.0	109.1	189.0

Model (A) gives an almost perfect fit, whereas the other models are not even close.

(b) Looking at the table, we see the following behavior as t increases:

(A) Leveling off; (B) Rising (C) Rising (they begin to fall after 7 months, however) (D) Rising

24.

c	\$2000	\$5000
h	1900	2050

Point: (2000, 1900)

Slope: $m = \frac{h_2 - h_1}{c_2 - c_1} = \frac{2050-1900}{5000-2000}$

$= \frac{150}{3000} = 0.05$

Intercept: $b = y_1 - mx_1$

$= 1900-(0.05)(2000)$

$= 1900 - 100 = 1800$

Thus, the equation is

$$h = mc + b$$

$$h = 0.05c + 1800$$

(b) A budget of \$6000 per month for banner ads corresponds to $c = 6000$

$$h(6000) = 0.05(6000) + 1800$$

$= 300 + 1800 = 2100$ hits per day

(c) We are given $h = 2500$ and want c.

$$2500 = 0.05c + 1800$$

$$0.05c - 1500-1800 = 700$$

$$c = \frac{700}{0.05} = \$14000 \text{ per month}$$

25. $h = -0.000005c^2 + 0.085c + 1750$

(a) Currently, $c = \$6000$, so

$h = -0.000005(6000)^2+0.085(6000)+1750$

$= -180 + 510 + 1750 = 2080$ hits per day

(b) Here is a portion of the graph of h:

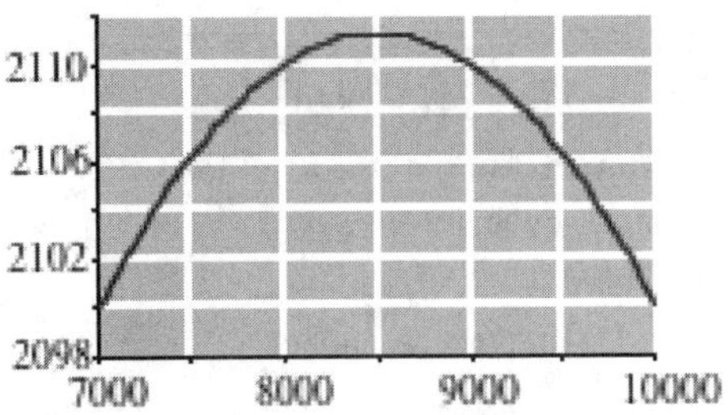

Technology formula:

`-0.000005*x^2+0.085*x+1750`

For $c = 8500$ or larger, we see that Web site traffic is projected to decrease as advertising increases, and then drop toward zero. Thus, the model does not appear to give a reasonable prediction of traffic at expenditures larger than \$8,500 per month.

26.

Cost function: $C = mx + b$

b = fixed cost = \$900 per month

m = marginal cost = \$4 per novel

Thus, the linear cost function is

$C = 4x + 900$

Revenue function: $R = mx + b$

b = fixed revenue = 0

m = marginal revenue = \$5.50 per novel

Thus, the linear revenue function is

$R = 5.50x$

Profit function: $P = R - C$

$$P = 5.50x - (4x + 900)$$
$$= 5.50x - 4x - 900$$
$$= 1.50x - 900$$

(b) For break-even, $P = 0$

$$1.50x - 900 = 0$$
$$1.50x = 900$$
$$x = \frac{900}{1.50} = 600 \text{ novels per month}$$

(c) $R = 5.00x$

$$P = 5.00x - (4x + 900)$$
$$= 5x - 4x - 900 = x - 900$$

For break-even,

$x - 900 = 0,$

$x = 900$ novels per month

27.

(a) Demand: We are given two points:

$(p, q) = (10, 350)$ and $(5.5, 620)$ Slope:

$$m = \frac{q_2 - q_1}{p_2 - p_1} = \frac{620-350}{5.5-10} = \frac{270}{-4.5} = -60$$

Intercept:

$$b = q_1 - mp_1 = 350-(-60)(10)$$
$$= 350 + 600 = 950$$

Thus, the demand equation is

$$q = mp + b$$
$$q = -60p + 950$$

(b) When $p = \$15$, the demand is

$$q = -60(15) + 950 = -900 + 950$$
$$= 50 \text{ novels per month}$$

(c) From Question 8, the cost function is

$C = 4q + 900$

We use q for the monthly sales rather than x

$= 4(-60p + 950) + 900$

We want everything expressed in terms of p, so we used the demand equation.

$= -240p + 3800 + 900$

$C = -240p + 4700$

To compute the profit in terms of price, we need the revenue as well:

$$R = pq = p(-60p + 950)$$
$$= -60p^2 + 950p$$

Profit: $P = R - C$

$$P = -60p^2 + 950p - (-240p + 4700)$$
$$= -60p^2 + 1190p - 4700$$

Now we compare profits for the three prices:

$$P(5.50) = -60(5.5)^2 + 1190(5.5) - 4700$$
$$= \$30$$
$$P(10) = -60(10)^2 + 1190(10) - 4700$$
$$= \$1200$$
$$P(15) = -60(15)^2 + 1190(15) - 4700$$
$$= -\$350 \text{ (loss)}$$

Thus, charging \$10 will result in the largest monthly profit of \$1200.

28.

p	5.50	10	12	15
q	620	350	300	100

(a) Calculation of the regression line:

	x	y	xy	x^2
	5.5	620	3410	30.25
	10	350	3500	100
	12	300	3600	144
	15	100	1500	225
Σ	42.5	1370	12010	499.25

$n = 4$ (number of data points)

Slope: $m = \dfrac{n(\Sigma xy) - (\Sigma x)(\Sigma y)}{n(\Sigma x^2) - (\Sigma x)^2}$

$= \dfrac{4(12010) - (42.5)(1370)}{4(499.25) - (42.5)^2}$

$= \dfrac{-10185}{190.75} \approx 53.3945$

Intercept: $b = \dfrac{\Sigma y - m(\Sigma x)}{n}$

$\approx \dfrac{1370 - (-53.394495)(42.5)}{4}$

$\approx \dfrac{3639.26606}{4} \approx 909.8165$

Thus, the regression equation is

$q = -53.3945p + 909.8165$

(q plays the role of y and p plays the role of x)

(b) We are given $p = \$8$, so the demand is

$q = -53.3945(8) + 909.8165$

≈ 483 novels per month (rounded)

Chapter 2

2.1

1. $f(x) = x^2 + 3x + 2$

$a = 1, b = 3, c = 2; -b/(2a) = -3/2;$

$f(-3/2) = -1/4$, so: vertex: $(-3/2, -1/4)$

y intercept $= c = 2$

$x^2 + 3x + 2 = (x + 1)(x + 2)$, so:

x intercepts: $-2, -1$

$a > 0$ so the parabola opens upward

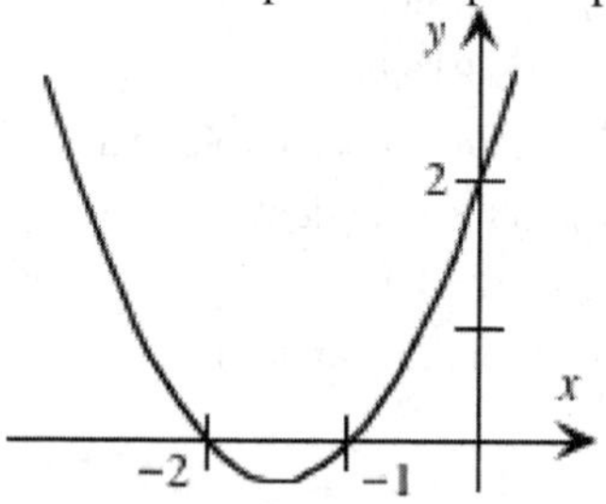

3. $f(x) = -x^2 + 4x - 4$

$a = -1, b = 4, c = -4; -b/(2a) = 2;$

$f(2) = 0$, so: vertex: $(2,0)$

y intercept $= c = -4$

$-x^2 + 4x - 4 = -(x - 2)^2$, so:

x intercept: 2

$a < 0$ so the parabola opens downward

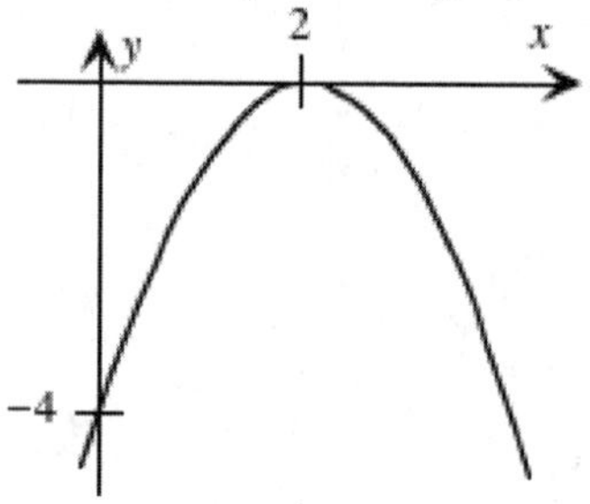

5. $f(x) = -x^2 - 40x + 500$

$a = -1, b = -40, c = 500; -b/(2a) = -20;$

$f(-20) = 900$, so: vertex: $(-20, 900)$

y intercept $= c = 500$

$-x^2 - 40x + 500 = -(x + 50)(x - 10)$, so:

x intercepts: $-50, 10$

$a < 0$ so the parabola opens downward

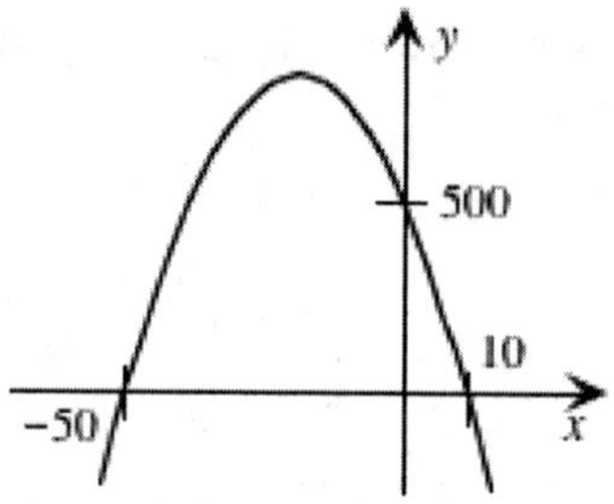

7. $f(x) = x^2 + x - 1$

$a = 1, b = 1, c = -1; -b/(2a) = -1/2;$

$f(-1/2) = -5/4$, so: vertex: $(-1/2, -5/4)$

y intercept $= c = -1$

from the quadratic formula:

x intercepts: $-1/2 \pm \sqrt{5}/2$

$a > 0$ so the parabola opens upward

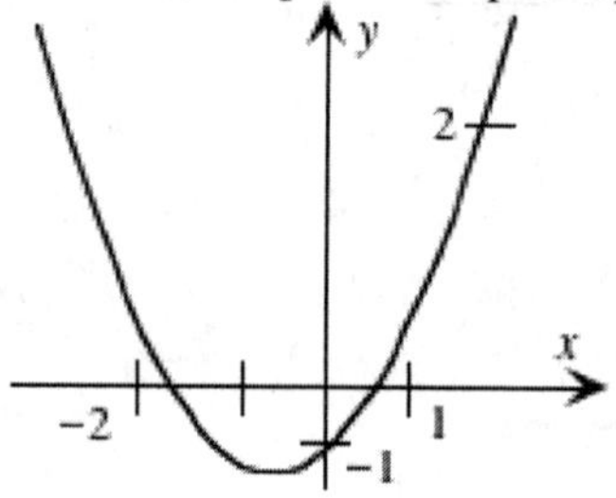

9. $f(x) = x^2 + 1$

$a = 1, b = 0, c = 1; -b/(2a) = 0;$

$f(0) = 1$, so: vertex: $(0, 1)$

y intercept $= c = 1$

$b^2 - 4ac = -4 < 0$, so no x intercept

$a > 0$ so the parabola opens upward

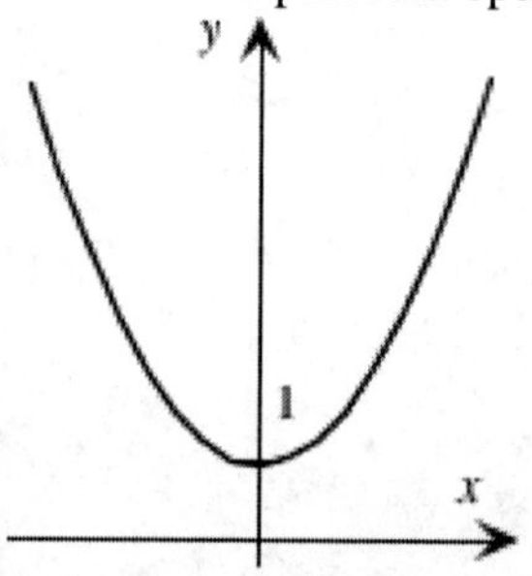

11. $q = -4p + 100$
$R = pq = p(-4p + 100) = -4p^2 + 100p$;
maximum revenue when $p = -b/(2a) = \$12.50$

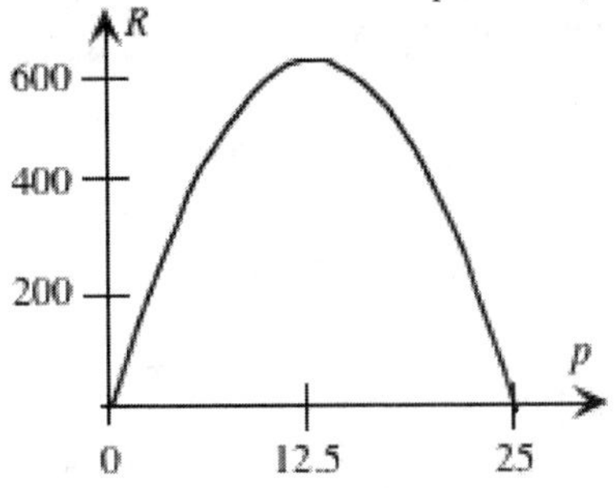

13. $q = -2p + 400$
$R = pq = p(-2p + 400) = -2p^2 + 400p$
maximum revenue when $p = -b/(2a) = \$100$

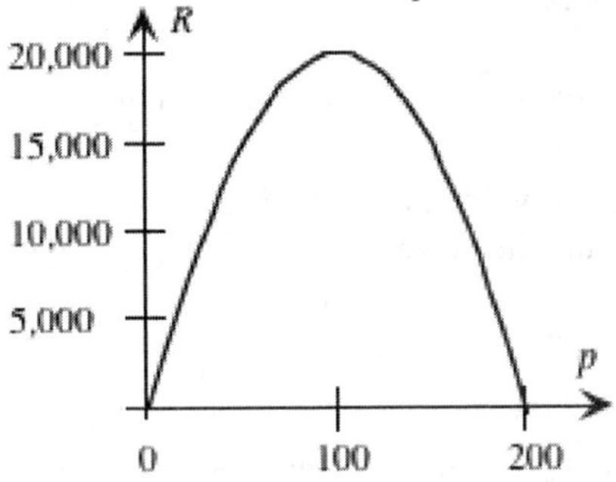

15. $y = -0.7955x^2 + 4.4591x - 1.6000$

17. $y = -1.1667x^2 - 6.1667x - 3$

19. **(a)** Positive because the data suggest a curve that is concave up.
(b) The data suggest a parabola that is concave up (a positive) and with y-intercept around 50. Only choice **(C)** has both properties.
(c) $-b/(2a) = 7/6 \approx 1.2$, which is in 1995. The parabola rises to the left of the vertex and thus predicts increasing trade as we go back in time, contradicting history.

21. $W = 3t^2 - 90t + 4200 (5 \le t \le 27)$

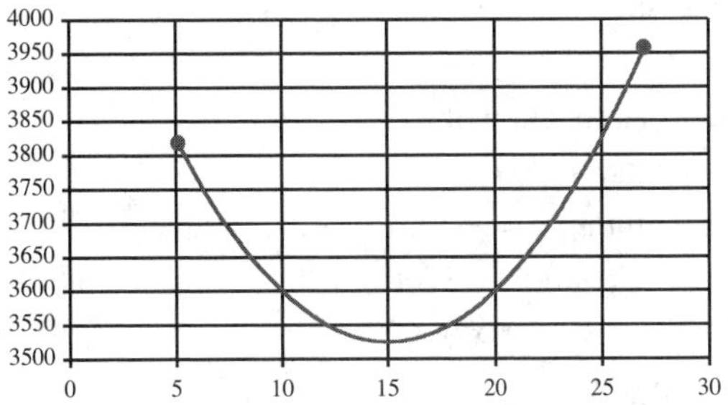

$-b/(2a) = 15$, which represents 1985. $W(15) = 3525$ pounds was the average weight that year.

23. $-b/(2a) = 5000$ pounds. The model is not trustworthy for vehicle weights larger than 5000 pounds because it predicts increasing fuel economy with increasing weight. Also, 5000 is close to the upper limit of the domain of the function.

25. $q = -0.5p + 140$.
Revenue is $R = pq = -0.5p^2 + 140p$. Maximum revenue occurs when
$p = -b/(2a) = \$140$; the corresponding revenue is $R = \$9800$.

27. The given data points are $(p, q) = (40, 200{,}000)$ and $(60, 160{,}000)$. The line passing through these points is $q = -2000p + 280{,}000$. Revenue is $R = pq = -2000p^2 + 280{,}000p$. Maximum revenue occurs when $p = -b/(2a) = 70$ houses; the corresponding revenue is $R = \$9{,}800{,}000$.

29. **(a)** The data points are
$(x, q) = (2, 280)$ and $(1.5, 560)$.
Thus, $q = -560x + 1400$ and
$R = xq = -560x^2 + 1400x$.
(b) $P = R - C = -560x^2 + 1400x - 30$. The largest monthly profit occurs when

$x = -b/(2a) = \$1.25$ and then
$P = \$845$ per month.

31. As a function of q,
$C = 0.5q + 20$.
Substituting $q = -400x + 1200$, we get
$C = 0.5(-400x + 1200) + 20$
$\quad = -200x + 620$.
The profit is
$P = R - C = xq - C$
$\quad = -400x^2 + 1400x - 620$.
The profit is largest when
$x = -b/(2a) = \$1.75$ per log on; the corresponding profit is
$P = \$605$ per month.

33. (a) The data points are
$(p, q) = (10, 300)$ and $(15, 250)$,
so $q = -10p + 400$.
(b) $R = pq = -10p^2 + 400p$.
(c) $C = 3q + 3000 = 3(-10p + 400) + 3000 = -30p + 4200$
(d) $P = R - C = -10p^2 + 430p - 4200$. The maximum profit occurs when $p = -b/(2a) = \$21.50$.

35. Here is the Excel tabulation of the data, together with the scatter plot and the quadratic (polynomial order 2) trendline (with the option "Display equation on chart" checked):

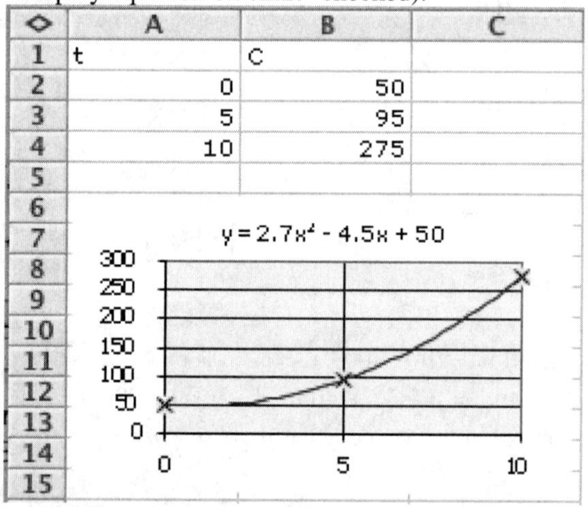

From the trendline, the quadratic model is

$$C(t) = 2.7t^2 - 4.5t + 50.$$

To estimate the value of US trade with China in 2000, substitute the corresponding value $t = 6$, to obtain

$$C(6) = 2.7(6)^2 - 4.5(6) + 50 = 120.2.$$

The actual figure (from Exercise 19) is \$120 billion, which agrees with the predicted value to the nearest \$1 billion.

37. (a) Here is the Excel tabulation of the data, together with the scatter plot and the quadratic (polynomial order 2) trendline (with the option "Display equation on chart" checked):

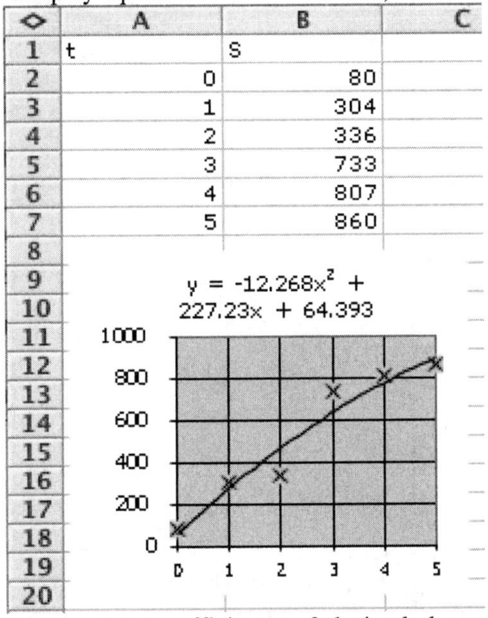

	A	B	C
1	t	s	
2	0	80	
3	1	304	
4	2	336	
5	3	733	
6	4	807	
7	5	860	

We round the coefficients to 2 decimal places:

$$S(t) = -12.27t^2 + 227.23t + 64.39$$

(b) To estimate sales in the fourth quarter of 2004 ($t = 6$) compute

$$S(7) = -12.27(6)^2+227.23(6)+64.39$$
$$= 986.05 \approx 986 \text{ thousand,}$$

or 986,000 units.

(c) Actual sales were more than double the quantity predicted by the regression equation. This fact suggest that mathematical regression cannot reliably be used to make predictions about sales.

39. The x coordinate of the vertex represents the unit price that leads to the maximum revenue, the y coordinate of the vertex gives the maximum possible revenue, the x intercepts give the unit prices that result in zero revenue, and the y intercept gives the revenue resulting from zero unit price (which is obviously zero).

41, Graph the data to see whether the points suggest a curve rather than a straight line. If the curve suggested by the graph is concave up or concave down, then a quadratic model would be a likely candidate.

43. If $q = mp + b$ (with $m < 0$), then the revenue is given by $R = pq = mp^2 + bp$. This is the equation of a parabola with $a = m < 0$, and so is concave down. Thus, the vertex is the highest point on the parabola, showing that there is a single highest value for R, namely the y coordinate of the vertex.

45. Because $R = pq$, the demand must be given by

$$q = \frac{R}{p} = \frac{-50p^2+60p}{p} = -50p + 60.$$

2.2

1. `4^x`

x	−3	−2	−1	0	1	2	3
$f(x)$	$\frac{1}{64}$	$\frac{1}{16}$	$\frac{1}{4}$	1	4	16	64

3. `3^(-x)`

x	−3	−2	−1	0	1	2	3
$f(x)$	27	9	3	1	$\frac{1}{3}$	$\frac{1}{9}$	$\frac{1}{27}$

5. `2*2^x` or `2*(2^x)`

x	−3	−2	−1	0	1	2	3
$f(x)$	$\frac{1}{4}$	$\frac{1}{2}$	1	2	4	8	16

7. `-3*2^(-x)`

x	−3	−2	−1	0	1	2	3
$f(x)$	−24	−12	−6	−3	$-\frac{3}{2}$	$-\frac{3}{4}$	$-\frac{3}{8}$

9. `2^x-1`

x	−3	−2	−1	0	1	2	3
$f(x)$	$-\frac{7}{8}$	$-\frac{3}{4}$	$-\frac{1}{2}$	0	1	3	7

11. `2^(x-1)`

x	−3	−2	−1	0	1	2	3
$f(x)$	$\frac{1}{16}$	$\frac{1}{8}$	$\frac{1}{4}$	$\frac{1}{2}$	1	2	4

13.

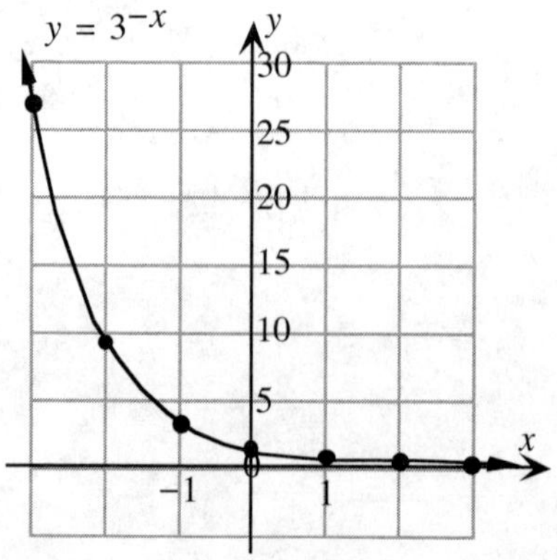

15.

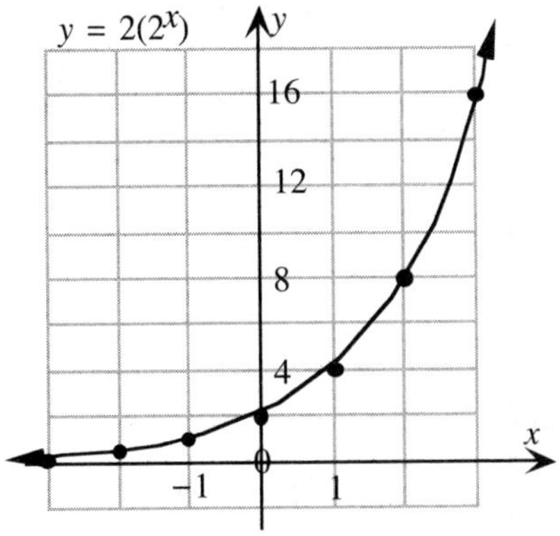

17.

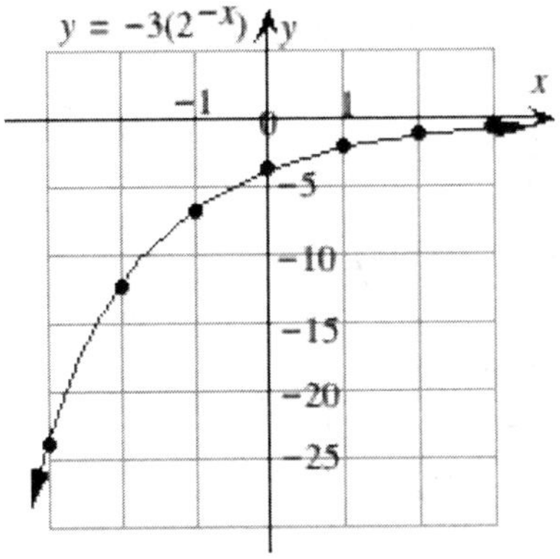

19.

x	−2	−1	0	1	2
$f(x)$	0.5	1.5	4.5	13.5	40.5
$g(x)$	8	4	2	1	$\frac{1}{2}$

For every increase in x by one unit, the value of f is multiplied by 3, so f is exponential.

Since $f(0) = 4.5$, the exponential model is $f(x) = 4.5(3^x)$.

For every increase in x by one unit, the value of g is multiplied by 1/2, so g is exponential.

Since $g(0) = 2$, the exponential model is $g(x) = 2(1/2)^x$, or $2(2^{-x})$

21.

x	−2	−1	0	1	2
$f(x)$	22.5	7.5	2.5	7.5	22.5
$g(x)$	0.3	0.9	2.7	8.1	16.2

When x increases from −1 to 0, the value of f is multiplied by 1/3, but when x is increased from 0 to 1, the value of f is multiplied by 3. So f is not exponential.

When x increases from −1 to −0, the value of g is multiplied by 3, but when x is increased from 1 to 2, the value of g is multiplied by 2. So g is not exponential.

23.

x	−2	−1	0	1	2
$f(x)$	100	200	400	600	800
$g(x)$	100	20	4	0.8	0.16

The values of $f(x)$ double for every one-unit increase in x except when x increases from 1 to 2, when the value of f is multiplied by 4/3. Hence f is not linear.

When g is multiplied by 0.2 for every increase by one unit in x, so g is exponential.

Since $g(0) = 4$, the exponential model is $g(x) = 4(0.2)^x$.

25. `e^(-2*x)` or `EXP(-2*x)`

x	−3	−2	−1	0	1	2	3
$f(x)$	403.4	54.60	7.389	1	0.1353	0.01832	0.002479

27. `1.01*2.02^(-4*x)`

x	−3	−2	−1	0	1	2	3
$f(x)$	4662	280.0	16.82	1.01	0.06066	0.003643	0.0002188

29. `50*(1+1/3.2)^(2*x)`

x	−3	−2	−1	0	1	2	3
$f(x)$	9.781	16.85	29.02	50	86.13	148.4	255.6

The following solutions also show some common errors you should avoid.

31. `2^(x-1)` *not* `2^x-1`

33. `2/(1-2^(-4*x))` *not* `2/1-2^-4*x` *and not* `2/1-2^(-4*x)`

35. `(3+x)^(3*x)/(x+1)` or `((3+x)^(3*x))/(x+1)` *not* `(3+x)^(3*x)/x+1` *and not* `(3+x^(3*x))/(x+1)`

37. `2*e^((1+x)/x)` or `2*EXP((1+x)/x)` *not* `2*e^1+x/x` *and not* `2*e^(1+x)/x` *and not* `2*EXP(1+x)/x`

In the following solutions, f_1 is black and f_2 is gray.

39.

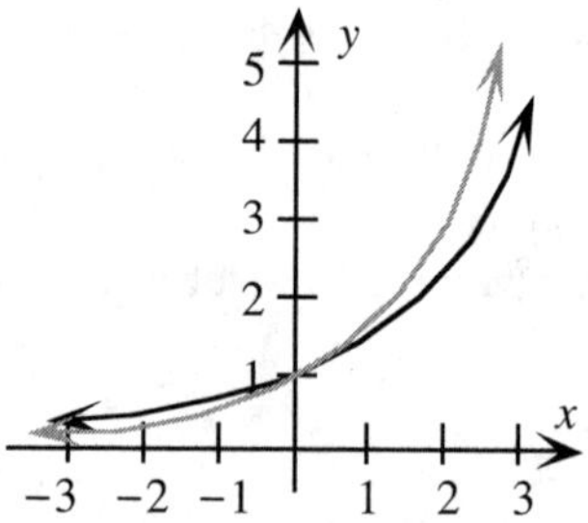

41. (Note that the x-axis shown here crosses at 200, not 0.)

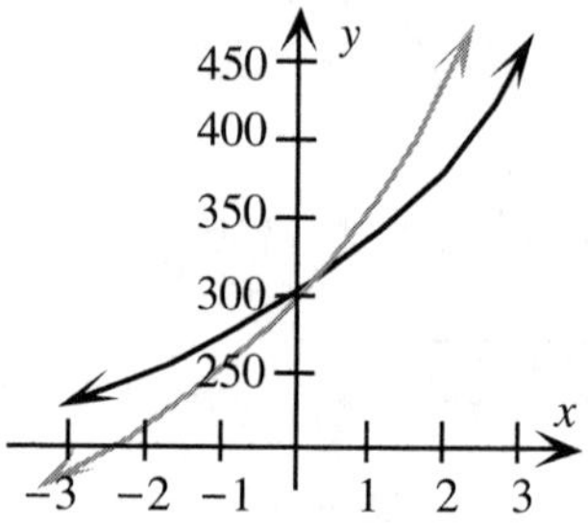

43.

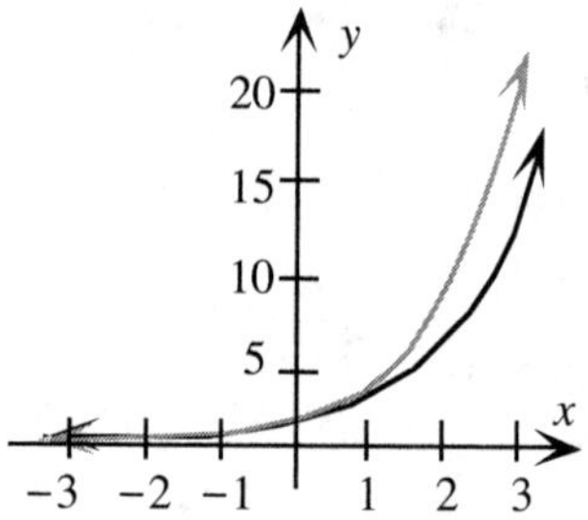

45. (Note that the x-axis shown here crosses at 900.)

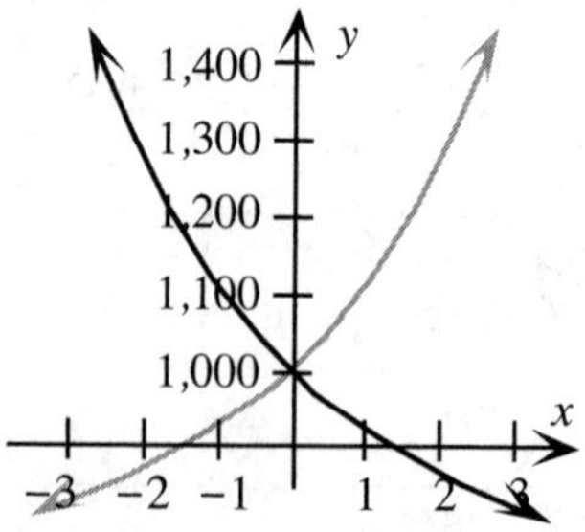

47. Each time x increases by 1, $f(x)$ is multiplied by 0.5. Also, $f(0) = 500$. So, $f(x) = 500(0.5)^x$.

49. Each time x increases by 1, $f(x)$ is multiplied by 3. Also, $f(0) = 10$. So, $f(x) = 10(3)^x$.

51. Each time x increases by 1, $f(x)$ is multiplied by $225/500 = 0.45$. Also, $f(0) = 500$. So, $f(x) = 500(0.45)^x$.

53. Write $f(x) = Ab^x$. We have $Ab^1 = -110$ and $Ab^2 = -121$. Dividing, $b = -121/(-110) = 1.1$. Substituting, $A(1.1) = -110$, so $A = -100$. Thus, $f(x) = -100(1.1)^x$.

55. We want an equation of the form $y = Ab^x$. Substituting the coordinates of the given points gives

$$36 = Ab^2$$
$$324 = Ab^4$$

Dividing the second equation by the first gives

$$324/36 = 9 = b^2$$

so $b = 3$.

Substituting into the first equation now gives

$$36 = A(3)^2 = 9A$$

so $A = 36/9 = 4$.

Hence the model is

$$y = Ab^x = 4(3^x)$$

57. We want an equation of the form $y = Ab^x$. Substituting the coordinates of the given points gives

$$-25 = Ab^{-2}$$
$$-0.2 = Ab$$

Dividing the second equation by the first gives

$$-0.2/(-25) = 0.008 = b^3$$

so $b = 0.2$.

Substituting into the second equation now gives

$$-0.2 = 0.2A$$

so $A = -1$.

Hence the model is

$$y = Ab^x = -1(0.2^x)$$

59. Write $f(x) = Ab^x$. We have $Ab^1 = 3$ and $Ab^3 = 6$. Dividing, $b^2 = 6/3 = 2$, so $b = \sqrt{2} \approx 1.4142$. Substituting, $A\sqrt{2} = 3$, so $A = 3/\sqrt{2} \approx 2.1213$. Thus, $y = 2.1213\,(1.4142^x)$.

61. Write $f(x) = Ab^x$. We have $Ab^2 = 3$ and $Ab^6 = 2$. Dividing, $b^4 = 2/3$, so $b = \sqrt[4]{2/3} \approx 0.9036$. Substituting, $A\left(\sqrt[4]{2/3}\right)^2 = 3$, so $A = 3/\left(\sqrt[4]{2/3}\right)^2 \approx 3.6742$. Thus, $y = 3.6742(0.9036^x)$.

63. Use the formula $f(t) = Pe^{rt}$ with $P = 5000$ and $r = 0.10$, giving $f(t) = 5000e^{0.10t}$.

65. Use the formula $f(t) = Pe^{rt}$ with $P = 1000$ and $r = -0.063$, giving $f(t) = 1000e^{-0.063t}$.

67. $y = 1.0442(1.7564)^x$

69. $y = 15.1735(1.4822)^x$

71. If y represents the size of the culture at time t, then $y = Ab^t$. We are told that the initial size is 1000, so $A = 1000$. We are told that the size doubles every 3 hours, so $b^3 = 2$, or $b = 2^{1/3}$. Thus, $y = 1000(2^{1/3})^t = 1000(2^{t/3})$. There will be $1000(2^{48/3}) = 65{,}536{,}000$ bacteria after 2 days.

73. Apply the formula

$$A(t) = P\left(1+\frac{r}{m}\right)^{mt}$$

with $P = 5000$, $r = 0.0439$, and $m = 1$. We get the model

$$A(t) = 5000(1+0.0439)^t$$
$$= 5000(1.0439)^t$$

At the end of 2008, the deposit would be worth $5000(1.0439)^5 \approx$ £6198.

75. From the answer to Exercise 73, the value of the investment after t years is $A(t) = 5000(1.0439)^t$.

TI83: Enter

`Y1 = 5000*(1.0439)^X,`

Press [2nd] [TBLSET], and set `Indpnt` to Ask. (You do this once and for all; it will permit you to specify values for x in the table screen.) Then, press [2nd] [TABLE], and you will be able to evaluate the function at several values of x. Here are some values of x and the resulting values of Y_1.

x	Y1
1	5219.5
2	5448.63605
3	5687.83117
4	5937.52696
5	6198.18439
6	6470.28469
7	6754.33019
8	7050.84528
9	7360.37739
10	7683.49796

Notice that Y_1 first exceeds 7500 when $x = 10$. Since $x = 0$ represents the beginning of 2004, $x = 10$ represents the start of 2014, so the investment will first exceed £7500 at the beginning of 2014.

Excel: You can obtain a similar table to the one above by setting up your spreadsheet as follows:

◇	A	B
1	t	A
2	1	=5000*(1.0439)^A2
3	2	
4	3	
5	4	
6	5	
7	6	
8	7	
9	8	
10	9	
11	10	

77. $C(t) = 104(0.999879)^t$, so $C(10{,}000) \approx 31.0$ grams, $C(20{,}000) \approx 9.25$ grams, and $C(30{,}000) \approx 2.76$ grams.

79. We are looking for t such that $4.06 = 46(0.999879)^t$. Among the values suggested we find that $C(15{,}000) \approx 7.5$, $C(20{,}000) \approx 4.09$ and $C(25{,}000) \approx 2.23$. Thus, the answer is 20,000 years to the nearest 5000 years.

81. Let y represent the amount of aspirin in the bloodstream at time t hours. Then $y = Ab^t$. The initial value is given to us as $A = 300$ mg. We are told that half the amount is removed every 2 hours, so $b^2 = 0.5$, giving $b = \sqrt{0.5} \approx 0.7071$. Thus, $y = 300(0.7071)^t$. After 5 hours the amount left is $300(0.7071)^5 \approx 53$ mg.

83. **(a)** The line through $(t, P) = (0, 360)$ and $(3, 480)$ is $P = 40t + 360$ million dollars. **(b)** We need to find the exponential curve $P = Ab^t$ through $(0, 360)$ and $(3, 480)$. We see that $A = 360$ and $b^3 = 480/360 = 4/3$, so $b = \sqrt[3]{4/3} \approx 1.1006$. Thus, $P = 360(1.1006)^t$. Neither model is applicable to the data: The data move erratically up and down rather than increasing steadily either linearly or exponentially.

85. **(a)** Let $P = Ab^t$ million people. We are given the data points $(0, 180)$ and $(44, 294)$. Substituting them into the equation $P = Ab^t$ gives

$$A = 180$$
$$294 = 180\, b^{44}$$

so

$$b^{44} = 294/180,$$

which gives

$$b = (294/180)^{1/44} \approx 1.01121.$$

Thus, the model is

$$P(t) = 180(1.01121)^t \text{ million people.}$$

(b) Rounding to 3 decimal places gives $180(1.011)^t$. Putting $t = 44$ *gives*

$$P(44) = 180(1.011)^{44} \approx 291 \neq 294.$$

Rounding to 4 decimal places gives $180(1.0112)^t$. Putting $t = 44$ *gives*

$$P(44) = 180(1.0112)^{44} \approx 294,$$

which is accurate to 3 significant digits. Therefore, we should round to 4 decimal places. **(c)** When $t = 60$, $P = 180(1.01121)^{60} \approx 351$ million people.

87. (a) Let y be the number of frogs in year t, with $t = 0$ representing two years ago; we seek a model of the form $y = Ab^t$. We are given the initial value of $A = 50{,}000$ and are told that $50{,}000b^2 = 75{,}000$. This gives $b = (75{,}000/50{,}000)^{1/2} = 1.5^{1/2}$. Thus, $y = 50{,}000(1.5^{1/2})^t = 50{,}000(1.5^{t/2})$. **(b)** When $t = 3$, $y = 50{,}000(1.5)^{3/2} = 91{,}856$ tags.

89. From the continuous compounding formula, $1000e^{0.04\times 10} = \$1491.82$, of which \$491.82 is interest.

91. (a)

year	1950	2000	2050	2100
C(t) parts per million	561	669	799	953

(b) Testing the decades between 2000 ($t = 250$) and 2050, we find that $C(260) \approx 694$ and $C(270) \approx 718$. Testing individual years between $t = 260$ and $t = 270$, we find that the level surpasses 700 parts per million for the first time when $t = 263$. Thus, to the nearest decade, the level passes 700 in 2010 ($t = 260$).

93. (a) Here is the Excel tabulation of the data, together with the scatter plot and the quadratic (polynomial order 2) trendline (with the option "Display equation on chart" checked):

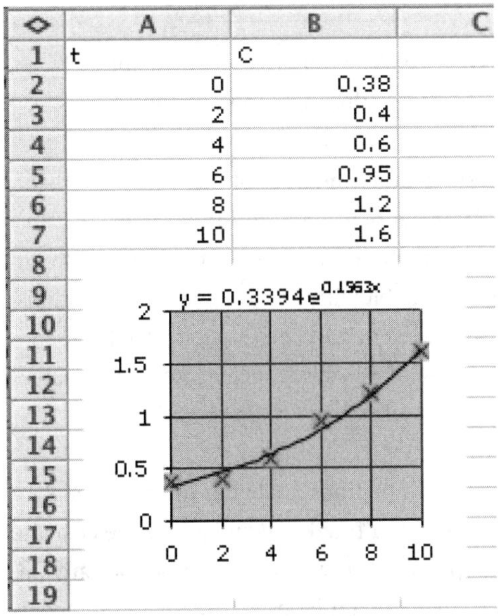

Note that the displayed model has the form $P(t) = 0.3394e^{0.1563t}$. To express this in the form $A(b)^t$ we write

$$0.339e^{0.1563t} = 0.339(e^{0.1563})^t$$
$$\approx 0.339(1.169)^t.$$

so the model is

$$P(t) = 0.339(1.169)^t.$$

(b) In 2005, $t = 11$, so the predicted cost is

$$P(15) = 0.339(1.169)^{11}$$
$$\approx \$1.9 \text{ million}$$

95. (a) $y = 5.4433(1.0609)^t$

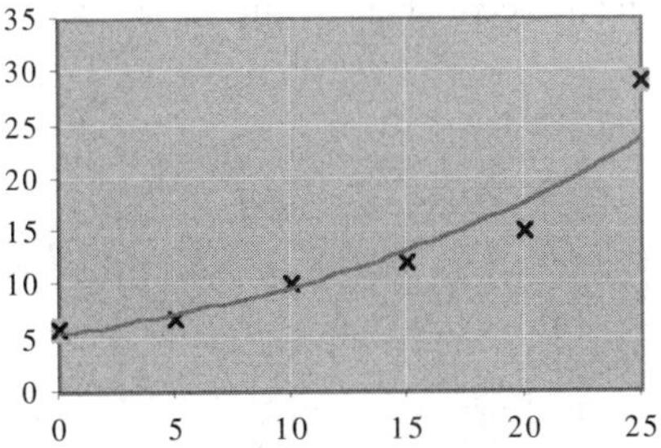

(b) Each year spending increases by a factor of 1.0609, which represents a 6.09% annual increase. **(c)** $y(18) \approx \$16$ billion

97. (B) An exponential function eventually becomes larger than any polynomial.

99. Exponential functions of the form $f(x) = Ab^x$ $(b > 0)$ increase rapidly for large values of x. In real-life situations, such as population growth, this model is reliable only for relatively short periods of growth. Eventually, population growth tapers off because of pressures such as limited resources and overcrowding.

101. Linear functions are better for cost models where there is a fixed cost and a variable cost and for simple interest, where interest is paid only on the original amount invested. Exponential models are better for compound interest and population growth. In both of these latter examples, the rate of growth depends on the current number of items, rather than on a fixed initial quantity.

103. Take the ratios y_2/y_1 and y_3/y_2. If they are the same, the points fit on an exponential curve.

105. This reasoning is suspect—the bank need not use its computer resources to update all the accounts every minute, but can instead use the continuous compounding formula to calculate the balance in any account at any time as needed.

2.3

1.

Exponential form	$10^4 = 10{,}000$	$4^2 = 16$	$3^3 = 27$	$5^1 = 5$	$7^0 = 1$	$4^{-2} = \frac{1}{16}$
Logarithmic form	$\log_{10}10{,}000 = 4$	$\log_4 16 = 2$	$\log_3 27 = 3$	$\log_5 5 = 1$	$\log_7 1 = 0$	$\log_4 \frac{1}{16} = -2$

3.

Exponential form	$(0.5)^2 = 0.25$	$5^0 = 1$	$10^{-1} = 0.1$	$4^3 = 64$	$2^8 = 256$	$2^{-2} = \frac{1}{4}$
Logarithmic form	$\log_{0.5}0.25 = 2$	$\log_5 1 = 0$	$\log_{10}0.1 = -1$	$\log_4 64 = 3$	$\log_2 256 = 8$	$\log_2 \frac{1}{4} = -2$

5. $x = \log_3 5 = \log 5/\log 3 = 1.4650$

7. $-2x = \log_5 40 = \log 40/\log 5$, so $x = -(\log 40/\log 5)/2 = -1.1460$

9. $e^x = 2/4.16$, so $x = \ln(2/4.16) = -0.7324$

11. $1.06^{2x+1} = 11/5$, so $2x + 1 = \log_{1.06}(11/5)$, $x = (\log(11/5)/\log(1.06) - 1)/2 = 6.2657$

13. $f(x) = \log_4 x$

We plot some points and draw the graph:

x	$y = \log_4(x)$
1/16	−2
1/4	−1
1/2	− 1/2
1	0
2	1/2
4	1

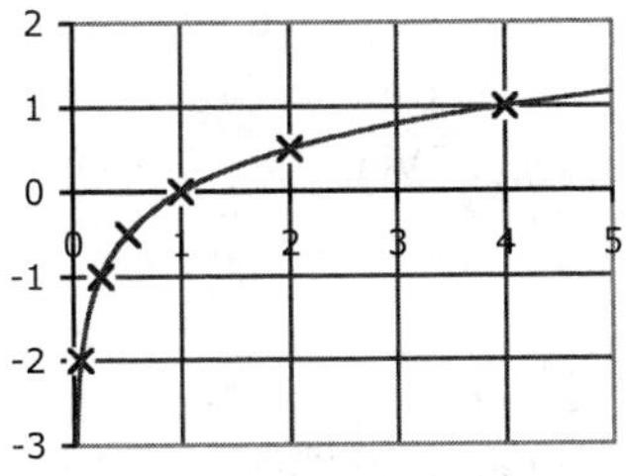

15. $f(x) = \log_4(x-1)$

We plot some points and draw the graph:

x	$y = \log_4(x-1)$
17/16	−2
5/4	−1
3/2	− 1/2
2	0
3	1/2
5	1

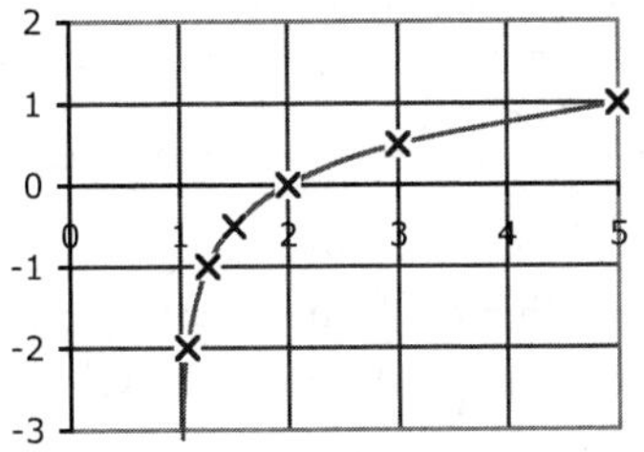

17. $f(x) = \log_{1/4}x$

We plot some points and draw the graph:

x	$y = \log_{1/4}(x)$
1/16	1
1/4	1
1/2	1/2
1	0
2	−1/2
4	−1

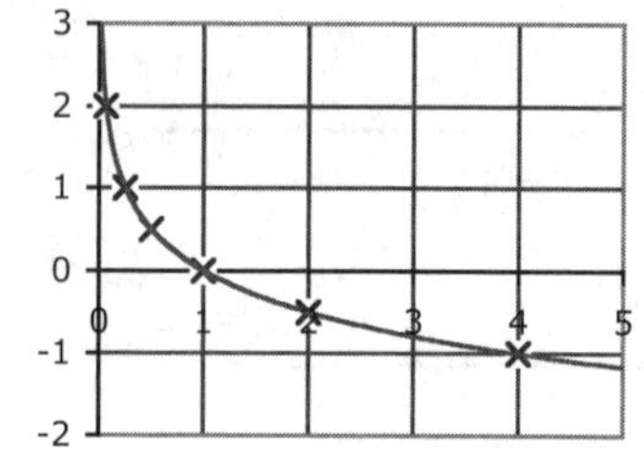

19. $Q = 1000$ when $t = 0$; Half-life = 1
We want a model of the form $Q = Q_0 e^{-kt}$ for suitable Q_0 and k. We are given $Q_0 = 1000$. For k, we use the formula $t_h k = \ln 2$ with t_h = half-life = 1, so $k = \ln 2$, and the model is

$$Q = Q_0 e^{-kt} = 1000e^{-t \ln 2}$$

21. $Q = 1000$ when $t = 0$; Doubling time = 2
We want a model of the form $Q = Q_0 e^{kt}$ for suitable Q_0 and k. We are given $Q_0 = 1000$. For k, we use the formula $t_d k = \ln 2$ with t_d = doubling time = 2, so $2k = \ln 2$, giving $k = (\ln 2)/2$ and the model is

$$Q = Q_0 e^{kt} = 1000e^{t\,(\ln 2)/2}$$

23. $Q = 1000e^{0.5t}$
Since the exponent is positive, the model represents exponential growth. We use the formula $t_d k = \ln 2$, giving $t_d(0.5) = \ln 2$, giving doubling time = $t_d = (\ln 2)/0.5 = 2 \ln 2$

25. $Q = Q_0 e^{-4t}$
Since the exponent is negative, the model represents exponential decay. We use the formula $t_h k = \ln 2$, giving $t_h(4) = \ln 2$, giving half-life = t_h = $(\ln 2)/4$

27. $f(x) = 4e^{2x} = 4(e^2)^x \approx 4(7.389)^x$

29. $f(t) = 2.1(1.001)^t = 2.1e^{kt}$

$$1.001^t = e^{kt}$$

$$t \ln(1.001) = kt$$

$$k = \ln(1.001) \approx 0.0009995$$

so $f(t) = 2.1e^{0.000\,9995t}$

31. $f(t) = 10(0.987)^t = 10e^{-kt}$

$$0.987^t = e^{-kt}$$

$$t \ln(0.987) = -kt$$

$$k = -\ln(0.987) \approx 0.01309$$

so $f(t) = 10e^{-0.013\,09t}$

33. Substitute $A = 700$, $P = 500$, and $r = 0.10$ in the continuous compounding formula: $700 = 500e^{0.10t}$. Solve for t: $e^{0.10t} = 700/500 = 7/5$, $0.10t = \ln(7/5)$, $t = 10\ln(7/5) \approx 3.36$ years.

35. Substitute $A = 3P$ and $r = 0.10$ in the continuous compounding formula: $3P = Pe^{0.10t}$. Solve for t: $e^{0.10t} = (3P)/P = 3$, $0.10t = \ln 3$, $t = 10\ln 3 \approx 11$ years

37. We want a model of the form $A = Pe^{rt}$. For the rate of interest r, we use the formula $t_d r = \ln 2$ with t_d = doubling time = 3, so $3r = \ln 2$, giving $r = (\ln 2)/3 \approx 0.231$, or 23.1%

39. If 99.95% has decayed, then 0.05% remains, so $C(t) = 0.0005A$. Therefore, $0.0005A = A(0.999879)^t$, $(0.999879)^t = 0.0005$, so $t = \log_{0.999879} 0.0005 = \log(0.0005)/\log(0.999879) \approx \approx$ 63,000 years old.

41. Substitute $A = 15{,}000$, $P = 10{,}000$, $r = 0.053$, and $m = 1$ into the compound interest formula: $15{,}000 = 10{,}000(1 + 0.053)^t$, $(1.053)^t = 1.5$, $t = \log_{1.053} 1.5 = \log(1.5)/\log(1.053) \approx 8$ years.

43. Substitute $A = 20{,}000$, $P = 10{,}400$, $r = 0.052$, and $m = 12$ into the compound interest formula: $20{,}000 = 10{,}400(1 + 0.052/12)^{12t}$,

$(1 + 0.052/12)^{12t} = 200/104$, $12t =$ $\log(200/104)/\log(1 + 0.052/12) \approx 151$ months.

45. Substitute $A = 2P$, $r = 0.06$, and $m = 2$ into the compound interest formula: $2P = P(1 + 0.06/2)^{2t}$, $(1.03)^{2t} = 2$, $2t = \log 2/\log 1.03$, $t = (\log 2/\log 1.03)/2 \approx 12$ years.

47. Substitute $A = 2P$ and $r = 0.053$ in the continuous compounding formula: $2P = Pe^{0.053t}$, $e^{0.053t} = 2$, $0.053t = \ln 2$, $t = (\ln 2)/0.053 \approx 13.08$ years.

49. We want to find the value of t for which $C(t)$ = the weight of undecayed radium-226 left = half the original weight = $0.5A$. Substituting, we get

$$0.5A = A(0.999567)^t$$

so

$$0.5 = 0.999567^t$$

$$t = \log_{0.999567} 0.5 \approx 1600 \text{ years}$$

51. (a) If $y = Ab^t$, substitute $y = 3A$ and $t = 6$ to get $3A = Ab^6$, so $b = 3^{1/6} \approx 1.20$.
(b) Now substitute $y = 2A$ and solve for t: $2A = A(1.20)^t$, so $t = \log 2/\log 1.20 \approx 3.8$ months.

53. (a) $t_h k = \ln 2$

$$5k = \ln 2$$

$$k = (\ln 2)/5 \approx 0.139$$

$$Q(t) = Q_0 e^{-kt} \approx Q_0 e^{-0.139t}$$

(b) One third has decayed when 2/3 is left:

$$\frac{2}{3}Q_0 = Q_0 e^{-0.139t}$$

$$\frac{2}{3} = e^{-0.139t}$$

$$\ln(2/3) = -0.139t$$

$$t = \ln(2/3)/(-0.139) \approx 3 \text{ years}$$

55. Let $Q(t)$ be the amount left after t million years. We first find a model of the form $Q(t) = Q_0 e^{-kt}$.

To find k use

$t_h k = \ln 2$

$$710k = \ln 2$$

$$k = (\ln 2)/710 \approx 0.000\,976\,3$$

$$Q(t) = Q_0 e^{-kt} \approx Q_0 e^{-0.000\,976\,3t}$$

We now answer the question: $Q_0 = 10$g, $Q(t) = 1$g. Substituting in the model gives

$$1 = 10e^{-0.000\,976\,3t}$$

$$e^{-0.000\,976\,3t} = 0.1$$

$$-0.000\,976\,3t = \ln(0.1)$$

$$t = -\ln 0.1/0.000\,976\,3$$

$$\approx 2360 \text{ million years}$$

(rounded to 3 significant digits).

57. Let $Q(t)$ be the amount left after t hours. We first find a model of the form $Q(t) = Q_0 e^{-kt}$.

To find k use

$t_h k = \ln 2$

$$2k = \ln 2$$

$$k = (\ln 2)/2 \approx 0.3466$$

$$Q(t) = Q_0 e^{-kt} \approx Q_0 e^{-0.3466t}$$

We now answer the question: $Q_0 = 300$mg, $Q(t) = 100$mg. Substituting in the model gives

$$100 = 300e^{-0.3466t}$$

$$e^{-0.3466t} = 1/3$$

$$-0.3466t = \ln(1/3)$$

$$t = -\ln(1/3)/0.3466 \approx 3.2 \text{ hours}$$

59. We first find a model of the form $A(t) = Pb^t$. We are given the data points (0, 1) and (2, 0.7), so $P = 1$ and $b^2 = 0.7$, so $b = (0.7)^{1/2}$; the model is $A(t) = (0.7)^{t/2}$. We wish to find when $A(t) = 0.5$:

$$0.5 = (0.7)^{t/2}$$

$$t/2 = \log_{0.7}(0.5)$$

$$t = 2\log(0.5)/\log(0.7) \approx 3.89 \text{ days.}$$

61. (a) To obtain the logarithmic regression equation for the data in Excel, do a scatter plot and add a Logarithmic trendline with the option "Display equation on chart" checked. On he TI-83, use STAT, select CALC, and choose the option `LnReg`. The resulting regression equation with coefficients rounded to 4 digits is

$P(t) = 6.591 \ln(t) - 17.69$

b. The year 1940 is represented by $t = 40$, so

$P(40) = 6.591 \ln(40) - 17.69 \approx 6.6,$

which is accurate only to one digit.

c. The logarithm increases without bound (choice (A)).

63. $M(t) = 11.622 \ln t - 7.1358$. The model is unsuitable for large values of t because, for sufficiently large values of t, $M(t)$ will eventually become larger than 100%.

65. (a) Substitute:

$8.2 = (2/3)(\log E - 11.8).$

Solve for $\log E$:

$\log E = (3/2)\times 8.2 + 11.8 = 24.1,$

$E = 10^{24.1} \approx 1.259\times 10^{24}$ ergs

(b) Compute the energy released as in (a):

$7.1 = (2/3)(\log E - 11.8)$

$\log E = (3/2)\times 7.1 + 11.8 = 22.45$

$E = 10^{22.4} \approx 2.818\times 10^{22}$ ergs.

Comparing, the energy released in 1989 was

$$\frac{2.818\times 10^{22}}{1.259\times 10^{24}}$$

$\approx 2.24\%$ of the energy released in 1906.

(c) Solving for E in terms of R we get

$E = 10^{1.5R + 11.8}.$

This gives

$$\frac{E_2}{E_1} = \frac{10^{1.5R_2 + 11/8}}{10^{1.5R_1 + 11.8}} = 10^{1.5(R_2 - R_1)}$$

(d) If $R_2 - R_1 = 2$, then $E_2/E_1 = 10^{1.5(2)} = 10^3 = 1000$.

67. (a) By substitution: 75 dB, 69 dB, 61 dB

(b) $D = 10 \log(320\times 10^7) - 10 \log r^2$

$\approx 95 - 20 \log r$

(c) Solve $0 = 95 - 20 \log r$: $\log r = 95/20$, $r = 10^{95/20} \approx 57{,}000$ feet (rounding up so that we're beyond the point where the decibel level is 0).

69. The logarithm of a negative number, were it defined, would be the power to which a base must be raised to give that negative number. But raising a base to a power never results in a negative number, so there can be no such number as the logarithm of a negative number.

71. $\log_4 y$

73. 8. Note that $b^{\log_b a} = a$ for any a and b, since $\log_b a$ is the power to which you raise b to get a.

75. x. To what power to you raise e to get e^x? x!

77. Any logarithmic curve $y = \log_b t + C$ will eventually surpass 100%, and hence not be suitable as a long-term predictor of market share.

79. Time is increasing logarithmically with population: Solving $P = Ab^t$ for t gives $t = \log_b(P/A) = \log_b P - \log_b A$, which is of the form $t = \log_b P + C$.

2.4

1. $N = 7$, $A = 6$, $b = 2$; `7/(1+6*2^-x)`

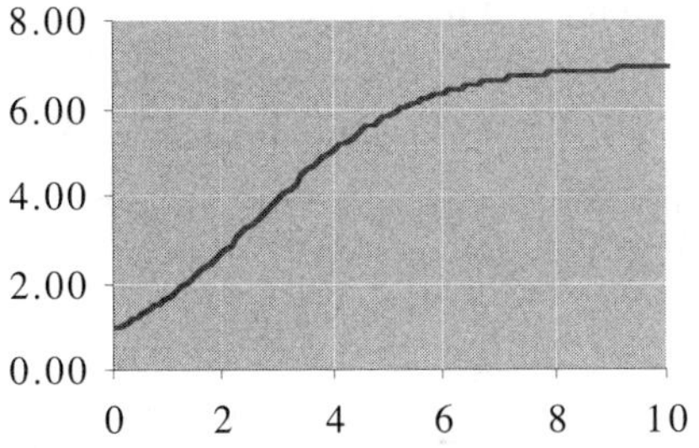

3. $N = 10$, $A = 4$, $b = 0.3$;
`10/(1+4*0.3^-x)`

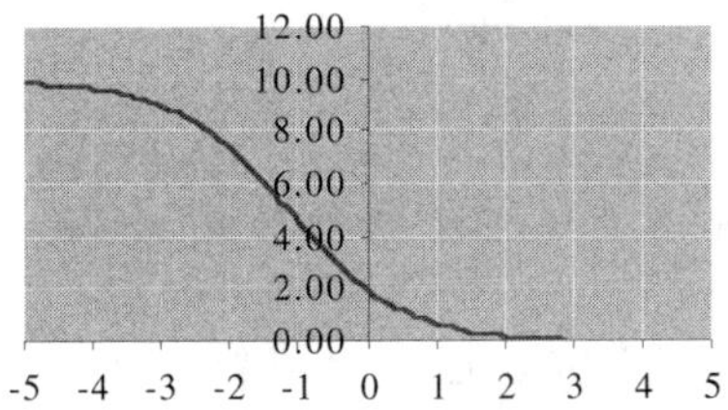

5. $N = 4$, $A = 7$, $b = 1.5$; `4/(1+7*1.5^-x)`

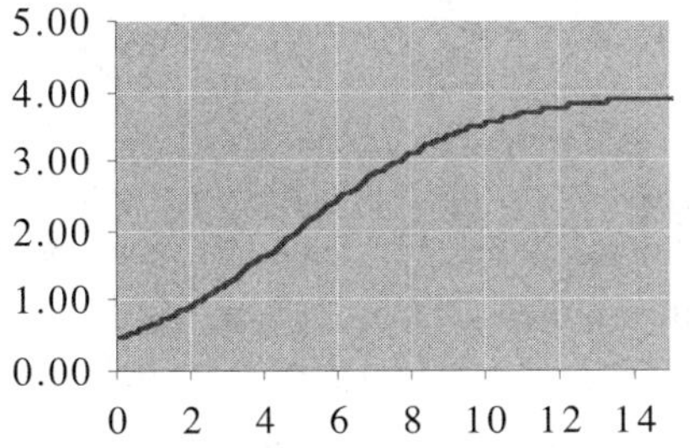

7. $N = 200$, the limiting value. $10 = f(0) = N/(1 + A) = 200/(1 + A)$, so $A = 19$. For small values of x, $f(x) \approx 10b^x$; to double with every increase of 1 in x we must therefore have $b = 2$. This gives $f(x) = \dfrac{200}{1 + 19\cdot(2^{-x})}$.

9. $N = 6$, the limiting value. $3 = f(0) = N/(1 + A) = 6/(1 + A)$, so $A = 1$. $4 = f(1) = 6/(1 + b^{-1})$, $1 + b^{-1} = 6/4 = 3/2$, $b^{-1} = 1/2$, $b = 2$. So, $f(x) = \dfrac{6}{1 + 2^{-x}}$.

11. B: The limiting value is 9, eliminating (A). The initial value is $2 = 9/(1 + 3.5)$, not $9/(1 + 0.5)$, so the answer is (B), not (C).

13. B: The graph decreases, so $b < 1$, eliminating (C). The y intercept is $2 = 8/(1 + 3)$ as in (B).

15. C: The initial value is 2, as in (A) or (C). If b were 5, an increase in 1 in x would multiply the value of $f(x)$ by approximately 5 when x is small. However, increasing x from 0 to 10 does not quite double the value. Hence b must be smaller, as in (C).

17. $y = \dfrac{7.2}{1 + 2.4\,(1.05)^{-x}}$

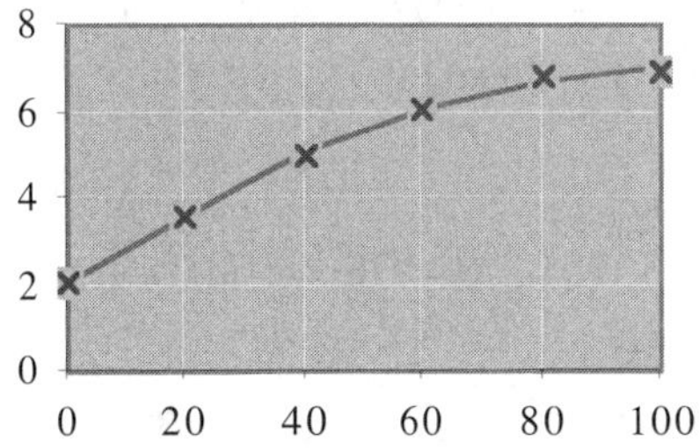

19. $y = \dfrac{97}{1 + 2.2(0.942)^{-x}}$

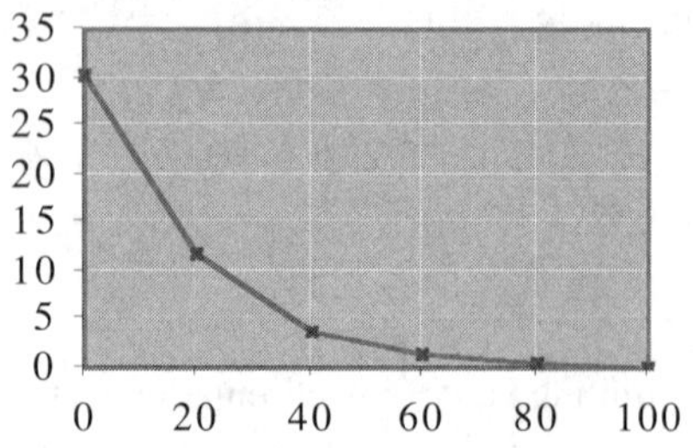

21. **(a)** We can eliminate (C) and (D) because $b = 0.8$ is less that 1, giving a decreasing function of t. The value $N = 4.0$ in (B) predicts a leveling-off value of around 4000 articles, which is clearly too small. Thus, we are left with choice (A).
(b) 1985 is in the initial period of the model, when the rate of growth is governed by exponential growth. Since $b = 1.2$, the rate of growth is 20% per year.

23. **(a)** For the extremely wealthy we look at large values of x, where $P(x)$ is close to $N = 91\%$. **(b)** $P(x) \approx [91/(1 + 5.35)](1.05)^x \approx 14.33(1.05)^x$. **(c)** Set $P(x) = 50$ and solve for x: $50 = 91/[1 + 5.35(1.05)^{-x}]$, $1 + 5.35(1.05)^{-x} = 91/50$, $x = -\log[41/(50\times5.35)]/\log(1.05) \approx \$38{,}000$.

25. $N = 10{,}000$, the susceptible population. $1000 = 10{,}000/(1 + A)$, so $A = 9$. In the initial stages, the rate of increase was 25% per day, so $b = 1.25$. Thus, $N(t) = \dfrac{10{,}000}{1 + 9(1.25)^{-t}}$. $N(7) \approx 3463$ cases.

27. $N = 3000$, the total available market. $100 = 3000/(1 + A)$, so $A = 29$. Sales are initially doubling every 5 days, so $b^5 = 2$, or $b = 2^{1/5}$. Thus, $N(t) = \dfrac{3000}{1 + 29(2^{1/5})^{-t}}$. Set $N(t) = 700$ and solve for t: $700 = 3000/[1 + 29(2^{1/5})^{-t}]$, $1 + 29(2^{1/5})^{-t} = 30/7$, $(2^{1/5})^{-t} = 23/(29\times7)$, $t = -\log[23/(29\times7)]/\log(2^{1/5}) \approx 16$ days.

29. **(a)** Following Example 2, if using Excel, start with initial rough estimates of N, A, and b. (Their exact value is not important.)

N = leveling off value ≈ 7

$b = 1.1$ (slightly larger than 1, since N is increasing with t)

A: Use y-intercept $= N/(1+A)$

$$1.2 = \frac{7}{1+A}$$

$$1+A = 7/1.2 \approx 6$$

So, $A \approx 5$.

Solver then gives the following solution:

$N \approx 6.3$, $A \approx 4.8$, $b \approx 1.2$.

Thus, the regression model is

$$A(t) = \frac{6.3}{1 + 4.8(1.2)^{-t}}.$$

The model predicts that the number will level off around $N = 6.3$ thousand, or 6300 articles.

(b) $A(17) = \dfrac{6.3}{1 + 4.8(1.2)^{-17}} \approx 5.2$, or 5200 articles.

31. **(a)** $N(t) = \dfrac{82.8}{1 + 21.8(7.14)^{-t}}$ The model predicts that books sales will level off at around 82.8 million books per year. **(b)** Not consistent; 15% of the market is represented by more than double the predicted value. This shows the difficulty in making long-term predictions from regression models obtained from a small amount of data. **(c)** Set $N(t) = 80$ and solve for t: $t = -\log[2.8/(80\times21.8)]/\log(7.14) \approx 3.3$, so book sales first exceed 80 million in 2001.

33. $N(t) = \dfrac{5}{1 + 1.081(1.056)^{-t}}$. Set $N(t) = 3.5$ and solve for t:

$t = -\log[1.5/(3.5\times1.081)]/\log(1.056) \approx 17$ which represents the year 2010.

35. Just as diseases are communicated via the spread of a pathogen (such as a virus), new technology is communicated via the spread of information (such as advertising and publicity). Further, just as the spread of a disease is ultimately limited by the number of susceptible individuals, so the spread of a new technology is ultimately limited by the size of the potential market.

37. It can be used to predict where the sales of a new commodity might level off.

Chapter 2 Review Exercises

1. $a = 1$, $b = 2$, $c = -3$, so $-b/(2a) = -1$; $f(-1) = -4$, so the vertex is $(-1, -4)$. y intercept: $c = -3$. $x^2 + 2x - 3 = (x + 3)(x - 1)$, so the x intercepts are -3 and 1. $a > 0$, so the parabola opens upward.

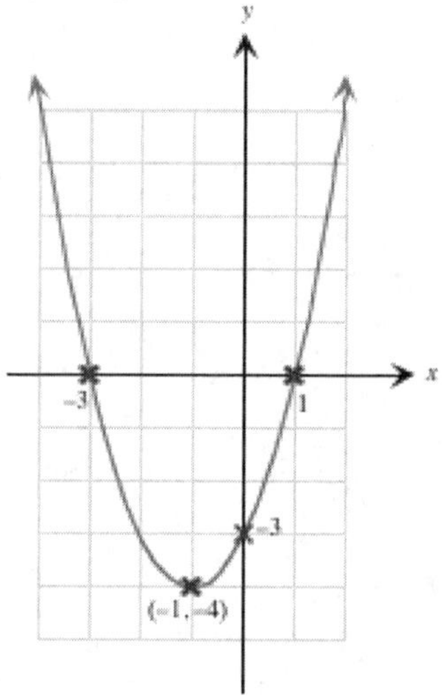

2. $a = -1$, $b = -1$, $c = -1$, so $-b/(2a) = -1/2$; $f(-1/2) = -3/4$, so the vertex is $(-1/2, -3/4)$. y intercept: $c = -1$. $b^2 - 4ac = -3 < 0$, so there are no x intercepts. $a < 0$, so the parabola opens downward.

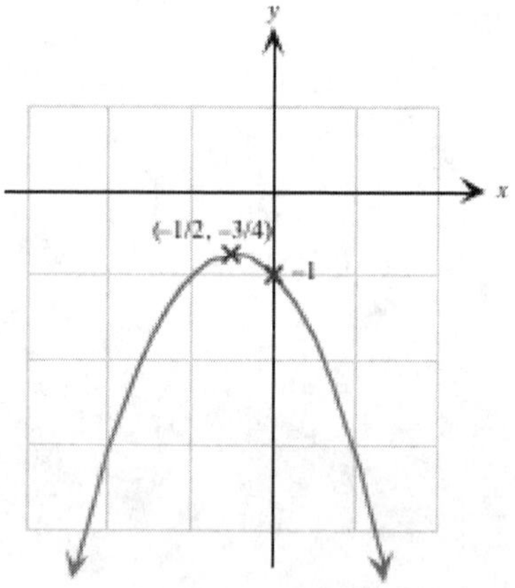

3. For every increase in x by one unit, the value of f is multiplied by 1/2, so f is exponential.
Since $f(0) = 5$, the exponential model is $f(x) = 5(1/2)^x$, or $5(2^{-x})$.
$g(2) = 0$, whereas exponential functions are never zero, so g is not exponential.

4. The values of f decrease by 2 for every increase in x by one unit, so f is linear (not exponential).
For every increase in x by one unit, the value of g is multiplied by 2, so g is exponential.
Since $g(0) = 3$, the exponential model is $g(x) = 3(2^x)$.

5. We use the following table of values:

x	$f(x)$	$g(x)$
–3	1/54	27/2
–2	1/18	9/2
–1	1/6	3/2
0	1/2	1/2
1	3/2	1/6
2	9/2	1/18
3	27/2	1/54

Graphing these gives the following curves:

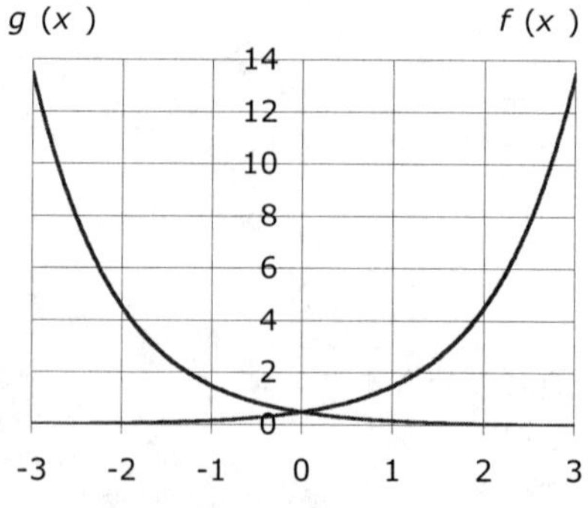

6. We use the following table of values:

x	$f(x)$	$g(x)$
–3	1/32	128
–2	1/8	32
–1	1/2	8
0	2	2
1	8	1/2
2	32	1/8
3	128	1/32

Graphing these gives the following curves:

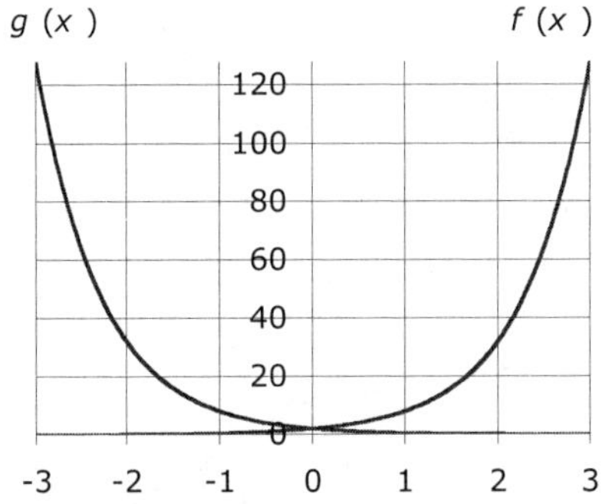

7. The technology formulas for the two functions are

TI-83: `e^x; e^(0.8*x)`

Excel: `=EXP(x); =EXP(0.8*x)`

Technology gives us the following graphs:

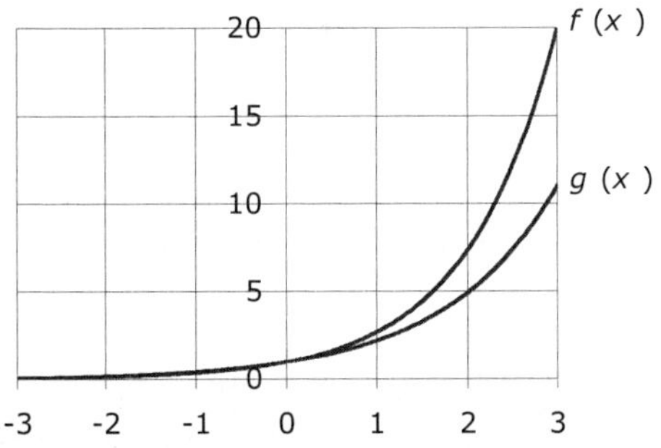

8. The technology formulas for the two functions are

TI-83 and Excel: `2*1.01^x; 2*0.99^x`

Technology gives us the following graphs:

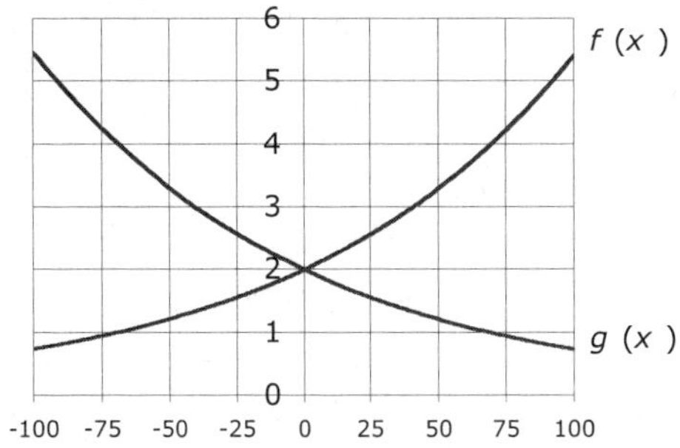

9. $A = P(1 + \frac{r}{m})^{mt}$;

$P = 3000, r = 0.03; m = 12; t = 5$

$A = 3000(1+0.03/12)^{60} \approx \$3484.85$

10. $A = P(1 + \frac{r}{m})^{mt}$;

$P = 10{,}000, r = 0.025; m = 4; t = 10$

$A = 10{,}000(1+0.025/4)^{40} \approx \$12{,}830.27$

11. $A = P(1 + \frac{r}{m})^{mt}$;

$A = 5000, r = 0.03; m = 12; t = 10$

$5000 = P(1+0.03/12)^{120}$

$P = 5000/(1+0.03/12)^{120} \approx \$3705.48$

12. $A = P(1 + \frac{r}{m})^{mt}$;

$A = 10{,}000, r = 0.025; m = 4; t = 10$

$10{,}000 = P(1+0.025/4)^{40}$

$P = 10{,}000/(1+0.025/4)^{40} \approx \$7794.07$

13. $A = Pe^{rt}$;

$P = 3000, r = 0.03; t = 5$

$A = 3000e^{0.15} \approx \$3485.50$

14. $A = Pe^{rt}$;

$P = 10{,}000, r = 0.025; t = 10$

$A = 10{,}000e^{0.25} \approx \$12{,}840.25$

15. Increasing x by 1/2-unit triples the value of f. Therefore, increasing x by 1-unit multiplies the value of f by 9, giving $b = 9$. $A = f(0) = 4.5$. Therefore,

$f(x) = Ab^x = 4.5(9^x)$

16. Increasing x by one unit decreases the value of f by 75%. That is, it reduces the value of f to 25% of its original value. Therefore, $b = 0.25$. $A = f(0) = 5$. Therefore,

$f(x) = Ab^x = 5(0.25^x)$

17. $2 = Ab^1$ and $18 = Ab^3$. Dividing, $b^2 = 9$, so $b = 3$. Then $2 = 3A$, so $A = 2/3$: $f(x) = \frac{2}{3}3^x$

18. $10 = Ab^1$ and $5 = Ab^3$. Dividing, $b^2 = 1/2$, so $b = 1/\sqrt{2}$. Then $10 = A/\sqrt{2}$, so $A = 10\sqrt{2}$: $f(x) = 10\sqrt{2}\left(\frac{1}{\sqrt{2}}\right)^x$

19. $-2x = \log_3 4$, so $x = -\frac{1}{2}\log_3 4$

20. $2x^2 - 1 = \log_2 2 = 1$, $x^2 = 1$, $x = \pm 1$

21. $10^{3x} = 315/300 = 1.05$, $3x = \log 1.05$, $x = \frac{1}{3}\log 1.05$

22. $(1 + i)^{mx} = A/P$, $mx = \log_{1+i}(A/P) = \ln(A/P)/\ln(1+i)$, $x = \dfrac{\ln(A/P)}{m\ln(1+i)}$

23. We use the following table of values:

x	$f(x)$	$g(x)$
1/27	–3	3
1/9	–2	2
1/3	–1	1
1	0	0
3	1	–1
9	2	–2
27	3	–3

Graphing these gives the following curves:

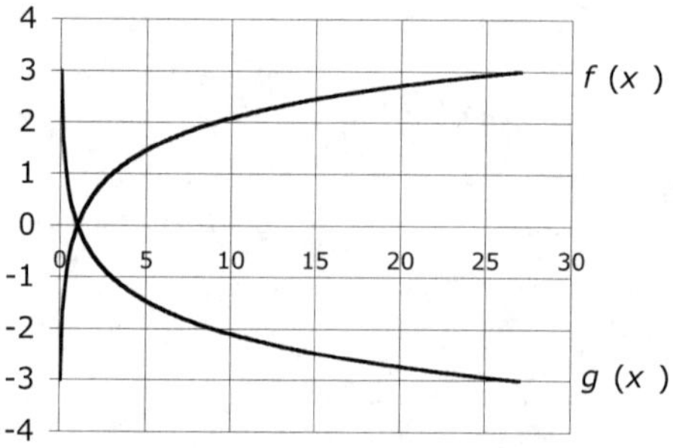

24. We use the following table of values:

x	$f(x)$	$g(x)$
1/1000	–3	3
1/100	–2	2
1/10	–1	1
1	0	0
10	1	–1
100	2	–2
1000	3	–3

Graphing these gives the following curves:

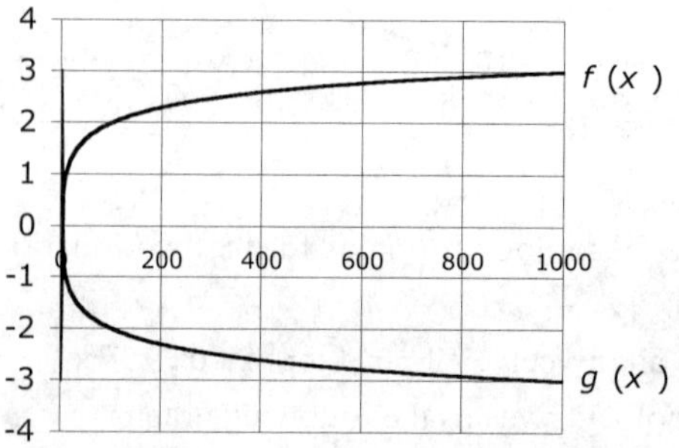

25. $Q_0 = 5$ (given)
$t_h k = \ln 2$
$100k = \ln 2$
$k = (\ln 2)/100 \approx 0.00693$
$Q = Q_0 e^{-kt}$
$Q = 5e^{-0.00693t}$

26 $Q_0 = 10{,}000$ (given)
$t_h k = \ln 2$
$5k = \ln 2$
$k = (\ln 2)/5 \approx 0.139$
$Q = Q_0 e^{-kt}$
$Q = 10{,}000e^{-0.139t}$

27. $Q_0 = 2.5$ (given)
$t_d k = \ln 2$
$2k = \ln 2$
$k = (\ln 2)/2 \approx 0.347$
$Q = Q_0 e^{kt}$
$Q = 2.5e^{0.347t}$

28. $Q_0 = 10{,}000$ (given)
$t_d k = \ln 2$
$15k = \ln 2$
$k = (\ln 2)/15 \approx 0.0462$
$Q = Q_0 e^{kt}$
$Q = 10{,}000e^{0.0462t}$

29. $3000 = 2000(1 + 0.04/12)^{12t}$
$t = [\log(3/2)/\log(1 + 0.04/12)]/12$
≈ 10.2 years

30. $3000 = 2000(1 + 0.0675/365)^{365t}$
$t = [\log(3/2)/\log(1 + 0.0675/365)]/365$
≈ 6 years

31. $3000 = 2000e^{0.0375t}$
$t = \ln(3/2)/0.0375$
≈ 10.8 years

32. $r = 1$, $m = 4$
$1200 = 1000(1 + 1/4)^{4t} = 1000(1.25^{4t})$
$t = [\log(1200/1000)/\log(1.25)]/4$
≈ 0.2 years

33. $N = 900$ is given, $100 = 900/(1 + A)$ gives $A = 8$, initially increasing 50% per unit increase in x gives $b = 1.5$: $f(x) = \dfrac{900}{1 + 8(1.5)^{-x}}$

34. $N = 25$ is given, $5 = 25/(1 + A)$ gives $A = 4$, $b = 1.1$ is given in the initial exponential form: $f(x) = \dfrac{25}{1 + 4(1.1)^{-x}}$

35. $N = 20$ is given, $20/(1 + A) = 5$, so $A = 3$, decreasing at a rate of 20% per unit of x near 0 means $b = 1 - 0.2 = 0.8$: $f(x) = \dfrac{20}{1 + 3(0.8)^{-x}}$

36. $N \approx 10$ and $b = 0.8$ are given, and for small x,
$N/(1+A) = 10/(1+A) = 2$,
so $A = 4$, giving
$$f(x) = \frac{10}{1 + 4(0.8)^{-x}}$$

37. (a) The largest volume will occur at the vertex: $-b/(2a) = 0.085/(2\times0.000005) \approx \$8500$ per month; substituting $c = 8500$ gives $h =$ an average of approximately 2100 hits per day. **(b)** Solve $h = 0$ using the quadratic formula: $c \approx \$29{,}049$ per month (the other solution given by the quadratic formula is negative). **(c)** The fact that -0.000005, the coefficient of c^2, is negative.

38. (a) The rate of growth is a minimum at the vertex:
$$t = -\frac{b}{2a} = -\frac{-2}{4} = \frac{1}{2},$$
midway through 2001.

(b) A zero rate of growth would correspond to an intercept of the t-axis. However, the discriminant

$b^2 - 4ac = 4 - 4(2)(8) = -60$

is negative, so there are no t-intercepts, and hence no zero rate of growth.

(c) The fact that the coefficient of t^2 is positive.

(d) No. What was decreasing was the number of *new* broadband users. Put another way, the number of broadband users was growing at a declining rate in the first half of 2001.

39. $R = pq = -60p^2 + 950p$. The maximum revenue occurs at the vertex: $p = -b/(2a) = 950/(2\times 60) =$ \$7.92 per novel. At that price the monthly revenue is $R =$ \$3760.42.

40. $C = 900 + 4q = 900 + 4(-60p + 950) = -240p + 4700$, so $P = R - C = -60p^2 + 950p - (-240p + 4700) = -60p^2 + 1190p - 4700$. The maximum monthly profit occurs at the vertex: $p = -b/(2a) = 1190/(2\times 60) =$ \$9.92 per novel. At that price, the monthly profit is $P =$ \$1200.42.

41. (a) 1997 corresponds to $t = 0$, and so the harvest was $n = 10(0.66^0) = 10$ million pounds. The value of the base $b = 0.66$ of the exponential model tells us that the value each year 0.66 times, or 66% of the value the previous year. In other words, the value decreases by 100–66 = 34% each year.

(b) 2005 corresponds to $t = 8$, and

$n(8) = 10(0.66^8) \approx 0.36,$

or about 360,000 pounds.

42. Let S be the stock price and t be the time, in hours since the IPO. Model the stock price by $S = Ab^t$. Then $A = 10{,}000$ and $b^3 = 2$ (since the stock is doubling in price every 3 hours), so $b = 2^{1/3}$. After eight hours, $S = 10{,}000(2^{1/3})^8 =$ \$63,496.04.

43. The model for the lobster harvest from Exercise 39 is $n(t) = 10(0.66^t)$ million pounds. We need to find the value of t when it first dips below 100,000, which is 0.1 million pounds:

$0.1 = 10(0.66^t)$

$0.01 = 0.66^t$

$\log(0.01) = t\log(0,66)$

$t = \log(0.01)/\log(0.66) \approx 11.08$

Since $t = 0$ corresponds to June, 1997, $t = 11$ corresponds to June, 2008.

44. Solve $50{,}000 = 10{,}000(2^{t/3})$: $t = 3\log 5/\log 2 \approx 7.0$ hours.

45. After 10 hours the stock is worth $10{,}000(2^{10/3}) =$ \$100,793.68. From this level it follows a new exponential curve with $A = 100{,}793.68$ and $b^4 = 2/3$ (it loses 1/3 of its value every 4 hours), hence $b = (2/3)^{1/4}$. Now solve $10{,}000 = 100{,}793.68(2/3)^{t/4}$ for $t = 4\log(10{,}000/100{,}793.68)/\log(2/3) \approx 22.8$. Adding the first 10 hours, the stock will be worth \$10,000 again 32.8 hours after the IPO.

46. We use the data shown in the following table:

t	n
0	8.2
1	7.5
2	6.4
3	2.6
4	1.8
5	1.4
6	0.8

The regression model is

Excel: $n(t) = 10.318e^{-0.4145t} = 10.318(e^{-0.4145})^t \approx 10.318(0.6607)^t$

TI-83 and Web Site: $10.3182(0.660642)^t$

Rounding to 2 digits gives $n(t) = 10(0.66^t)$ million pounds of lobster.

47. C: (A) is true because $L_1 = L_0e^{-1/t} < L_0$. (B) is true because $L_{3t} = L_0e^{-3} > 0$. (D) is true because increasing t decreases x/t, which increases $e^{-x/t}$ (makes it closer to 1), hence increases L_x for every x. (E) is true because $e^{-x/t}$ is never 0. On the other hand, (C) is false because $L_t = L_0e^{-1} \approx 0.37L_0 < L_0/2$.

48. (a) The given function has the form Constant + logistic function. The logistic function part levels off at $N = 4470$. Therefore, adding the constant gives a leveling of at 4470 + 6050 = 10,520.

(b) We solve

$$10{,}000 = 6050 + \frac{4470}{1 + 14(1.73^{-t})}$$

$$3950 = \frac{4470}{1 + 14(1.73^{-t})}$$

$$1+14(1.73^{-t}) = 4470/3950$$

$$14(1.73^{-t}) = 4470/3950 - 1$$

$$1.73^{-t} = (4470/3950 - 1)/14$$

$$-t \log(1.73) = \log((4470/3950 - 1)/14)$$

$$t = -\log((4470/3950 - 1)/14)/\log(1.73)$$

$$\approx 8.5,$$

abound 8.5 weeks.

Chapter 3

3.1

1. 0:

x	$f(x)$
−0.1	0.01111111
−0.01	0.00010101
−0.001	1.001×10^{-6}
−0.0001	1.0001×10^{-6}
0	
0.0001	9.999×10^{-9}
0.001	9.99×10^{-7}
0.01	9.901×10^{-5}
0.1	0.00909091

3. 4:

x	$f(x)$
1.9	3.9
1.99	3.99
1.999	3.999
1.9999	3.9999
2	
2.0001	4.0001
2.001	4.001
2.01	4.01
2.1	4.1

5. Does not exist:

x	$f(x)$
−1.1	−22.1
−1.01	−202.01
−1.001	−2002.001
−1.0001	−20002
−1	
−0.9999	19998.0001
−0.999	1998.001
−0.99	198.01
−0.9	18.1

7. 1.5:

x	$f(x)$
10	2.66
100	1.58969231
1000	1.50877143
10,000	1.50087521
100,000	1.5000875

9. 0.5:

x	$f(x)$
−10	53.1578947
−1000	1
−100,000	0.505
−10,000,000	0.50005
−1,000,000,000	0.5000005

11. Diverges to $+\infty$:

x	$f(x)$
10	76.9423077
100	258.966135
1000	2059.17653
10000	20059.1977
100000	200059.2

13. 0:

x	$f(x)$
10	0.79988005
100	0.02600019
1000	0.00206
10000	0.0002006
100000	2.0006×10^{-5}

15. 1:

x	$f(x)$
1.9	0.90483742
1.99	0.99004983
1.999	0.9990005
1.9999	0.9999
2	
2.0001	1.00010001
2.001	1.0010005
2.01	1.01005017
2.1	1.10517092

17. 0:

x	$f(x)$
10	0.000453999
100	3.72008×10^{-42}
1000	0
10000	0
100000	0

(The last three values of e^{-x}, while not mathematically 0, are too small to be represented in Excel, which just gives the values as 0.)

19. **(a)** −2 **(b)** −1

21. **(a)** 2 **(b)** 1 **(c)** 0 **(d)** +∞

23. **(a)** 0 **(b)** 2: As x approaches 0 from the right, $f(x)$ approaches the solid dot at height 2. **(c)** −1: As x approaches 0 from the left, $f(x)$ approaches the open dot at height −1. The fact that $f(0) = 2$ is irrelevant. **(d)** Does not exist: Parts (b) and (c) show that the one-sided limits, though they both exist, do not agree. **(e)** 2: The solid dot indicates the actual value of $f(0)$. **(f)** +∞

25. **(a)** 1 **(b)** 1: Similar to Exercise 23. **(c)** 2 **(d)** Does not exist **(e)** 1 **(f)** 2

27. **(a)** 1 **(b)** +∞ **(c)** +∞ **(d)** +∞ **(e)** Not defined **(f)** −1

29. **(a)** −1: Approaching from the left or the right, the value of $f(x)$ approaches the height of the open dot, −1. **(b)** +∞ **(c)** −∞ **(d)** Does not exist **(e)** 2 **(f)** 1: The value of the function is given by the closed dot on the graph.

31. Here is a table of values for $A(t)$

Technology formula:

```
7.0/(1+5.4*1.2^(-x))
```

t	A(t)
0	1.09375
10	3.7390563
20	6.135755
30	6.8443011
40	6.9743759
50	6.9958488
60	6.9993292
70	6.9998917
80	6.9999825
90	6.9999972
100	6.9999995

At t gets larger and larger, the values of $A(t)$ are getting closer and closer to 7.0. Thus, we estimate

$$\lim_{t \to +\infty} A(t) = 7.0$$

Since $A(t)$ represents the number of thousands of research articles per year in *Physics Review* written by researchers in Europe, we interpret the result as follows: In the long term, the number of research articles in *Physics Review* written by researchers in Europe approaches 7000 per year.

33. Here is a table of values for $S(x)$

Technology formula:

```
470-136*0.974^x
```

x	S(x)
0	334
50	433.567686
100	460.240342
150	467.385537
200	469.299626
250	469.81238
300	469.94974
350	469.986536
400	469.996393
450	469.999034
500	469.999741

At t gets larger and larger, the values of $S(x)$ are getting closer and closer to 470. Thus, we estimate

$$\lim_{x\to+\infty} S(x) = 470$$

Since $S(x)$ represents the average SAT verbal score of a student whose income is x thousand dollars per year, we interpret the result as follows: Students whose parents earn an exceptionally large income score an average of 470 on the SAT verbal test.

35. $\lim_{t\to1^-} C(t) = 0.06$, $\lim_{t\to1^+} C(t) = 0.08$, so that $\lim_{t\to1} C(t)$ does not exist.

37. $\lim_{t\to+\infty} I(t) = +\infty$, $\lim_{t\to+\infty} (I(t)/E(t)) \approx 2.5$. In the long term, U.S. imports from China will rise without bound and be 2.5 times U.S. exports to China. In the real world, imports and exports cannot rise without bound. Thus, the given models should not be extrapolated far into the future.

39. $\lim_{t\to+\infty} n(t) \approx 80$. On-line book sales can be expected to level off at 80 million per year in the long term.

41. To approximate $\lim_{x\to a} f(x)$ numerically, choose values of x closer and closer to and on either side of $x = a$, and evaluate $f(x)$ for each of them. The limit (if it exists) is then the number that these values of $f(x)$ approach. A disadvantage of this method is that it may never give the exact value of the limit, but only an approximation. (However, we can make the approximation as accurate as we like.)

43. It is possible for $\lim_{x\to a} f(x)$ to exist even though $f(a)$ is not defined. An example is

$$\lim_{x\to1} \frac{x^2-3x+2}{x-1}.$$

45. Any situation in which there is a sudden change can be modeled by a function in which $\lim_{t\to a^+} f(t)$ is not the same as $\lim_{t\to a^-} f(t)$ One example is the value of a stock market index before and after a crash: $\lim_{t\to a^-} f(t)$ is the value immediately before the crash at time $t = a$, while $\lim_{t\to a^+} f(t)$ is the value immediately after the crash. Another example might be the price of a commodity that is suddenly increased from one level to another.

47. An example is $f(x) = (x-1)(x-2)$.

3.2

1. Continuous on its domain

3. Continuous on its domain

5. Discontinuous at $x = 0$:
$\lim_{x\to 0^+} f(x) \neq \lim_{x\to 0^-} f(x)$, so $\lim_{x\to 0} f(x)$ does not exist.

7. Discontinuous at $x = -1$:
$\lim_{x\to -1^+} f(x) \neq \lim_{x\to -1^-} f(x)$, so $\lim_{x\to -1} f(x)$ does not exist.

9. Continuous on its domain:
Note that $f(0)$ is not defined, so 0 is not in the domain of f.

11. Discontinuous at $x = -1$ and 0:
$\lim_{x\to -1} f(x) = -1 \neq f(-1)$
and $\lim_{x\to 0} f(x)$ does not exist.
Note that $f(0)$ is defined $[f(0) = 2]$ so 0 is in the domain of f.

13. (A), (B), (D), (E) are continuous on their domains:
Note that 1 is not in the domain in (B) and (D) and that the domain in (E) is $(-\infty, -1] \cup (1, +\infty)$; the "horizontal break" in the graph in (E) does not make the function discontinuous. (C) is discontinuous at 1 because the limit there does not equal the function's value.

In Exercises 15–22, either a graph of the function or a table of values can be used to compute the indicated limit. We show a graph drawn with technology in each case. Note: The vertical lines near discontinuities in some of the graphs below is typical behavior of graphing technology.

15. Technology Formula:
`(x^2-2*x+1)/(x-1)`
Graph drawn with technology:

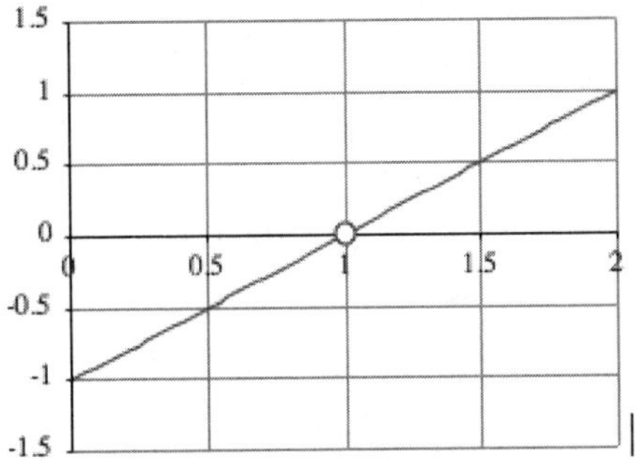

From the graph we see that
$$\lim_{x\to 1} \frac{x^2 - 2x + 1}{x - 1} = 0,$$
so setting $f(1) = 0$ makes it continuous at 1.

17. Technology Formula:
`x/(3*x^2-x)`
Graph drawn with technology:

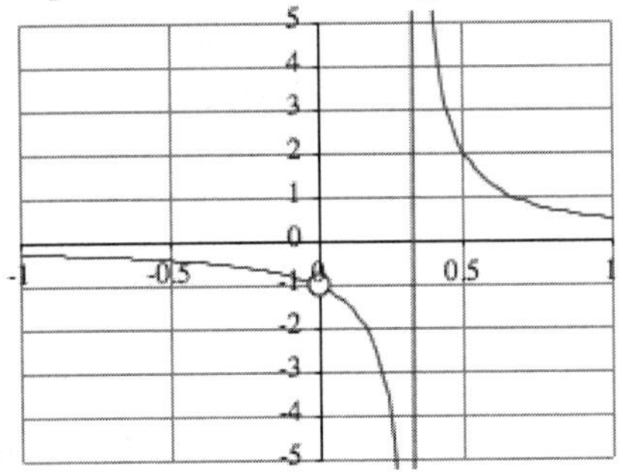

From the graph we see that
$$\lim_{x\to 0} \frac{x}{3x^2 - x} = -1,$$
so setting $f(0) = -1$ makes it continuous at 0.

19. Technology Formula:

`3/(3*x^2-x)`

Graph drawn with technology:

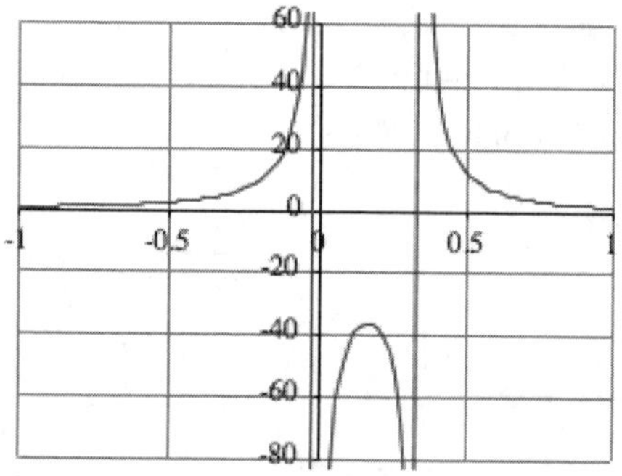

From the graph we see that

$$\lim_{x\to 0} \frac{3}{3x^2 - x} \text{ is undefined,}$$

so no value of $f(0)$ will make it continuous at 0.

21. Technology Formula:

TI-83/84: `(1-e^x)/x`

Excel: `(1-exp(x))/x`

Graph drawn with technology:

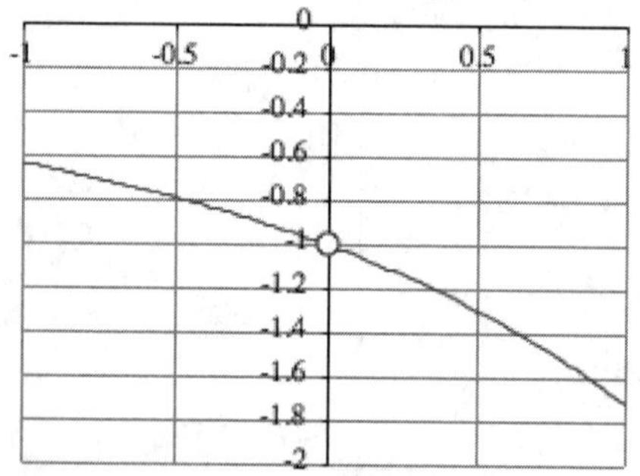

From the graph we see that

$$\lim_{x\to 0} \frac{1 - e^x}{x} = -1,$$

so setting $f(0) = -1$ will make it continuous at 0.

Note: The vertical lines near discontinuities in some of the graphs below is typical behavior of graphing technology.

23. Continuous on its domain:

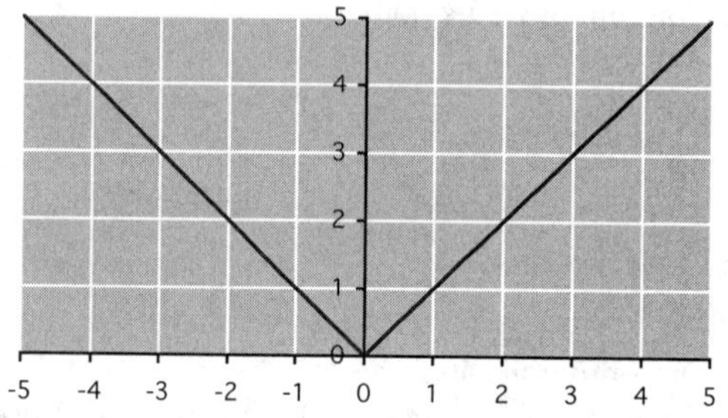

25. Continuous on its domain: Note that 1 and -1 are not in the domain of g.

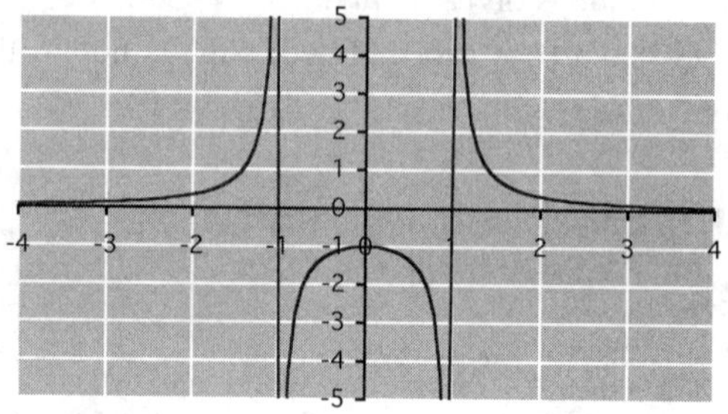

27. Discontinuity at $x = 0$:

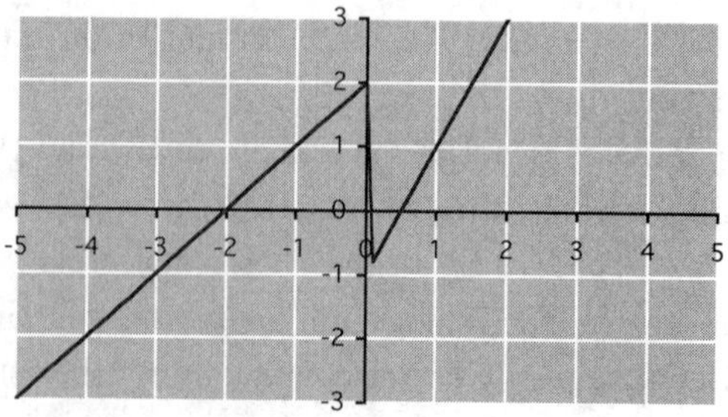

29. Discontinuity at $x = 0$: The limit of $h(x)$ as $x \longrightarrow 0$ does not exist, but 0 is now in the domain of h.

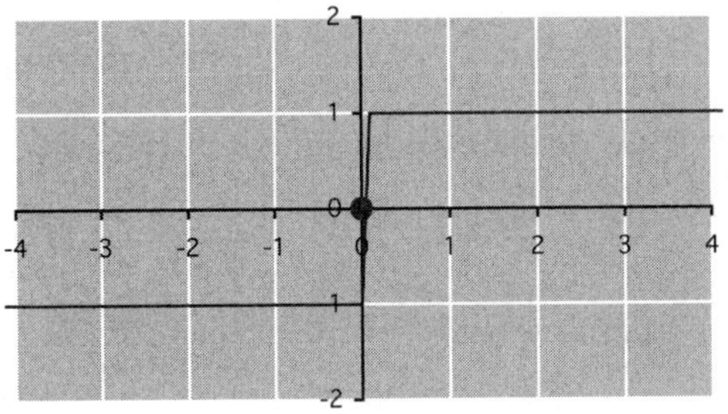

31. Continuous on its domain:

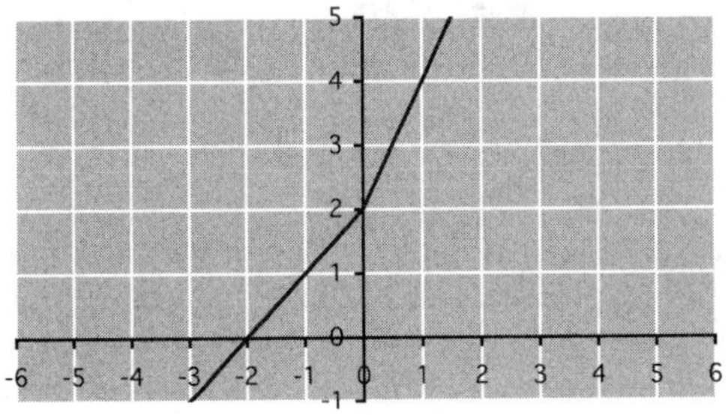

33. Not unless the domain of the function consists of all real numbers. (It is impossible for a function to be continuous at points not in its domain.) For example, $f(x) = 1/x$ is continuous on its domain—the set of nonzero real numbers—but not at $x = 0$.

35. True. If the graph of a function has a break in its graph at any point a, then it cannot be continuous at the point a.

37. Answers may vary.

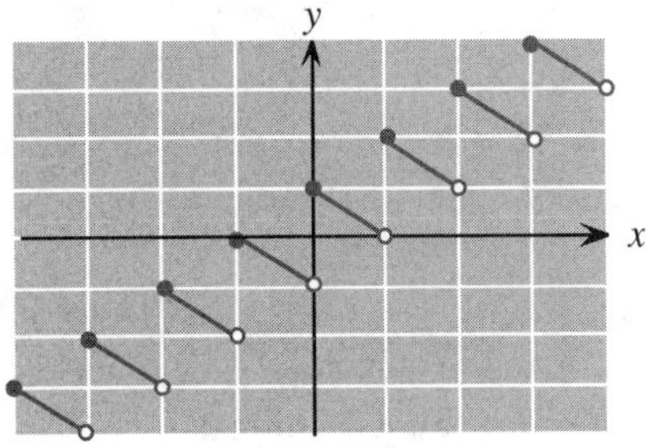

39. Answers may vary. The price of OHaganBooks.com stock suddenly drops by \$10 as news spreads of a government investigation. Let $f(x)$ = the price of OHaganBooks.com stock.

3.3

1. The natural domain of $f(x) = \dfrac{1}{x-1}$ consists of all real numbers except $x = 1$. Therefore, f is continuous for all x except $\underline{x = 1.}$

3. Since $x = 2$ is in the domain of $f(x) = \sqrt{x+1}$, we can obtain the limit by substituting:

$$\lim_{x\to3} \sqrt{x+1} = \sqrt{3+1} = \sqrt{4} = 2.$$

5. $\lim_{x\to0} (x + 1) = 0 + 1 = 1$

7. $\lim_{x\to2} \dfrac{2 + x}{x} = \dfrac{2 + 2}{2} = 2$

9. $\lim_{x\to-1} \dfrac{x + 1}{x} = \dfrac{-1 + 1}{-1} = 0$

11. $\lim_{x\to8} (x - \sqrt[3]{x}) = 8 - \sqrt[3]{8} = 6$

13. $\lim_{h\to1} (h^2 + 2h + 1) = 1^2 + 2(1) + 1 = 4$

15. $\lim_{h\to3} 2 = 2$

17. $\lim_{h\to0} \dfrac{h^2}{h + h^2} = \lim_{h\to0} \dfrac{h}{1 + h} = \dfrac{0}{1 + 0} = 0$

19. $\lim_{x\to1} \dfrac{x^2 - 2x + 1}{x^2 - x} = \lim_{x\to1} \dfrac{(x-1)^2}{x(x-1)} = \lim_{x\to1} \dfrac{x - 1}{x} = \dfrac{1 - 1}{1} = 0$

21. $\lim_{x\to2} \dfrac{x^3 - 8}{x - 2} = \lim_{x\to2} \dfrac{(x - 2)(x^2 + 2x + 4)}{x - 2}$

$= \lim_{x\to2} (x^2 + 2x + 4) = 2^2 + 2(2) + 4 = 12$

23. $\lim_{x\to0^+} \dfrac{1}{x^2}$ diverges to $+\infty$ because it has the form "$k/0$" and $1/x^2$ is positive.

25. $\lim_{x\to-1} \dfrac{x^2 + 1}{x + 1}$ does not exist: The limit from the left is $-\infty$ whereas the limit from the right is $+\infty$.

27. $\lim_{x\to+\infty} \dfrac{3x^2 + 10x - 1}{2x^2 - 5x} = \lim_{x\to+\infty} \dfrac{3x^2}{2x^2} = \lim_{x\to+\infty} \dfrac{3}{2} = 3/2$

29. $\lim_{x\to+\infty} \dfrac{x^5 - 1000x^4}{2x^5 + 10{,}000} = \lim_{x\to+\infty} \dfrac{x^5}{2x^5} = \lim_{x\to+\infty} \dfrac{1}{2} = 1/2$

31. $\lim_{x\to+\infty} \dfrac{10x^2 + 300x + 1}{5x + 2} = \lim_{x\to+\infty} \dfrac{10x^2}{5x} = \lim_{x\to+\infty} (2x)$ diverges to $+\infty$

33. $\lim_{x\to+\infty} \dfrac{10x^2 + 300x + 1}{5x^3 + 2} = \lim_{x\to+\infty} \dfrac{10x^2}{5x^3} = \lim_{x\to+\infty} \dfrac{2}{x} = 0$

35. $\lim_{x\to-\infty} \dfrac{3x^2 + 10x - 1}{2x^2 - 5x} = \lim_{x\to-\infty} \dfrac{3x^2}{2x^2} = \lim_{x\to-\infty} \dfrac{3}{2} = 3/2$

37. $\lim_{x\to-\infty} \dfrac{x^5 - 1000x^4}{2x^5 + 10{,}000} = \lim_{x\to-\infty} \dfrac{x^5}{2x^5} = \lim_{x\to-\infty} \dfrac{1}{2} = 1/2$

39. $\lim_{x \to -\infty} \frac{10x^2 + 300x + 1}{5x + 2} = \lim_{x \to -\infty} \frac{10x^2}{5x} = \lim_{x \to -\infty} \frac{10x}{5}$ diverges to $-\infty$

41. $\lim_{x \to -\infty} \frac{10x^2 + 300x + 1}{5x^3 + 2} = \lim_{x \to -\infty} \frac{10x^2}{5x^3} = \lim_{x \to -\infty} \frac{10}{5x} = 0$

43. The only possible discontinuity is at $x = 0$. There, $\lim_{x \to 0^-} f(x) = 0 + 2 = 2$ whereas $\lim_{x \to 0^+} f(x) = 2(0) - 1 = -1$. Since these disagree, there is a discontinuity at $x = 0$.

45. The only possible discontinuities are at $x = 0$ and $x = 2$. We have $\lim_{x \to 0^-} g(x) = 0 + 2 = 2$ and $\lim_{x \to 0^+} g(x) = 2(0) + 2 = 2 = g(0)$, hence g is continuous at 0. We have $\lim_{x \to 2^-} g(x) = 2(2) + 2 = 8$ and $\lim_{x \to 2^+} g(x) = 2^2 + 2 = 8 = g(2)$. Hence, g is continuous everywhere.

47. The only possible discontinuity is at $x = 0$. We have $\lim_{x \to 0^-} h(x) = 0 + 2 = 2 \neq h(0) = 0$, so there is a discontinuity at $x = 0$. (Note that we don't have to bother computing the limit from the right, which also equals 2.)

49. The only possible discontinuities are at $x = 0$ and $x = 2$. We have $\lim_{x \to 0^-} f(x) = \lim_{x \to 0^-} (1/x) = -\infty$, so there is a discontinuity at $x = 0$. On the other hand, $\lim_{x \to 2^-} f(x) = 2 = f(2)$ and $\lim_{x \to 2^+} f(x) = 2^{2-1} = 2$ also, so f is continuous at 2.

51. (a) Since $f(t) = 0.04t + 0.33$ when $t < 4$ and is a closed-form function in this range, we compute the limit $\lim_{t \to 4^-} f(t)$ by substituting $t = 4$ to get

$$\lim_{t \to 4^-} f(t) = \lim_{t \to 4^-} 0.04t + 0.33$$
$$= 0.04(4) + 0.33 = 0.49$$

Thus, shortly before 1999 ($t = 4$), annual advertising expenditures were close to \$0.49 billion.

Since $f(t) = -0.01t + 1.2$ when $t > 4$ and is a closed-form function in this range, we compute the limit $\lim_{t \to 4^+} f(t)$ by substituting $t = 4$ to get

$$\lim_{t \to 4^+} f(t) = \lim_{t \to 4^+} -0.01t + 1.2$$
$$= -0.01(4) + 1.2 = 1.16$$

Thus, shortly after 1999 ($t = 4$), annual advertising expenditures were close to \$1.16 billion.

(b) By part (a), $\lim_{t \to 4^-} f(t) \neq \lim_{t \to 4^+} f(t)$, and so f is not continuous at $t = 4$. Interpretation: Movie advertising expenditures jumped suddenly in 1999.

53. $\lim_{t \to +\infty} \frac{P(t)}{C(t)} = \lim_{t \to +\infty} \frac{1.745t + 29.84}{1.097t + 10.65} = \lim_{t \to +\infty} \frac{1.745t}{1.097t} = \lim_{t \to +\infty} \frac{1.745}{1.097} \approx 1.59$

If the trend continues indefinitely, the annual spending on police will be 1.59 times the annual spending on courts in the long run.

55. $\lim_{t \to +\infty} I(t) = \lim_{t \to +\infty} (t^2 + 3.5t + 50) = +\infty$

$\lim_{t \to +\infty} \frac{I(t)}{E(t)} = \lim_{t \to +\infty} \frac{t^2 + 3.5t + 50}{0.4t^2 - 1.6t + 14} = \lim_{t \to +\infty} \frac{t^2}{0.4t^2}$

$= \lim_{t \to +\infty} \frac{1}{0.4} = 2.5.$

In the long term, U.S. imports from China will rise without bound and be 2.5 times U.S. exports to China. In the real world, imports and exports cannot rise without bound. Thus, the given models should not be extrapolated far into the future.

57. $\lim_{t\to+\infty} p(t) = \lim_{t\to+\infty} 100\left(1 - \frac{12{,}200}{t^{4.48}}\right) =$ $100(1 - 0) = 100$. The percentage of children who learn to speak approaches 100% as their age increases.

59. Yes: $\lim_{t\to 8^-} C(t) = \lim_{t\to 8^+} C(t) = 1.24$.

61. To evaluate $\lim_{x\to a} f(x)$ algebraically, first check whether $f(x)$ is a closed-form function. Then check whether $x = a$ is in its domain. If so, the limit is just $f(a)$; that is, it is obtained by substituting $x = a$. If not, then try to first simplify $f(x)$ in such a way as to transform it into a new function such that $x = a$ is in its domain, and then substitute. A disadvantage of this method is that it is sometimes extremely difficult to evaluate limits algebraically, and rather sophisticated methods are often needed.

63. She is wrong. Closed-form functions are continuous only at points in their domains, and $x = 2$ is not in the domain of the closed-form function $f(x) = 1/(x-2)^2$.

65. The statement may not be true. For example, if $f(x) = \begin{cases} x+2 & \text{if } x < 0 \\ 2x-1 & \text{if } x \geq 0, \end{cases}$

then $f(0)$ is defined and equals -1, and yet $\lim_{x\to 0} f(x)$ does not exist. The statement can be corrected by requiring that f be a closed-form function: "If f is a closed form function, and $f(a)$ is defined, then $\lim_{x\to a} f(x)$ exists and equals $f(a)$."

67. Answers may vary. For example,

$$f(x) = \begin{cases} 0 & \text{if } x \text{ is any number other than 1 or 2} \\ 1 & \text{if } x = 1 \text{ or } 2 \end{cases}$$

3.4

1. $[f(3) - f(1)]/(3 - 1) = (-1 - 5)/2 = -3$

3. $[f(-1) - f(-1)]/[-1 - (-3)] = [-1.5 - (-2.1)]/2 = 0.3$

5. $[R(6) - R(2)]/(6 - 2) = (20.1 - 20.2)/4 = -\$25{,}000$ per month

7. $[q(5.5) - q(5)]/(5.5 - 5) = (300 - 400)/0.5 = -200$ items per dollar

9. $[S(5) - S(2)]/(5 - 2) = (27 - 23)/3 \approx \$1.33$ per month

11. $[U(4) - U(0)]/(4 - 0) = (8 - 5)/4 = 0.75$ percentage point increase in unemployment per 1 percentage point increase in the deficit

13. $[f(3) - f(1)]/(3 - 1) = [6 - (-2)]/2 = 4$

15. $[f(0) - f(-2)]/[0 - (-2)] = (4 - 0)/2 = 2$

17. $[f(3) - f(2)]/(3 - 2) = [9/2 + 1/3 - (2 + 1/2)]/1 = 7/3$

19. $f(x) = 2x^2$

Average rate of change

$$= \frac{f(a+h) - f(a)}{h} = \frac{f(h) - f(0)}{h}$$

because $a = 0$.

$$h = 1: \frac{f(h) - f(0)}{h} = \frac{f(1) - f(0)}{1} = \frac{2 - 0}{1} = 2$$

$$h = 0.1: \frac{f(h) - f(0)}{h} = \frac{f(0.1) - f(0)}{0.1} = \frac{0.02 - 0}{0.1} = 0.2$$

Technology can be used to compute the remaining cases. All the values are shown in the following table:

h	Ave. Rate of Change
1	2
0.1	0.2
0.01	0.02
0.001	0.002
0.0001	0.0002

21. $f(x) = 1/x$

Average rate of change

$$= \frac{f(a+h) - f(a)}{h} = \frac{f(2+h) - f(2)}{h}$$

because $a = 2$.

$$h = 1: \frac{f(2+h) - f(2)}{h} = \frac{f(3) - f(2)}{1} = \frac{1/3 - 1/2}{1} = -1/6$$

≈ -0.1667

$$h = 0.1: \frac{f(2+h) - f(2)}{h} = \frac{f(2.1) - f(2)}{0.1} = \frac{1/2.1 - 1/2}{0.1}$$

≈ -0.2381

Technology can be used to compute the remaining cases. All the values are shown in the following table:

h	Ave. Rate of Change
1	−0.1667
0.1	−0.2381
0.01	−0.2488
0.001	−0.2499
0.0001	−0.24999

23. $f(x) = x^2 + 2x$

Average rate of change

$$= \frac{f(a+h) - f(a)}{h} = \frac{f(3+h) - f(3)}{h}$$

because $a = 3$.

$$h = 1: \frac{f(3+h) - f(3)}{h} = \frac{f(4) - f(3)}{1} = \frac{24 - 15}{1} = 9$$

$$h = 0.1: \frac{f(3+h) - f(3)}{h} = \frac{f(3.1) - f(3)}{0.1} = \frac{15.81 - 15}{0.1}$$

$= 8.1$

Technology can be used to compute the remaining cases. All the values are shown in the following table:

h	Ave. Rate of Change
1	9
0.1	8.1
0.01	8.01
0.001	8.001
0.0001	8.0001

25. (a) $[P(4) - P(0)]/(4 - 0)$
$= (131 - 132)/4$
$= -0.25$ million people per year.
During the period 2000–2004, employment in the U.S. decreased at an average rate of 0.25 million people per year.
(b) $[P(2) - P(-1)]/(2 - (-1))$
$= (130 - 120)/3$
$= 0$ people per year.
During the period 1999–2002 the average rate of change of employment in the U.S. was zero people per year.

27. (a) Look for the largest increase in N over a period of two years; it occurs over the period 1998–2000. The number of companies that invested in venture capital each year was increasing most rapidly during the period 1998–2000, when it grew at an average rate of $[N(10) - N(8)]/(10 - 8) = (1700 - 400)/2 =$ 650 companies per year. **(b)** Look for the least increase (or greatest decrease) in N over a period of two years; it occurs over the period 1999–2001. The number of companies that invested in venture capital each year was decreasing most rapidly during the period 1999–2001, when its rate of increase was $[N(11) - N(9)]/(11 - 9) = (900 - 1000)/2 = -50$, i.e., it decreased at an average rate of 50 companies per year.

29. (a) Check each interval ([3, 5], [3, 7], and so on). You will find the most negative average drop, in [3, 5], is $(4.6–5.1)/(5–3) = –0.25$ thousand articles per year. So, during the period 1993–1995, the number of articles authored by U.S. researchers decreased at an average rate of 250 articles per year.
(b) The percentage rate of change is
$[N(13) - N(3)]/N(3)$
$= (4.2–5.1)/5.1$
≈ -0.1765;
the average rate of change is
$[N(13) - N(3)]/(13 - 3)$
$= (4.2–5.1)/10$
$= -0.09$ thousand articles per year.
Over the period 1993–2003, the number of articles authored by U.S. researchers decreased at an average rate of 90 per year, representing an 17.65% decrease over that period.

31. (a) $(680 - 380)/4 = 75$ teams per year
(b) It decreased: The slope of the graph from $t = 16$ to $t = 20$ is clearly less than the slope from $t = 14$ to $t = 18$.

33. (a) $[N(1) - N(0)]/(1 - 0)$
$= (600 - 350)/1$
$= 250$ million transactions per year;
$[N(2) - N(1)]/(2 - 1)$
$= (450 - 600)/1$
$= -150$ million transactions per year;
$[N(2) - N(0)]/(2 - 0)$
$= (450 - 350)/2$
$= 50$ million transactions per year.
Over the period January 2000–January 2001, the (annual) number of on-line shopping transactions

in the U.S. increased at an average rate of 250 million per year. From January 2001 to January 2002, this number decreased at an average rate of 150 million per year. From January 2000 to January 2002, this number increased at an average rate of 50 million per year.

(b) The average rate of change of $N(t)$ over [0, 2] is the average of the rates of change over [0, 1] and [1, 2]: In this case 50 is the average of 250 and −150; in general the average of

$[N(1) - N(0)]/(1 - 0)$ and

$[N(2) - N(1)]/(2 - 1)$ equals

$[N(2) - N(0)]/(2 - 0)$.

35. We can estimate the slope of the regression line using two grid points it passes through: (2, 1.25) and (6, 1.3).

Slope of regression line

$\approx (1.3-1.25)/(6-2)$

$= 0.0125$ billion dollars per year

(a) (C): On the interval [0, 4], the average rate of change of government funding was approximately

$(1.3-1.24)/(4-0) = 0.0125$,

the same as the estimated slope of the regression line.

(b) (A): On the interval [4, 8], the average rate of change of government funding was approximately

$(1.2-1.3)/(8-4) = -0.025$,

less than the estimated slope of the regression line.

(c) (B): On the interval [3, 6], the average rate of change of government funding was approximately

$(1.4-1.22)/(6-3) = 0.06$,

greater than the estimated slope of the regression line.

(d) Over the interval [0, 8] the average rate of change of government funding was approximately

$(1.2 - 1.25)/(8 - 0)$

$= -0.00625 \approx -0.0063$

(to two significant digits) billion dollars per year, (−\$6,300,000 per year). This is much less than the (positive) slope of the regression line, $0.0125 \approx 0.013$ billion dollars per year, (\$13,000,000 per year).

37. From 1991 to 1995 the volatility decreased at an average rate of 0.2 points per year, so decreased a total of $4\times0.2 = 0.8$ points. Since its value in 1995 was 1.1, its value in 1991 must have been $1.1 + 0.8 = 1.9$. Similarly, from 1995 to 1999 the volatility increased a total of $4\times0.3 = 1.2$ to end at $1.1 + 1.2 = 2.3$. In between these points we've found almost anything could happen, but the graph might look something like the following:

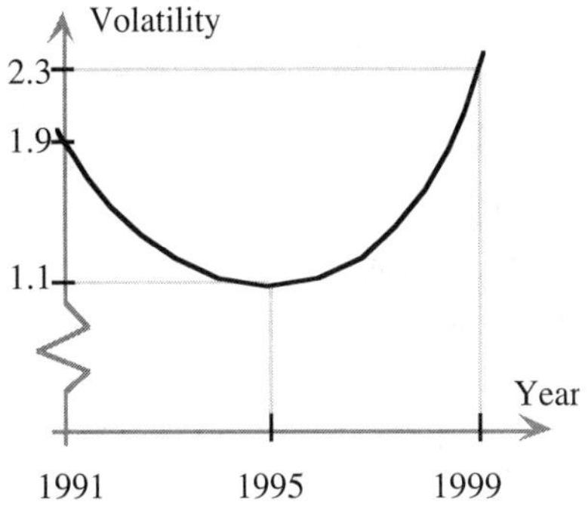

39. $[I(2) - I(0)]/(2 - 0)$

$= (1600 - 1000)/2 = 300$.

The index was increasing at an average rate of 300 points per day.

41. Average rate of change of e is

$$\frac{e(4.5) - e(0.5)}{4.5 - 0.5}$$

$$= \frac{1.279-0.959}{4} = \$\ 0.08 \text{ per year.}$$

The value of the euro in U.S. dollars was growing at an average rate of about \$0.08 per year over the period June 2000 – June 2004.

43. (a) $[f(6) - f(5)]/(6 - 5)$
$= (27.6 - 18.75)/1$
$= 8.85$ manatee deaths per 100,000 boats;
$[f(8) - f(7)]/(8 - 7)$
$= (66.6 - 43.55)/1$
$= 23.05$ manatee deaths per 100,000 boats

(b) More boats result in more manatee deaths per additional boat.

45. (a) $[R(2) - R(0)]/(2 - 0)$
$= (760 - 150)/2$
$= \$305$ million per year.

Over the period 1997–1999, annual advertising revenues increased at an average rate of \$305 million per year.

(b) (A): The average rate of change from 1998 to 1999 was larger than the average rate of change from 1997 to 1998.

(c) $[R(3) - R(2)]/(2 - 0)$
$= (1350 - 760)/1$
$= \$590$ million per year.

The model projects annual advertising revenues to increase by \$590 million per year in 2000.

47. (a) Here is an Excel worksheet that computes the successive rates of change:

◇	A	B	C
1	x	p(x)	Rate of Change
2	39	=0.092*A2^2-8.1*A2+190	
3	39.5		=(B3-B2)/0.5
4	40		
5	40.5		
6	41		
7	41.5		
8	42		

This worksheet leads to the following values, showing the desired rates of change in the rightmost column.

◇	A	B	C
1	x	p(x)	Rate of Change
2	39	14.032	
3	39.5	13.593	-0.88
4	40	13.2	-0.79
5	40.5	12.853	-0.69
6	41	12.552	-0.60
7	41.5	12.297	-0.51
8	42	12.088	-0.42

(b) From the table, the average rate of change of p over [40, 40.5] is -0.69. The unites of measurement are units of p per unit ot x: percentage points per \$1000 of household income. Thus, for household incomes between \$40,000 and \$40,500, the poverty rate decreases at an average rate of 0.69 percentage points per \$1000 increase in the median household income.

(c) All the rates of change in obtained in part (a) are negative, showing that the poverty rate decreases as the median houshold income increases (Choice B).

(d) Although all the rates of change in the table are negative, they become less so as the household income increases. Thus, the effect is decreasing in magnitude (Choice B).

49. The average rate of change of f over an interval $[a, b]$ can be determined numerically; using a table of values; graphically, by measuring the slope of the corresponding line segment through two points on the graph; or algebraically, using an algebraic formula for the function. Of these, the least precise is the graphical method, because it relies on reading coordinates of points on a graph.

51. Answers will vary. Here is one possibility:

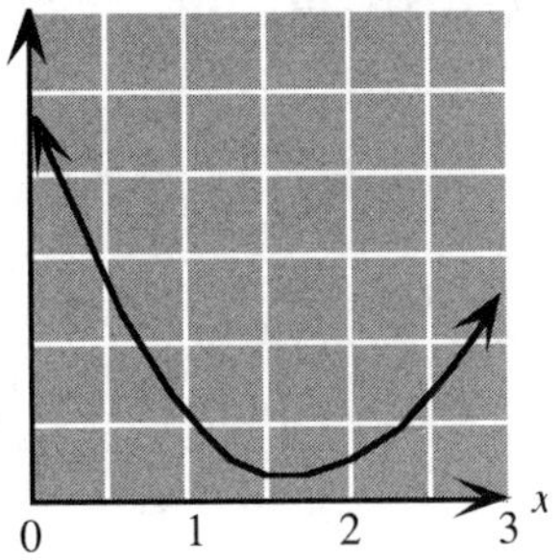

53. For every change of 1 in C, B changes by 3, so A changes by $2\times3 = 6$ units of quantity A per unit of quantity C

55. (A): The secant line given by $x = 1$ and $x = 1 + h$ is steeper for smaller values of h.

57. Yes. Here is an example, in which the average rate of growth for 2000–2003 is negative, but the average rates of growth for 2000–2001 and 2001–2002 are positive:

Year	2000	2001	2002	2003
Revenue (billion)	$10	$20	$30	$5

59. **(A)**: This can be checked by algebra:
$\{[f(2) - f(1)]/(2 - 1) + [f(3) - f(2)]/(3 - 2)\}/2 = [f(3) - f(1)]/(3 - 1)$.

3.5

1. 6: The average rates of change are approaching 6 for both positive and negative values of h approaching 0.

3. −5.5: The average rates of change are approaching −5.5 for both positive and negative values of h approaching 0.

5. The average rate of change is $\frac{R(a+h) - R(a)}{h}$.

Here, $a = 5$.

$h = 1$:

$$\frac{R(5+1) - R(5)}{1} = \frac{R(6) - R(5)}{1} = \frac{39}{1} = 39$$

$h = 0.1$:

$$\frac{R(5+0.1) - R(5)}{0.1} = \frac{R(5.1) - R(5)}{0.1} = \frac{3.99}{0.1} = 39.9$$

$h = 0.01$:

$$\frac{R(5+0.01) - R(5)}{0.01} = \frac{R(5.01) - R(5)}{0.01} = \frac{0.3999}{0.01} = 39.99$$

Table:

h	1	0.1	0.01
Ave. rate	39	39.9	39.99

The average rates are approaching an instantaneous rate of 40 rupees per day.

7. The average rate of change is $\frac{R(a+h) - R(a)}{h}$.

Here, $a = 1$.

$h = 1$:

$$\frac{R(1+1) - R(1)}{1} = \frac{R(2) - R(1)}{1} = \frac{140}{1} = 140$$

$h = 0.1$:

$$\frac{R(1+0.1) - R(1)}{0.1} = \frac{R(1.1) - R(1)}{0.1} = \frac{6.62}{0.1} = 66.2$$

$h = 0.01$:

$$\frac{R(1+0.01) - R(1)}{0.01} = \frac{R(1.01) - R(1)}{0.01} = \frac{0.60602}{0.01} = 60.602$$

Table:

h	1	0.1	0.01
Ave. rate	140	66.2	60.602

The average rates are approaching an instantaneous rate of 60 rupees per day.

9.

The average cost to manufacture h more items is the average rate of change: $\frac{C(a+h) - C(a)}{h}$.

Here, $a = 1000$.

$h = 10$:

$$\frac{C(1000+10) - C(1000)}{10} = \frac{C(1010) - C(1000)}{10} = \frac{47.99}{10} = 4.799$$

$h = 1$:

$$\frac{C(1000+1) - C(1000)}{1} = \frac{C(1001) - C(1000)}{1} = \frac{4.7999}{1} = 4.7999$$

Table:

h	10	1
C_{ave}	4.799	4.7999

$C'(1{,}000) = \$4.8$ per item

11. The average cost to manufacture h more items is the average rate of change: $\frac{C(a+h) - C(a)}{h}$.

Here, $a = 100$.

$h = 10$:

$$\frac{C(100+10) - C(100)}{10} = \frac{C(110) - C(100)}{10}$$

$$= \frac{999.0909091}{10} \approx 99.91$$

$h = 1$:

$$\frac{C(100+1) - C(100)}{1} = \frac{C(101) - C(100)}{1}$$

$$= \frac{99.9009901}{1} \approx 99.90$$

Table:

h	10	1
C_{ave}	99.91	99.90

$C'(100) = \$99.9$ per item

In each of 13–18, the answer to (a) is the point at which the graph is rising with the steepest slope or falling with the shallowest slope; the answer to (b) is the point at which the graph is rising with the shallowest slope or falling with the steepest slope.

13. **(a)** R **(b)** P

15. **(a)** P **(b)** R

17. **(a)** Q **(b)** P

In each of 19–22, the answer is obtained by estimating the slope of the tangent line shown.

19. In the graph, the tangent line passes through (0, 2) and (6, 5). Therefore its slope is (5−2)/(6−0) = 1/2.

21. In the graph, the tangent is horizontal. Therefore its slope is 0.

In each of 23–26, the answer may be obtained by estimating the slopes of the tangent lines to the given points and comparing these slopes to the given numbers.

23. **(a)** Q **(b)** R **(c)** P

25. **(a)** R **(b)** Q **(c)** P

27. **(a)** The only point where the tangent line has slope 0 is (1, 0).

(b) None; the graph never rises.

(c) The only point where the tangent line has slope −1 is (−2, 1).

29. **(a)** The points where the tangent line has slope 0 are (−2, 0.3), (0, 0), and (2, −0.3).

(b) None; the graph never rises that steeply.

(c) None; the graph never falls that steeply.

31. $(a, f(a))$; $f'(a)$.

33. (B): The derivative is the slope of the tangent line. It is not any particular average rate of change or difference quotient; these only *approximate* the derivative.

35. **(a)** (A): The graph rises above the tangent line at $x = 2$.

(b) (C): The secant line is roughly parallel to the tangent line at $x = 0$.

(c) (B): The slopes of the tangent lines are decreasing.

(d) (B): The slopes of the tangent lines decrease to 0 then increase again.

(e) (C): The height of the graph is approximately 0.7 while the slope of the tangent line at $x = 4$ is approximately 1.

37. $[f(2 + 0.0001) - f(2 - 0.0001)]/0.0002 = -2$

39. $[f(-1 + 0.0001) - f(-1 - 0.0001)]/0.0002 \approx -1.5$

In each of 41–48 we use the "quick approximation" method of estimating the derivative using the balanced difference quotient. You could also use the ordinary difference quotient or a table of average rates of change with values of h approaching 0.

41. $[g(t + 0.0001) - g(t - 0.0001)]/0.0002 \approx -5$

43. $[y(2 + 0.0001) - y(2 - 0.0001)]/0.0002 = 16$

45. $[s(-2 + 0.0001) - s(-2 - 0.0001)]/0.0002 = 0$

47. $[R(20 + 0.0001) - R(20 - 0.0001)]/0.0002 \approx -0.0025$

49. (a) $[f(-1 + 0.0001) - f(-1 - 0.0001)]/0.0002 \approx 3$ **(b)** The equation of the line through $(-1, -1)$ with slope 3 is $y = 3x + 2$.

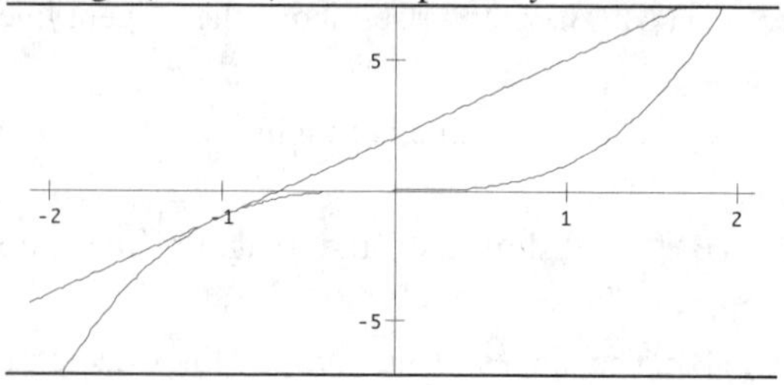

51. (a) $[f(2 + 0.0001) - f(2 - 0.0001)]/0.0002 \approx \frac{3}{4}$ **(b)** The equation of the line through (2, 2.5) with slope $\frac{3}{4}$ is $y = \frac{3}{4}x + 1$.

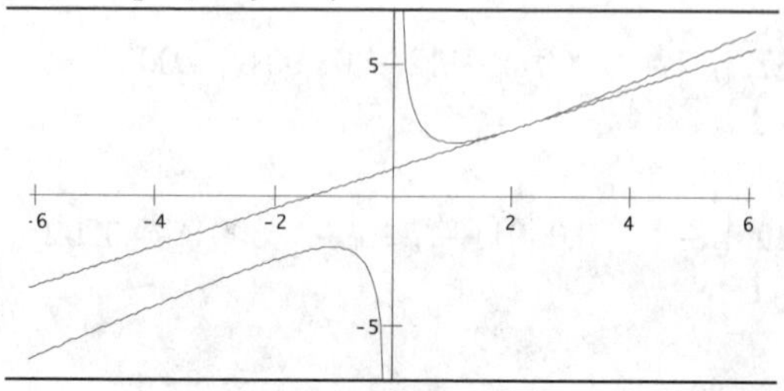

53. (a) $[f(4 + 0.0001) - f(4 - 0.0001)]/0.0002 \approx \frac{1}{4}$ **(b)** The equation of the line through (4, 2) with slope $\frac{1}{4}$ is $y = \frac{1}{4}x + 1$.

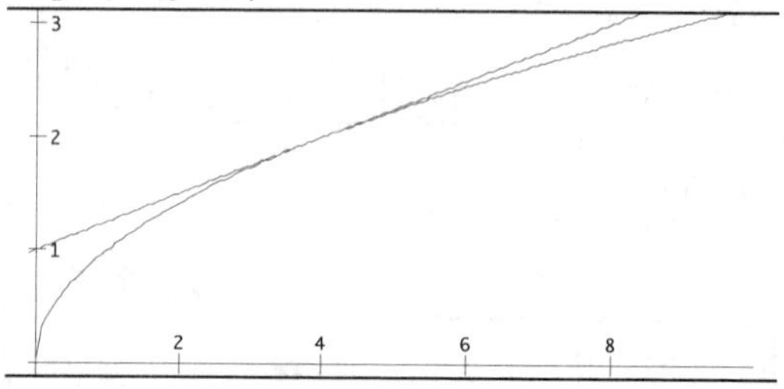

55. $[e^{0.0001} - e^{-0.0001}]/0.0002 \approx 1.000$

57. $[\ln(1 + 0.0001) - \ln(1 - 0.0001)]/0.0002 \approx 1.000$

59. (C): The graph of f is a falling straight line, so must have the same negative slope (derivative) at every point; f' must be a negative constant.

61. (A): The function f decreases until $x = 0$, where it turns around and starts to increase. Its derivative must be negative until $x = 0$, where the derivative is 0; past that the derivative becomes positive. This is exactly what (A) illustrates.

63. (F): The function f increases slowly at first, it becomes steeper around $x = 0$, then it returns to slowly rising. Its derivative starts as a small positive number, increases to become largest around $x = 0$, then decreases back toward 0. This is the behavior seen in (F).

65. The derivative:

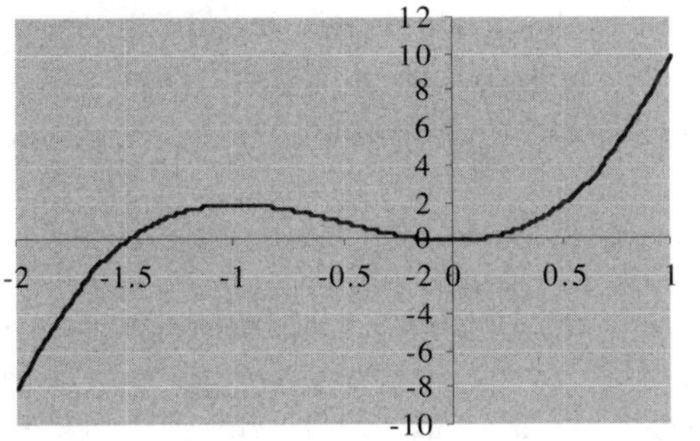

$x = -1.5$, $x = 0$: These are the points where the derivative is 0 (crosses the x axis).

67. Set up the spreadsheet as in Example 4:

◇	A	B	C	D	E
1	x	f(x)	f'(x)	Xmin	4
2	=$E$1	=(A2+2)/(A2-3)	=(B3-B2)/$E$2	h	0.1
3	=A2+$E$2				
4					
5					
6					
7					
8					
9					
10					
11					
12					

↓

◇	A	B	C	D	E
1	x	f(x)	f'(x)	Xmin	4
2	4	6	-4.545454545	h	0.1
3	4.1	5.545454545	-3.787878788		
4	4.2	5.166666667	-3.205128205		
5	4.3	4.846153846	-2.747252747		
6	4.4	4.571428571	-2.380952381		
7	4.5	4.333333333	-2.083333333		
8	4.6	4.125	-1.838235294		
9	4.7	3.941176471	-1.633986928		
10	4.8	3.777777778	-1.461988304		
11	4.9	3.631578947	-1.315789474		
12	5	3.5			

To graph $f(x)$ and $f'(x)$, highlight columns A through C and use the Chart Wizard to create the following Scatter Plot:

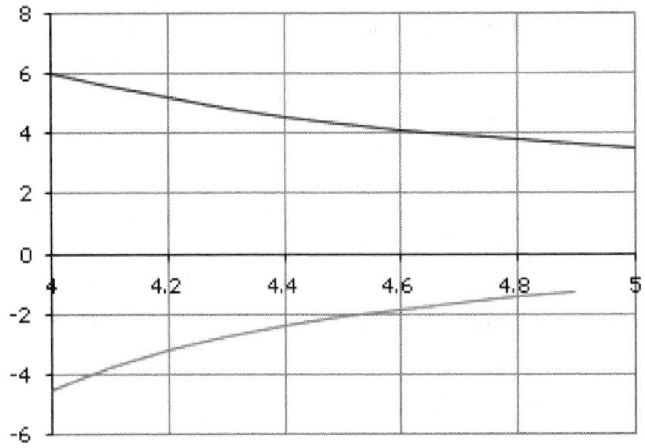

The top curve is $y = f(x)$; the bottom curve is $y = f'(x)$.

69. $q(100) = 50{,}000$, $q'(100) = -500$ (use one of the quick approximations). A total of 50,000 pairs of sneakers can be sold at a price of \$100, but the demand is decreasing at a rate of 500 pairs per \$1 increase in the price.

71. (a) Sales in 2000 were approximately 160,000 pools per year and were increasing at a rate of 6000 per year.
(b) Decreasing, since the slope is decreasing.

73. (a) (B): The graph is getting less steep.
(b) (B): The graph goes from above the tangent line to below it, so the slope of the tangent line is greater than the average rate of change.
(c) (A): From 0 to about 12 the graph is getting steeper, so the instantaneous rate of change is increasing; from that point on the graph is getting less steep, so the instantaneous rate of change is decreasing.
(d) 1992: This is the point ($t = 12$) where the graph is steepest.
(e) Reading values from the graph, we get the approximation $(1.2 - 0.8)/8 = 0.05$: In 1996, the total number of state prisoners was increasing at a rate of approximately 50,000 prisoners per year.

75. **(a)** $[s(4) - s(2)]/(4 - 2) = -96$ ft/sec
(b) $[s(4 + 0.0001) - s(4 - 0.0001)]/0.0002 = -128$ ft/sec

77. **(a)** $\frac{e(4) - e(0)}{4 - 0} = \frac{1.176 - 1}{4 - 0} = \$0.044$ per year. The value of the euro was increasing at an average rate of about \$0.044 per year over the period January 2000 – January 2004.
(b) $\frac{e(0 + 0.0001) - e(0 - 0.0001)}{0.0002} = \frac{0.99999 - 1.00001}{0.0002} = -\$0.10$ per year. In January, 2000, the value of the euro was decreasing at a rate of about \$0.10 per year. **(c)** The value of the euro was decreasing in January 2000, and then began to increase (making the average rate of change in part (a) positive).

79. **(a)** $[R(1 + 0.0001) - R(1 - 0.0001)]/0.0002 = \$305$ million per year.
(b) (A): If we estimate the instantaneous rates of change at 1997, 1998, and 1999, we get \$115 million per year, \$305 million per year, and \$495 million per year, respectively. We also know that the graph of R is a parabola opening upward, so it rises more and more steeply.
(c) $[R(3 + 0.0001) - R(3 - 0.0001)]/0.0002 = \$685$ million/year. In December 2000, AOL's advertising revenue was projected to be increasing at a rate of \$685 million per year.

81. $A(0) = 4.5$ million because A gives the number of subscribers; $A'(0) = 60{,}000$ because A' gives the rate at which the number of subscribers is changing.

83. **(a)** 60% of children can speak at the age of 10 months. At the age of 10 months, this percentage is increasing by 18.2 percentage points per month.
(b) As t increases, p approaches 100 percentage points (almost all children eventually learn to speak), and dp/dt approaches zero because the percentage stops increasing.

85. $S(5) \approx 109$, $\frac{dS}{dt}\Big|_{t=5} \approx [S(5 + 0.0001) - S(5 - 0.0001)]/0.0002 \approx 9.1$. After 5 weeks, sales are 109 pairs of sneakers per week, and sales are increasing at a rate of 9.1 pairs per week each week.

87. **(a)** $P(50) \approx 62$, $P'(50) \approx [P(50 + 0.0001) - P(50 - 0.0001)]/0.0002 \approx 0.96$; 62% of U.S. households with an income of \$50,000 have a computer. This percentage is increasing at a rate of 0.96 percentage points per \$1000 increase in household income.
(b) Using the techniques of Example 4, we obtain the following graphs of P and its derivative:

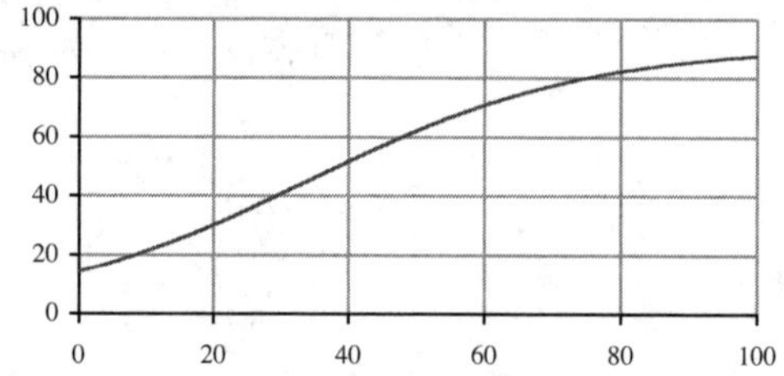

Graph of P

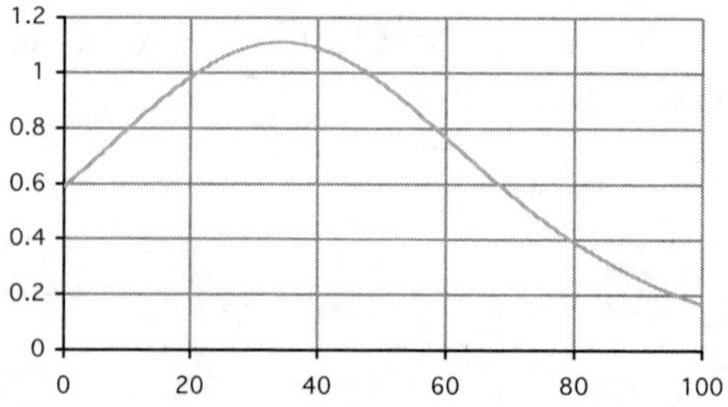

Graph of P'

As with any logistic function, the graph levels off for large values of x, so P' decreases toward zero. This can also be seen in the graph of P': notice that it decreases toward zero as x gets large.

89.

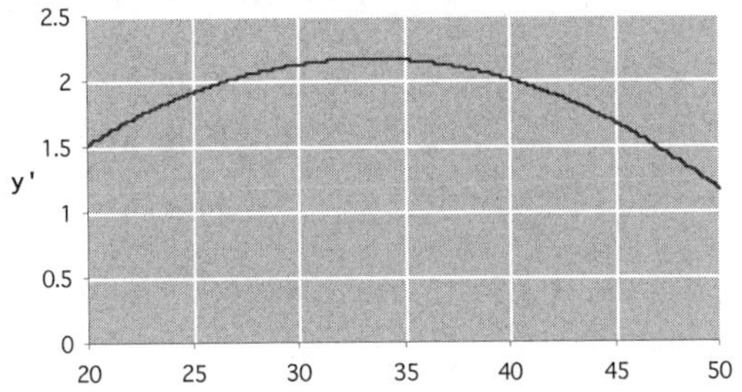

(a) (D): The graph of the derivative is rising.
(b) 33 days after the egg was laid: That is where the graph of the derivative is highest.
(c) 50 days after the egg was laid: That is where the graph of the derivative is lowest in the range $20 \le t \le 50$.

91. $L(0.95) \approx 31.2$ meters and $L'(0.95) \approx [L(0.95 + 0.0001) - L(0.95 - 0.0001)]/0.0002 \approx -304.2$ meters/warp. Thus, at a speed of warp 0.95, the spaceship has an observed length of 31.2 meters and its length is decreasing at a rate of 304.2 meters per unit warp, or 3.042 meters per increase in speed of 0.01 warp.

93. The difference quotient is not defined when $h = 0$ because there is no such number as 0/0.

95. The derivative is positive (sales are still increasing) and decreasing toward zero (sales are leveling off).

97. Company B. Although the company is currently losing money, the derivative is positive, showing that the profit is increasing. Company A, on the other hand, has profits that are declining.

99. (C) is the only graph in which the instantaneous rate of change on January 1 is greater than the one-month average rate of change.

101. The tangent to the graph is horizontal at that point, and so the graph is almost horizontal near that point.

103. Various graphs are possible.

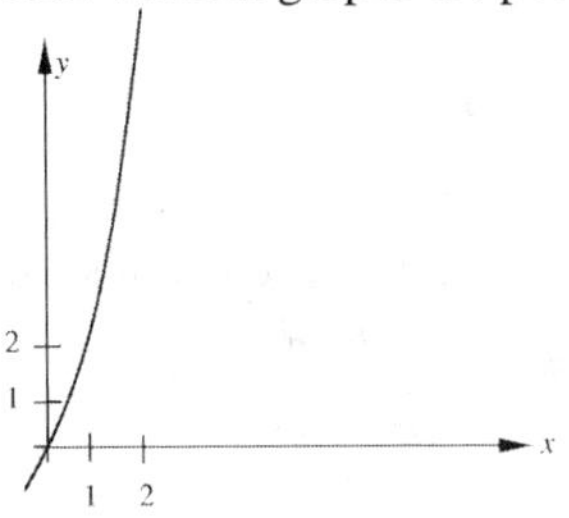

105. If $f(x) = mx + b$, then its average rate of change over any interval $[x, x+h]$ is

$$\frac{m(x+h) + b - (mx + b)}{h} = m.$$

Because this does not depend on h, the instantaneous rate is also equal to m.

107. Increasing, because the average rate of change appears to be rising as we get closer to 5 from the left (see the bottom row).

109. Answers may vary.

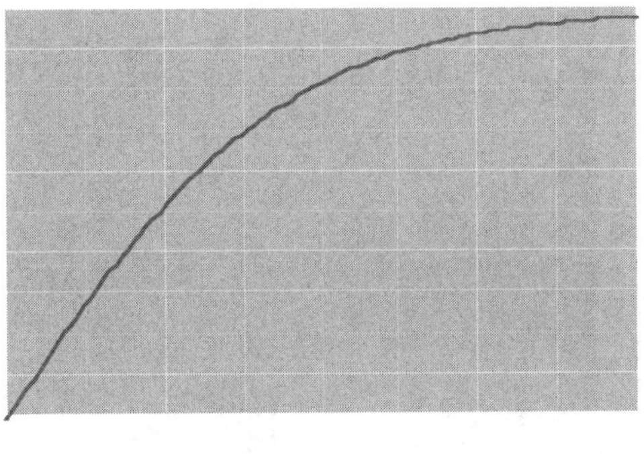

111.

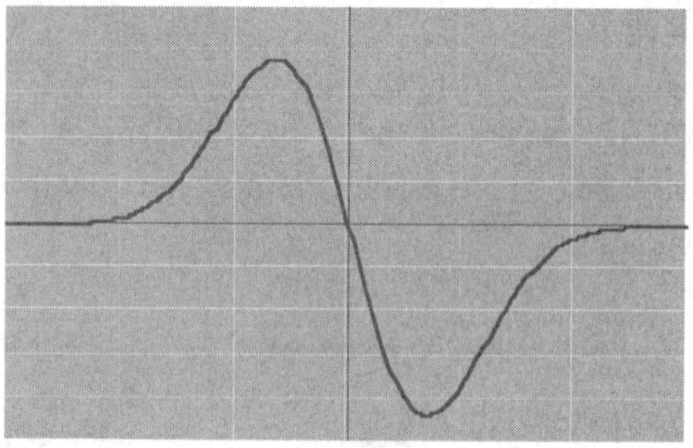

113. (B): His average speed was 60 miles per hour. If his instantaneous speed was always 55 mph or less, he could not have averaged more than 55 mph.

115. Answers will vary. Graph:

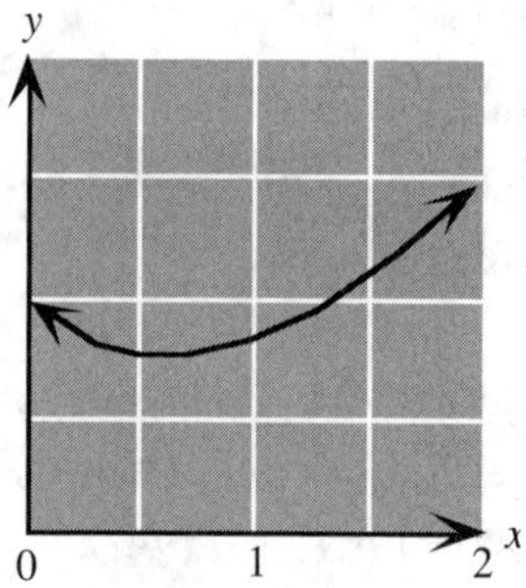

3.6

1. $f'(2) = \lim_{h\to 0} \frac{f(2+h) - f(2)}{h} = \lim_{h\to 0} \frac{(2+h)^2 + 1 - (2^2 + 1)}{h} = \lim_{h\to 0} \frac{4 + 4h + h^2 + 1 - 5}{h} =$

$\lim_{h\to 0} \frac{4h + h^2}{h} = \lim_{h\to 0} (4 + h) = 4$

3. $f'(-1) = \lim_{h\to 0} \frac{f(-1+h) - f(-1)}{h} = \lim_{h\to 0} \frac{3(-1 + h) - 4 - (-3 - 4)}{h} = \lim_{h\to 0} \frac{-3 + 3h - 4 + 7}{h} =$

$\lim_{h\to 0} \frac{3h}{h} = \lim_{h\to 0} 3 = 3$

5. $f'(1) = \lim_{h\to 0} \frac{f(1+h) - f(1)}{h} = \lim_{h\to 0} \frac{3(1+h)^2 + (1+h) - (3 + 1)}{h} = \lim_{h\to 0} \frac{3 + 6h + 3h^2 + 1 + h - 4}{h}$

$= \lim_{h\to 0} \frac{7h + 3h^2}{h} = \lim_{h\to 0} (7 + 3h) = 7$

7. $f'(-1) = \lim_{h\to 0} \frac{f(-1+h) - f(-1)}{h} = \lim_{h\to 0} \frac{2(-1+h) - (-1+h)^2 - (-2 - 1)}{h} =$

$\lim_{h\to 0} \frac{-2 + 2h - 1 + 2h - h^2 + 3}{h} = \lim_{h\to 0} \frac{4h - h^2}{h} = \lim_{h\to 0} (4 - h) = 4$

9. $f'(2) = \lim_{h\to 0} \frac{f(2+h) - f(2)}{h} = \lim_{h\to 0} \frac{(2+h)^3 + 2(2+h) - (8 + 4)}{h} =$

$\lim_{h\to 0} \frac{8 + 12h + 6h^2 + h^3 + 4 + 2h - 12}{h} = \lim_{h\to 0} \frac{14h + 6h^2 + h^3}{h} = \lim_{h\to 0} (14 + 6h + h^2) = 14$

11. $f'(1) = \lim_{h\to 0} \frac{f(1+h) - f(1)}{h} = \lim_{h\to 0} \frac{-1/(1+h) - (-1)}{h} = \lim_{h\to 0} \frac{-1 + (1 + h)}{h(1 + h)} = \lim_{h\to 0} \frac{h}{h(1 + h)} =$

$\lim_{h\to 0} \frac{1}{1 + h} = 1$

13. $f'(43) = \lim_{h\to 0} \frac{f(43+h) - f(43)}{h} = \lim_{h\to 0} \frac{m(43+h) + b - (43m + b)}{h} =$

$\lim_{h\to 0} \frac{43m + mh + b - 43m - b}{h} = \lim_{h\to 0} \frac{mh}{h} = \lim_{h\to 0} m = m$

15. $f'(x) = \lim_{h\to 0} \frac{f(x+h) - f(x)}{h} = \lim_{h\to 0} \frac{(x+h)^2 + 1 - (x^2 + 1)}{h} \quad \lim_{h\to 0} \frac{x^2 + 2xh + h^2 + 1 - x^2 - 1}{h} =$

$\lim_{h\to 0} \frac{2xh + h^2}{h} = \lim_{h\to 0} (2x + h) = 2x$

17, $f'(x) = \lim_{h\to 0} \frac{f(x+h) - f(x)}{h} = \lim_{h\to 0} \frac{3(x + h) - 4 - (3x - 4)}{h} = \lim_{h\to 0} \frac{3x + 3h - 4 - 3x + 4}{h} =$

$\lim_{h\to 0} \frac{3h}{h} = \lim_{h\to 0} 3 = 3$

19. $f'(x) = \lim_{h\to 0} \frac{f(x+h) - f(x)}{h} = \lim_{h\to 0} \frac{3(x+h)^2 + (x+h) - (3x^2 + x)}{h}$

$\lim_{h\to 0} \frac{3x^2 + 6xh + 3h^2 + x + h - 3x^2 - x}{h} = \lim_{h\to 0} \frac{6xh + 3h^2 + h}{h} = \lim_{h\to 0} (6x + 3h + 1) = 6x + 1$

21. $f'(x) = \lim_{h\to 0} \frac{f(x+h) - f(x)}{h} = \lim_{h\to 0} \frac{2(x+h) - (x+h)^2 - (2x - x^2)}{h}$

$\lim_{h\to 0} \frac{2x + 2h - x^2 - 2xh - h^2 - 2x + x^2}{h} = \lim_{h\to 0} \frac{2h - 2xh - h^2}{h} = \lim_{h\to 0} (2 - 2x - h) = 2 - 2x$

23. $f'(x) = \lim_{h\to 0} \frac{f(x+h) - f(x)}{h} = \lim_{h\to 0} \frac{(x+h)^3 + 2(x+h) - (x^3 + 2x)}{h} =$

$\lim_{h\to 0} \frac{x^3 + 3x^2h + 3xh^2 + h^3 + 2x + 2h - x^3 - 2x}{h} = \lim_{h\to 0} \frac{3x^2h + 3xh^2 + h^3 + 2h}{h} =$

$\lim_{h\to 0} (3x^2 + 3xh + h^2 + 2) = 3x^2 + 2$

25. $f'(x) = \lim_{h\to 0} \frac{f(x+h) - f(x)}{h} = \lim_{h\to 0} \frac{-1/(x+h) - (-1/x)}{h} = \lim_{h\to 0} \frac{-x + (x + h)}{hx(x + h)} = \lim_{h\to 0} \frac{h}{hx(x + h)}$

$= \lim_{h\to 0} \frac{1}{x(x + h)} = \frac{1}{x^2}$

27. $f'(x) = \lim_{h\to 0} \frac{f(x+h) - f(x)}{h} = \lim_{h\to 0} \frac{m(x+h) + b - (mx + b)}{h} = \lim_{h\to 0} \frac{mx + mh + b - mx - b}{h} =$

$\lim_{h\to 0} \frac{mh}{h} = \lim_{h\to 0} m = m$

29. $R'(2) = \lim_{h\to 0} \frac{R(2+h) - R(2)}{h} = \lim_{h\to 0} \frac{-0.3(2+h)^2 - (-0.3\times 2^2)}{h} =$

$\lim_{h\to 0} \frac{-1.2 - 1.2h - 0.3h^2 + 1.2}{h} = \lim_{h\to 0} \frac{-1.2h - 0.3h^2}{h} = \lim_{h\to 0} (-1.2 - 0.3h) = -1.2$

31. $U'(3) = \lim_{h\to 0} \frac{U(3+h) - U(3)}{h} = \lim_{h\to 0} \frac{5.1(3+h)^2 + 5.1 - (5.1\times 9 + 5.1)}{h} =$

$\lim_{h\to 0} \frac{45.9 + 30.6h + 5.1h^2 + 5.1 - 51}{h} = \lim_{h\to 0} \frac{30.6h + 5.1h^2}{h} = \lim_{h\to 0} (30.6 + 5.1h) = 30.6$

33. $U'(1) = \lim_{h\to 0} \frac{U(1+h) - U(1)}{h} = \lim_{h\to 0} \frac{-1.3(1+h)^2 - 4.5(1+h) - (-1.3 - 4.5)}{h} =$

$\lim_{h\to 0} \frac{-1.3 - 2.6h - 1.3h^2 - 4.5 - 4.5h + 5.8}{h} = \lim_{h\to 0} \frac{-7.1h - 1.3h^2}{h} = \lim_{h\to 0} (-7.1 - 1.3h) = -7.1$

35. $L'(1.2) = \lim_{h\to 0} \frac{L(1.2+h) - L(1.2)}{h} = \lim_{h\to 0} \frac{4.25(1.2+h) - 5.01 - (4.25\times 1.2 - 5.01)}{h} =$

$\lim_{h\to 0} \frac{5.1 + 4.25h - 5.01 - 5.1 + 5.01}{h} = \lim_{h\to 0} \frac{4.25h}{h} = \lim_{h\to 0} 4.25 = 4.25$

37. $q'(2) = \lim_{h\to 0} \frac{q(2+h) - q(2)}{h} = \lim_{h\to 0} \frac{2.4/(2+h) - 2.4/2}{h} = \lim_{h\to 0} \frac{4.8 - 2.4(2 + h)}{2h(2 + h)} =$

$\lim_{h\to 0} \frac{-2.4h}{2h(2 + h)} = \lim_{h\to 0} \frac{-1.2}{h(2 + h)} = -0.6$

39. Find the slope by finding the derivative:

$m = f'(2) = \lim_{h\to 0} \frac{f(2+h) - f(2)}{h} = \lim_{h\to 0} \frac{(2+h)^2 - 3 - (2^2 - 3)}{h} = \lim_{h\to 0} \frac{4 + 4h + h^2 - 3 - 1}{h} =$

$\lim_{h\to 0} \frac{4h + h^2}{h} = \lim_{h\to 0} (4 + h) = 4$. The tangent line has slope 4 and goes through $(2, f(2)) = (2, 1)$, so has equation $y = 4x - 7$.

41. Find the slope by finding the derivative:

$m = f'(3) = \lim_{h\to 0} \frac{f(3+h) - f(3)}{h} = \lim_{h\to 0} \frac{-2(3+h) - 4 - (-2\times 3 - 4)}{h} = \lim_{h\to 0} \frac{-6 - 2h - 4 + 10}{h} =$

$\lim_{h\to 0} \frac{-2h}{h} = \lim_{h\to 0} (-2) = -2$. The tangent line has slope -2 and goes through $(3, f(3)) = (3, -10)$, so has equation $y = -2x - 4$.

43. Find the slope by finding the derivative:

$$m = f'(-1) = \lim_{h\to 0} \frac{f(-1+h) - f(-1)}{h} = \lim_{h\to 0} \frac{(-1+h)^2 - (-1+h) - [(-1)^2 - (-1)]}{h} =$$

$$\lim_{h\to 0} \frac{1 - 2h + h^2 + 1 - h - 2}{h} = \lim_{h\to 0} \frac{-3h + h^2}{h} = \lim_{h\to 0} (-3 + h) = -3.$$ The tangent line has slope -3 and goes through $(-1, f(-1)) = (-1, 2)$, so has equation $y = -3x - 1$.

45. $$s'(4) = \lim_{h\to 0} \frac{s(4+h) - s(4)}{h} = \lim_{h\to 0} \frac{400 - 16(4+h)^2 - (400 - 16(4)^2)}{h} = \lim_{h\to 0} \frac{-128h - 16h^2}{h} =$$

-128 ft/sec

47. $$I'(5) = \lim_{h\to 0} \frac{I(5+h) - I(5)}{h}$$

$$= \lim_{h\to 0} \frac{(5+h)^2 + 3.5(5+h) + 50 - (5^2 + 3.5\times 5 + 50)}{h}$$

$$= \lim_{h\to 0} \frac{13.5h + h^2}{h} = 13.5;$$ annual U.S. imports from China were increasing by \$13.5 billion per year in 2000.

49. $$R'(t) = \lim_{h\to 0} \frac{R(t+h) - R(t)}{h}$$

$$= \lim_{h\to 0} \frac{17(t+h)^2 + 100(t+h) + 2300 - (17t^2 + 100t + 2300)}{h}$$

$$= \lim_{h\to 0} \frac{34th + 17h^2 + 100h}{h} = 34t + 100;$$ $R'(10) = 440$; annual U.S. sales of bottled water were increasing by 440 million gallons per year in 2000.

51. $$f'(8) = \lim_{h\to 0} \frac{f(8+h) - f(8)}{h} = \lim_{h\to 0} \frac{3.55(8+h)^2 - 30.2(8+h) + 81 - (3.55\times 8^2 - 30.2\times 8 + 81)}{h} =$$

$$\lim_{h\to 0} \frac{26.6h + 3.55h^2}{h} = 26.6$$ manatee deaths per 100,000 additional boats. At a level of 800,000 boats, the number of manatee deaths is increasing at a rate of 26.6 manatees per 100,000 additional boats.

53. The algebraic method, because it gives the exact value of the derivative. The other two approaches give only approximate values (except in some special cases).

55. Because the algebraic computation of $f'(a)$ is exact, and not an approximation, it makes no difference whether one uses the balanced difference quotient or the ordinary difference quotient in the algebraic computation.

57. The computation results in a limit that cannot be evaluated.

3.7

1. $5x^4$

3. $2(-2x^{-3}) = -4x^{-3}$

5. $-0.25x^{-0.75}$

7. $4(2x^3) + 3(x^2) - 0 = 8x^3 + 9x^2$

9. $-1 - 1/x^2$

11. $\frac{dy}{dx} = 10(0) = 0$ (constant multiple and power rule)

13. $\frac{dy}{dx} = \frac{d}{dx}(x^2) + \frac{d}{dx}(x)$ (sum rule) $= 2x + 1$ (power rule)

15. $\frac{dy}{dx} = \frac{d}{dx}(4x^3) + \frac{d}{dx}(2x) - \frac{d}{dx}(1)$ (sum and difference) $= 4\frac{d}{dx}(x^3) + 2\frac{d}{dx}(x) - \frac{d}{dx}(1)$ (constant multiples) $= 12x^2 + 2$ (power rule)

17. $f'(x) = 2x - 3$

19. $f'(x) = 1 + 0.5x^{-0.5}$

21. $g'(x) = -2x^{-3} + 3x^{-2}$

23. $g'(x) = \frac{d}{dx}(x^{-1} - x^{-2}) = -x^{-2} + 2x^{-3} = -\frac{1}{x^2} + \frac{2}{x^3}$

25. $h'(x) = \frac{d}{dx}(2x^{-0.4}) = -0.8x^{-1.4} = -\frac{0.8}{x^{1.4}}$

27. $h'(x) = \frac{d}{dx}(x^{-2} + 2x^{-3}) = -2x^{-3} - 6x^{-4} = -\frac{2}{x^3} - \frac{6}{x^4}$

29. $r'(x) = \frac{d}{dx}\left(\frac{2}{3}x^{-1} - \frac{1}{2}x^{-0.1}\right) = -\frac{2}{3}x^{-2} + \frac{0.1}{2}x^{-1.1} = -\frac{2}{3x^2} + \frac{0.1}{2x^{1.1}}$

31. $r'(x) = \frac{d}{dx}\left(\frac{2}{3}x - \frac{1}{2}x^{0.1} + \frac{4}{3}x^{-1.1} - 2\right) = \frac{2}{3} - \frac{0.1}{2}x^{-0.9} - \frac{4.4}{3}x^{-2.1} = \frac{2}{3} - \frac{0.1}{2x^{0.9}} - \frac{4.4}{3x^{2.1}}$

33. $t'(x) = \frac{d}{dx}(|x| + x^{-1}) = |x|/x - x^{-2} = |x|/x - 1/x^2$

35. $s'(x) = \frac{d}{dx}(x^{1/2} + x^{-1/2}) = \frac{1}{2}x^{-1/2} - \frac{1}{2}x^{-3/2} = \frac{1}{2\sqrt{x}} - \frac{1}{2x\sqrt{x}}$

37. $s'(x) = \frac{d}{dx}(x^3 - 1) = 3x^2$

39. $t'(x) = \frac{d}{dx}(x - 2x^2) = 1 - 4x$

41. $2.6x^{0.3} + 1.2x^{-2.2}$

43. $1.2(1 - |x|/x)$

45. $3at^2 - 4a$ (Remember to treat a as a constant, i.e., a number.)

47. $5.15x^{9.3} - 99x^{-2}$

49. $\frac{ds}{dt} = \frac{d}{dt}(2.3 + 2.1t^{-1.1} - \frac{1}{2}t^{0.6}) = -2.31t^{-2.1} + 0.3t^{-0.4} = -\frac{2.31}{t^{2.1}} - \frac{0.3}{t^{0.4}}$

51. $4\pi r^2$

In 53–58, we need to find the derivative at the indicated value of x or t.

53. $f'(x) = 3x^2$, so $f'(-1) = 3$

55. $f'(x) = -2$, so $f'(2) = -2$

57. $g'(t) = \frac{d}{dt}t^{-5} = -5t^{-6} = -\frac{5}{t^6}$, so $g'(1) = -5$

59. $f'(x) = 3x^2$, so $f'(-1) = 3$. The line with slope 3 passing through $(-1, f(-1)) = (-1, -1)$ is $y = 3x + 2$.

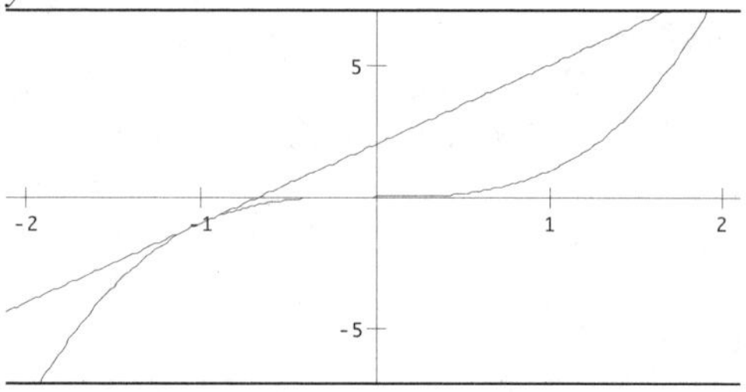

61. $f(x) = x + x^{-1}$, so $f'(x) = 1 - x^{-2} = 1 - \frac{1}{x^2}$; $f'(2) = 1 - 1/4 = 3/4$. The line with slope 3/4 passing through $(2, f(2)) = (2, 5/2)$ is $y = \frac{3}{4}x + 1$

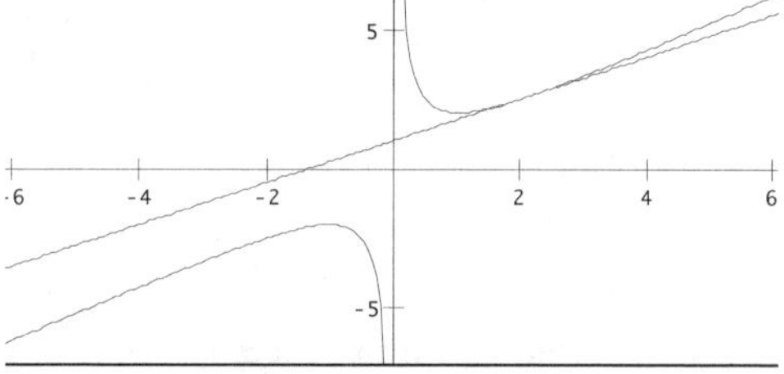

63. $f(x) = x^{1/2}$, so $f'(x) = \frac{1}{2}x^{-1/2} = \frac{1}{2\sqrt{x}}$; $f'(4) = 1/4$. The line with slope 1/4 passing through $(4, f(4)) = (4, 2)$ is $y = \frac{1}{4}x + 1$.

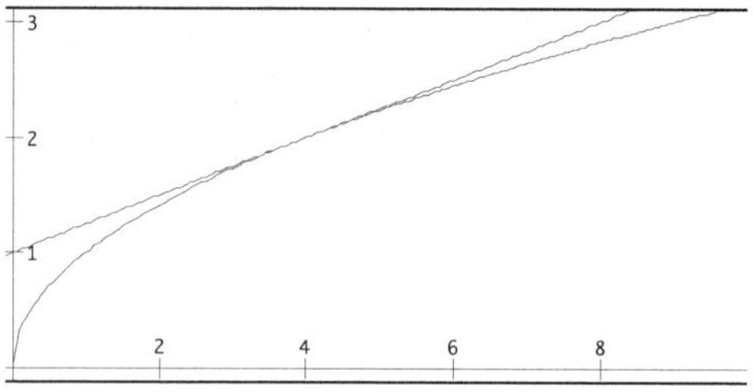

In 65–70 we need to find all values of x (if any) where the derivative is 0.

65. $y' = 4x + 3 = 0$ when $x = -3/4$

67. $y' = 2$ is never 0, so there are no such values of x

69. $y' = 1 - x^{-2} = 1 - 1/x^2 = 0$ when $x^2 = 1$, so $x = 1$ or -1

71. $\frac{d}{dx}x^4 = \lim_{h\to 0} \frac{(x+h)^4 - x^4}{h}$

$= \lim_{h\to 0} \frac{x^4 + 4x^3h + 6x^2h^2 + 4xh^3 + h^4 - x^4}{h}$

$= \lim_{h\to 0} \frac{4x^3h + 6x^2h^2 + 4xh^3 + h^4}{h} =$

$\lim_{h\to 0} (4x^3 + 6x^2h + 4xh^2 + h^3) = 4x^3$

73. The derivative:

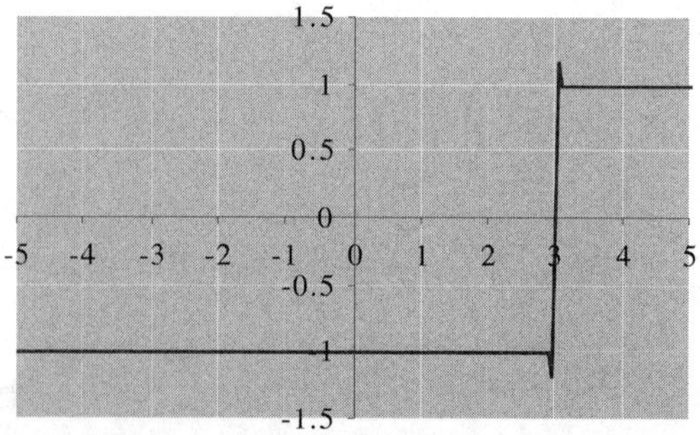

(a) $x = 3$: The sudden jump in value is a discontinuity and the derivative is not defined at $x = 3$.

(b) None: The derivative is never 0.

75. The derivative:

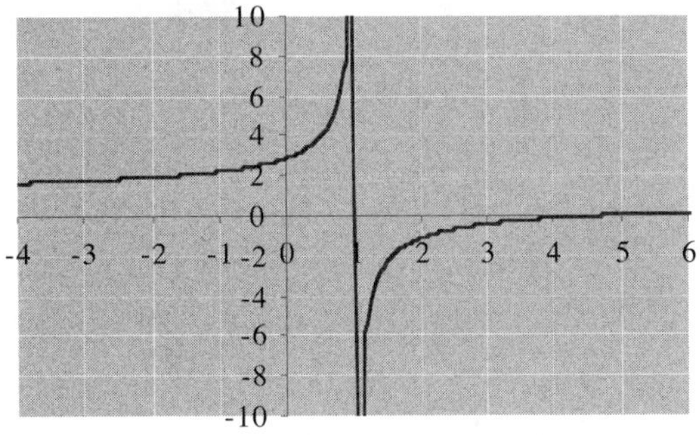

(a) $x = 1$: The sudden jump in value is a discontinuity and the derivative is not defined at $x = 1$.

(b) $x = 4.2$: The derivative is 0 at approximately 4.2. The derivative is not 0 at $x = 1$; that is just a defect of the graphing technology.

77. (a) $f'(1) = 1/3$:

h	$\frac{f(a+h) - f(a)}{h}$
−1	1
−0.1	0.34510615
−0.01	0.33445066
−0.001	0.33344451
−0.0001	0.33334445
1	0.25992105
0.1	0.32280115
0.01	0.33222835
0.001	0.33322228
0.0001	0.33332222

(b) f is not differentiable at 0:

h	$\frac{f(a+h) - f(a)}{h}$
−1	1
−0.1	4.64158883
−0.01	21.5443469
−0.001	100
−0.0001	464.158883
1	1
0.1	4.64158883
0.01	21.5443469
0.001	100
0.0001	464.158883

79. (a) Not differentiable at 1:

h	$\frac{f(a+h) - f(a)}{h}$
−1	0
−0.1	−4.4814047
−0.01	−21.472292
−0.001	−99.966656
−0.0001	−464.14341
1	−1.259921
0.1	−4.7914199
0.01	−21.615923
0.001	−100.03332
0.0001	−464.17435

(b) Not differentiable at 0:

h	$\frac{f(a+h) - f(a)}{h}$
−1	1.25992105
−0.1	4.79141986
−0.01	21.6159233
−0.001	100.033322
−0.0001	464.174355
1	0
0.1	4.48140475
0.01	21.4722917
0.001	99.9666555
0.0001	464.143411

81. Since putting $x = 0$ yields 0/0, L'Hospital's rule applies.

$$\lim_{x \to 1} \frac{x^2 - 2x + 1}{x^2 - x} = \lim_{x \to 1} \frac{2x - 2}{2x - 1} = \frac{0}{1} = 0$$

83. Since putting $x = 1$ yields 0/0, L'Hospital's rule applies.

$$\lim_{x \to 2} \frac{x^3 - 8}{x - 2} = \lim_{x \to 2} \frac{3x^2}{1} = \frac{12}{1} = 12$$

85. Since putting $x = 1$ yields 6/2 = 3. Since this is not an indeterminate form, L'Hospital's rule does not apply, and the limits is 3 (closed-form function).

87. Since putting $x = -\infty$ yields ∞/∞, L'Hospital's rule applies.

$$\lim_{x \to -\infty} \frac{3x^2 + 10x - 1}{2x^2 - 5x} = \lim_{x \to -\infty} \frac{6x + 10}{4x - 5} = \lim_{x \to -\infty} \frac{6}{4} = 3/2$$

89. Since putting $x = -\infty$ yields ∞/∞, L'Hospital's rule applies.

$$\lim_{x \to -\infty} \frac{10x^2 + 300x + 1}{5x + 2} = \lim_{x \to -\infty} \frac{20x + 300}{5}$$

diverges to $-\infty$

91. Since putting $x = -\infty$ yields ∞/∞, L'Hospital's rule applies.

$$\lim_{x \to -\infty} \frac{x^3 - 100}{2x^2 + 500} = \lim_{x \to -\infty} \frac{3x^2}{4x}$$

Since this still has the form ∞/∞, we can use L'Hospital's rule again to get

$$\lim_{x \to -\infty} \frac{3x^2}{4x} = \lim_{x \to -\infty} \frac{6x}{4} = -\infty/4 = -\infty$$

93. (a) $s(t) = 1.52t^2 + 9.45t + 82.7$, so $s'(t) = 3.04t + 9.45$

(b) The 1994–1995 academic year is represented by $t = 14$.

$s'(14) = 3.04(14) + 9.45 = 52.01$ teams/year

The number of women's soccer teams was increasing at a rate of about 52 teams per year.

95. $P'(t) = -5.2t + 13$. In January 2002, $P'(2) = 2.6$, so the percentage of people who had purchased anything on-line was increasing at a rate of 2.6 percentage points per year.

97. To find the rate of change of spending on food (y) with rrespect to x we take the derivative using the power rule:

$$y = \frac{35}{x^{0.35}} = 35x^{-0.35}$$

$$\frac{dy}{dx} = (35)(-0.35)x^{-0.35-1} = 12.25x^{-1.35}$$

We evaluate this at $x = 10\%$:

$$\left.\frac{dy}{dx}\right|_{x=10} = 12.25(10)^{-1.35} \approx 0.5472$$

≈ 0.55 percentage points per one percentage point increase in spending on education.

99. (a) $s'(t) = -32t$; $s'(0) = 0$, $s'(1) = -32$, $s'(2) = -64$, $s'(3) = -96$, $s'(4) = -128$ ft/sec
(b) $s(t) = 0$ when $400 - 16t^2 = 0$, so $t^2 = 400/16 = 25$, so at $t = 5$ seconds; the stone is traveling at the velocity $s'(5) = -160$, so downward at 160 ft/sec.

101. (a) $E(t) = 0.036t^2 - 0.10t + 1.0$

$E'(t) = (2)0.036t - 0.10 + 0$
$= 0.072t - 0.10$

January 2004 is represented by $t = 4$, and

$E'(4) = 0.072(4) - 0.10 = \$0.188$ per year.

Thus, the value of the euro was increasing at a rate of \$0.188 per year.

(b) One can answer this question either algebraically or graphically.

Algebraically: The range January 2000–January 2001 is represented by $0 \le t \le 1$.

$E'(0) = 0.072(0) - 0.10 = -0.10$

which is negative, so that E was decreasing at $t = 0$. Further, as t increases from the value 0, the derivative remains negative, but is less negative (has smaller absolute value). To summarize:

$E'(0)$ is negative $\Rightarrow$ e was decreasing at time $t = 0$.

$E'(t)$ remains negative but has smaller absolute value as t goes from 0 to 1 $\Rightarrow$ E decreases at a slower and slower rate.

Therefore, the value of e decreased at a slower and slower rate (Choice (D)).

Graphically: Here is the graph of E

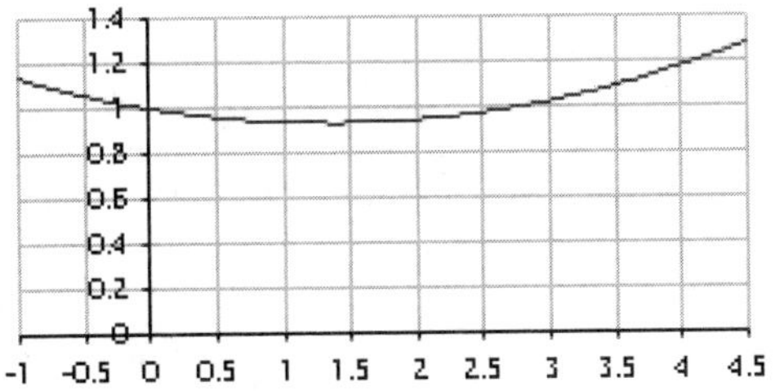

Notice that the graph has negative slope at $t = 0$ and and that the slope remains negative, but less so (has smaller absolute value) as t increases from 0 to 1.

Negative slope at $t = 0 \Rightarrow E$ was decreasing at time $t = 0$.

Negative, but less negative slope as t goes from 0 to 1 $\Rightarrow E$ decreases at a slower and slower rate.

Therefore, the value of E decreased at a slower and slower rate (Choice (D)).

103. (a) $f'(x) = 7.1x - 30.2$ manatees per 100,000 boats.

(b) $f'(x)$ is increasing; the number of manatees killed per additional 100,000 boats increases as the number of boats increases.

(c) $f'(8) = 26.6$ manatees per 100,000 additional boats. At a level of 800,000 boats, the number of manatee deaths is increasing at a rate of 26.6 manatees per 100,000 additional boats.

105. (a) $c(t)$ measures the combined market share (including MSN), while $m(t)$ measures the share due to MSN. Therefore, $c(t)-m(t)$ measures the combined market share of the other three providers (Comcast, earthlink, and AOL). Similarly, $c'(t)-m'(t)$ measures the rate of change of the combined market share of the other three providers.

(b) We can visualize $c(t) - m(t)$ on the graph as the vertical distance from the lower curve to the higher one:

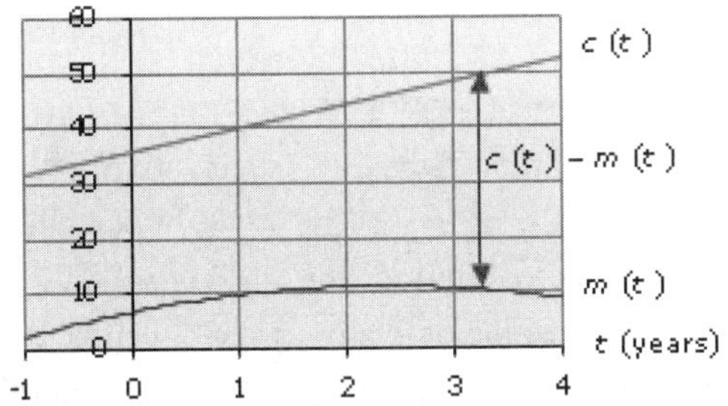

As we move from $t = 3$ to $t = 4$, that distance is increasing (choice A).

(c) From part (b), $c(t) - m(t)$ measures the vertical distance from the lower curve to the higher one, since this distance is increasing on [3, 4], its rate of change (derivative) $c'(t) - m'(t)$ is positive (choice A).

(d) $m(t) = -0.83t^2 + 3.8t + 6.8$

so $m'(t) = -1.66t + 3.8$

and $m'(2) = -1.66(2) + 3.8 = 0.48\%$ per year.

$c(t) = 4.2t + 36$,

so $c'(t) = c'(2) = 4.2\%$ per year.

Therefore, $c'(2) - m'(2) = 4.2 - 0.48 = 3.72\%$ per year. In 1992, the combined market share of the other three providers was increasing at a rate of about 3.72 percentage points per year.

107. After graphing the curve $y = 3x^2$, draw the line passing through $(-1, 3)$ with slope -6.

109. The slope of the tangent line of g is twice the slope of the tangent line of f because $g(x) = 2f(x)$, so $g'(x) = 2f'(x)$.

111. $g'(x) = -f'(x)$

113. The left-hand side is not equal to the right-hand side. The *derivative* of the left-hand side is equal to the right-hand side, so your friend should have written

$$\frac{d}{dx}(3x^4+11x^5) = 12x^3 + 55x^4.$$

115. The derivative of a constant times a function is the constant times the derivative of the function, so $f'(x) = (2)(2x) = 4x$. Your enemy mistakenly computed the *derivative* of the constant times the derivative of the function. (The derivative of a product of two functions is not the product of the derivative of the two functions. The rule for taking the derivative of a product is discussed in the next chapter.).

117. Answers may vary; here is one possibility. At one point its derivative is not defined but it has a tangent line: The tangent line at that point is vertical so has undefined slope.

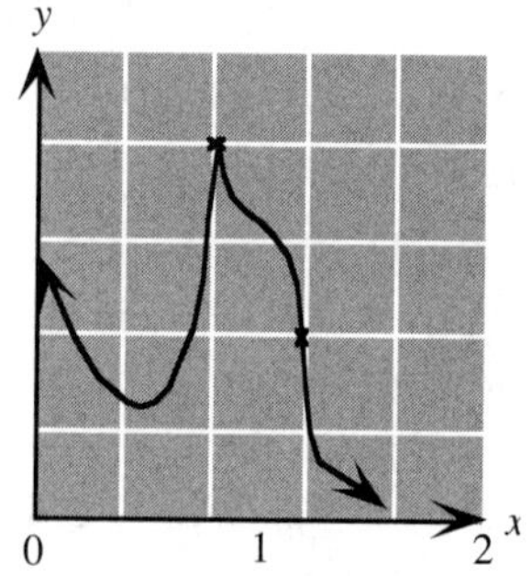

3.8

1. $C'(x) = 5 - 0.0002x$; $C'(1000) = \$4.80$ per item

3. $C'(x) = 100 - 1000/x^2$; $C'(100) = \$99.90$ per item

5. $C(x) = 4x$, so $C'(x) = 4$;
$R(x) = 8x - 0.001x^2$, so $R'(x) = 8 - x/500$;
$P(x) = R(x) - C(x) = R(x) = 8x - 0.001x^2 - 4x$
$= R(x) = 4x - 0.001x^2$, so $P'(x) = 4 - 0.002x$;
$P'(x) = 0$ when $x = 2000$. Thus, at a production level of 2000, the profit is stationary (neither increasing nor decreasing) with respect to the production level. This may indicate a maximum profit at a production level of 2000.

7. **(a)** (B): The slope of the graph decreases and then increases.
(b) (C): This is where the slope of the graph is least.
(c) (C): At $x = 50$, the height of the graph is about 3000, so the cost is \$3000. The tangent line at that point passes roughly through (100, 4000), so has a slope of roughly 20; hence the cost is increasing at a rate of about \$20 per item.

9. **(a)** $C(x) = 150 + 2250x - 0.02x^2$, so

$C'(x) = 2250 - 0.04x$.
$C'(4) = 2250 - 0.04(4)$
$= 2249.84$ thousand dollars;

that is, \$2,249,840. The cost is going up at a rate of \$2,249,840 per television commercial. The exact cost of airing the fifth television commercial is

$C(5) - C(4) = 11399.5 - 9149.68$
$= 2249.82$ thousand dollars, or \$2,249,820.

(b) $\bar{C}(x) = 150/x + 2250 - 0.02x$; $\bar{C}(4) =$ \$2,287,420 per television commercial. The average cost of airing the first four television commercials is \$2,287,420.

11. **(a)** $R(x) = 0.90x$
Marginal revenue $= R'(x) = 0.90$

$P(x) = R(x) - C(x)$
$= 0.90x - (70 + 0.10x + 0.001x^2)$
$= -70 + 0.80x - 0.001x^2$

Marginal Profit $= P'(x) = 0.80 - 0.002x$
(b) $R(x) = 0.90x$

$R(500) = 0.90(500) = \$450$

The total revenue from the sale of 500 copies is \$450.

$P(x) = -70 + 0.80x - 0.001x^2$
$P(500) = -70+0.80(500)-0.001(500)^2$
$= \$80$

The profit from the production and sale of 500 copies is \$80.

$R'(x) = 0.90$,
so $R'(500) = 0.90$

Approximate revenue from the sale of the 501st copy is 90¢.

$P'(x) = 0.80 - 0.002x$
$P'(500) = 0.80-0.002(500) = -0.2$

Approximate loss from the sale of the 501st copy is 20¢. (Negative marginal profit indicates a loss.)
(c) The marginal profit $P'(x)$ is zero when

$0.80 - 0.002x = 0$
$x = 0.80/0.002 = 400$ copies.

The graph of the profit function is a parabola with a vertex at $x = 400$, so the profit is a maximum when you produce and sell 400 copies.

13. $P(1000)$ represents the profit on the sale of 1000 DVDs. $P(1000) = 3000$, so the profit on the sale of 1000 DVDs is \$3000. $P'(1000)$ represents the rate of increase of the profit as a function of x.

$P'(1000) = -3$, so the profit is decreasing at a rate of \$3 per additional DVD sold.

15. $P(x) = 5x + \sqrt{x}$. Your current profit is

$P(50) = 5(50) + \sqrt{50} \approx \$257.07$.

The marginal profit is

$P'(x) = dP/dx = 5 + 1/(2\sqrt{x})$

$P'(50) = 5 + 1/(2\sqrt{50}) \approx 5.07$

The derivative is measured in dollars per additional magazine sold. Thus, your current profit is \$257.07 per month and this would increase at a rate of \$5.07 per additional magazine in sales.

17. (a) $q = 20{,}000/q^{1.5}$. When $q = 400$, $p = 20{,}000/(400)^{1.5} \approx \$2.50$ per pound.

(b) $R(q) = pq = (20{,}000/q^{1.5})q$

$= 20{,}000/q^{0.5}$

(c) $R(400) = \$1000$. This is the monthly revenue that will result from setting the price at \$2.50 per pound.

$R'(q) = -10{,}000/q^{1.5}$,

so $R'(400) = -10{,}000/(400)^{1.5}$

$= -\$1.25$ per pound of tuna.

Thus, at a demand level of 400 pounds per month, the revenue is decreasing at a rate of \$1.25 per pound.

(d) Since the revenue goes down with increasing demand, the fishery should raise the price to reduce the demand and hence increase revenue.

19. $P'(n) = 400 - n$, so $P'(50) = \$350$. This means that, at an employment level of 50 workers, the firm's daily profit will increase at a rate of \$350 per additional worker it hires.

21. (a) (B): $C'(x) = -0.002x + 0.3$ decreases as x increases.

(b) (B): $\bar{C}(x) = -0.001x + 0.3 + 500/x$ decreases as x increases.

(c) (C): $C'(100) = 0.1$, $\bar{C}(100) = 5.2$.

23. (a) $C(x) = 500{,}000 + 1{,}600{,}000x - 100{,}000\sqrt{x}$

$$C'(x) = 1{,}600{,}000 - \frac{50{,}000}{\sqrt{x}}$$

$$\bar{C}(x) = C(x)/x = \frac{500{,}000}{x} + 1{,}600{,}000 - \frac{100{,}000}{\sqrt{x}}$$

(b) $C'(3) \approx \$1{,}570{,}000$ per spot, $\bar{C}(3) \approx \$1{,}710{,}000$ per spot. Since the marginal cost is less than the average cost, the cost of the fourth ad is lower than the average cost of the first three, so the average cost will decrease as x increases.

25. (a) $C'(q) = 200q$ so $C'(10) = \$2000$ per one-pound reduction in emissions.

(b) $S'(q) = 500$. Thus $S'(q) = C'(q)$ when $500 = 200q$, or $q = 2.5$ pounds per day reduction.

(c) $N(q) = C(q) - S(q) = 100q^2 - 500q + 4000$. This is a parabola with lowest point (vertex) given by $q = 2.5$. The net cost at this production level is $N(2.5) = \$3375$ per day. The value of q is the same as that for part (b). The net cost to the firm is minimized at the reduction level for which the cost of controlling emissions begins to increase faster than the subsidy. This is why we get the answer by setting these two rates of increase equal to each other.

27. $M'(x) = \dfrac{3600x^{-2} - 1}{(3600x^{-1} + x)^2}$. So,

$$M'(10) = \frac{3600(10)^{-2} - 1}{(3600(10)^{-1} + 10)^2}$$

≈ 0.0002557 mpg/mph.

This means that, at a speed of 10 mph, the fuel economy is increasing at a rate of 0.0002557 miles per gallon per 1-mph increase in speed.

$$M'(60) = \frac{3600(60)^{-2} - 1}{(3600(60)^{-1} + 60)^2} = 0 \text{ mpg/mph.}$$

This means that, at a speed of 60 mph, the fuel economy is neither increasing nor decreasing with increasing speed.

$$M'(70) = \frac{3600(70)^{-2} - 1}{(3600(70)^{-1} + 70)^2}$$
$$\approx -0.00001799.$$

This means that, at 70 mph, the fuel economy is decreasing at a rate of 0.00001799 miles per gallon per 1-mph increase in speed. Thus 60 mph is the most fuel-efficient speed for the car.

29. (C): If the marginal cost were lower in one plant than another, moving some production from the higher cost plant to the lower would result in a lower cost for the same production level.

31. (D): The marginal product per dollar of salary is $2/1.5 \approx 1.33$ times as high for a junior professor as compared to a senior professor. Therefore, discharging senior professors and hiring more junior professors will result in a higher quantity of output for the same amount of money.

33. (B): (In most cases) This is why we use the marginal cost as an estimate of the actual cost of the item.

35. Cost is often measured as a function of the number of items x. Thus, $C(x)$ is the cost of producing (or purchasing, as the case may be) x items.

(a) The average cost function $\bar{C}(x)$ is given by $\bar{C}(x) = C(x)/x$. The marginal cost function is the derivative, $C'(x)$, of the cost function.

(b) The average cost $\bar{C}(r)$ is the slope of the line through the origin and the point on the graph where $x = r$. The marginal cost of the rth unit is the slope of the tangent to the graph of the cost function at the point where $x = r$.

(c) The average cost function $\bar{C}(x)$ gives the average cost of producing the first x items. The marginal cost function $C'(x)$ is the rate at which cost is changing with respect to the number of items x, or the incremental cost per item, and approximates the cost of producing the $(x+1)$st item.

37. The marginal cost: If the average cost is rising, then the cost of the next piano must be larger than the average cost of the pianos already built.

39. Not necessarily. For example, it may be the case that the marginal cost of the 101st item is larger than the average cost of the first 100 items (even though the marginal cost is decreasing). Thus, adding this additional item will *raise* the average cost.

41. The circumstances described suggest that the average cost function is at a relatively low point at the current production level, and so it would be appropriate to advise the company to maintain current production levels; raising or lowering the production level will result in increasing average costs.

Chapter 3 Review Exercises

1. 5:

x	$f(x)$
2.9	4.9
2.99	4.99
2.999	4.999
2.9999	4.9999
3	
3.0001	5.0001
3.001	5.001
3.01	5.01
3.1	5.1

2. Does not exist:

x	$f(x)$
2.9	33.9
2.99	303.99
2.999	3003.999
2.9999	30003.9999
3	
3.0001	−29996
3.001	−2995.999
3.01	−295.99
3.1	−25.9

3. Does Not Exist:

x	$f(x)$
−1.1	0.3226
−1.01	0.3322
−1.001	0.3332
−1.0001	0.3333
−1	
−0.9999	−0.3333
−0.999	−0.3334
−0.99	−0.3344
−0.9	−0.3448

4. 0:

x	$f(x)$
−1.1	−0.0529
−1.01	−0.0050
−1.001	−0.0005
−1.0001	-5×10^{-5}
−1	
−0.9999	5×10^{-5}
−0.999	−0.0005
−0.99	−0.0050
−0.9	−0.0478

5. (a) −1: As x approaches 0 from the left or right, $f(x)$ approaches the open dot at height −1. The fact that $f(0) = 3$ is irrelevant.
(b) 3: As x approaches 1 from the left or right, $f(x)$ approaches the point on the graph corresponding to $x = 1$, whose y-coordinate is 3.
(c) Does not exist: As x approaches 2 from the left, $f(x)$ approaches the open dot at height 2. As x approaches 2 from the right, $f(x)$ approaches the solid dot at height 1. Thus, the one-sided limits, though they both exist, do not agree.

6. (a) Does not exist: As x approaches 0 from the left, $f(x)$ approaches the open dot at height −1. As x approaches 0 from the right, $f(x)$ approaches the solid dot at height 3. Thus, the one-sided limits, though they both exist, do not agree.
(b) 2: As x approaches −2 from the left or right, $f(x)$ approaches the point on the graph corresponding to $x = -2$, whose y-coordinate is 2.
(c) 1: As x approaches 2 from the left or right, $f(x)$ approaches the open dot at height 1. The fact that $f(2) = 3$ is irrelevant.

7. $f(x) = \frac{x^2}{x-3}$ is a closed-form function whose domain includes $x = -2$. Therefore

$$\lim_{x \to -2} \frac{x^2}{x-3} = \frac{(-2)^2}{-2-3} = -4/5$$

8. $f(x) = \frac{x^2-9}{2x-6}$ is a closed-form function but its domain does not include $x = 3$, but we can simplify:

$$\lim_{x \to 3} \frac{x^2-9}{2x-6} = \lim_{x \to 3} \frac{(x-3)(x+3)}{2(x-3)}$$

$$\lim_{x \to 3} \frac{(x+3)}{2} = \frac{(3+3)}{2} = 3$$

9. $f(x) = \frac{x}{2x^2-x}$ is a closed-form function but its domain does not include $x = 0$, but we can simplify:

$$\lim_{x \to 0} \frac{x}{2x^2-x} = \lim_{x \to 0} \frac{x}{x(2x-1)}$$

$$\lim_{x \to 0} \frac{1}{(2x-1)} = \frac{1}{(0-1)} = -1$$

10. $f(x) = \frac{x^2-9}{x-1}$ is a closed-form function but its domain does not include $x = 1$, nor can we simplify to cancel the $(x-1)$ term. So instead, numerically evaluate the left- and right- limits.

$$\lim_{x \to 1^-} \frac{x^2-9}{x-1} = +\infty$$

$$\lim_{x \to 1^+} \frac{x^2-9}{x-1} = -\infty$$

Since both left-and right limits diverge, the overall limit does not exist.

11. Ignoring all the highest terms in numerator and denominator, we get:

$$\lim_{x \to -\infty} \frac{x^2-x-6}{x-3} = \lim_{x \to -\infty} \frac{x^2}{x}$$

$$= \lim_{x \to -\infty} x = -\infty,$$

So the limit diverges to $-\infty$.

12. Ignoring all the highest terms in numerator and denominator, we get:

$$\lim_{x \to \infty} \frac{x^2-x-6}{4x^2-3} = \lim_{x \to \infty} \frac{x^2}{4x^2}$$

$$= \lim_{x \to -\infty} 1/4 = 1/4$$

13. $f(x) = \frac{1}{x+1}$; $a = 0$

The average rate of change is $\frac{f(a+h) - f(a)}{h}$.

$h = 1$:

$$\frac{f(0+1) - f(0)}{1} = \frac{f(1) - f(0)}{1}$$

$$= \frac{-0.5}{1} = -0.5$$

$h = 0.01$:

$$\frac{f(0+0.01) - f(0)}{0.01} = \frac{f(0.01) - f(0)}{0.01}$$

$$= \frac{-0.009901}{0.01} = -0.99001$$

$h = 0.001$:

$$\frac{f(0+0.001) - f(0)}{0.001} = \frac{f(0.001) - f(0)}{0.001}$$

$$= \frac{-0.000999}{0.001} = -0.9990$$

Table:

h	Ave. Rate of Change
1	−0.5
0.01	−0.9901
0.001	−0.9990

The slope of the tangent is the limit as $h \to 0$, which appears to be -1.

Slope ≈ -1

14. $f(x) = x^x$; $a = 2$

The average rate of change is $\frac{f(a+h) - f(a)}{h}$.

$h = 1$:

$$\frac{f(2+1) - f(2)}{1} = \frac{f(3) - f(2)}{1} = \frac{23}{1} = 23$$

$h = 0.01$:

$$\frac{f(2+0.01) - f(2)}{0.01} = \frac{f(2.01) - f(2)}{0.01}$$

$$\approx \frac{0.068404}{0.01} = 6.8404$$

$h = 0.001$:

$$\frac{f(2+0.001) - f(2)}{0.001} = \frac{f(2.001) - f(2)}{0.001}$$

$$\approx \frac{0.0067793}{0.001} = 6.7793$$

Table:

h	Ave. Rate of Change
1	23
0.01	6.8404
0.001	6.7793

The slope of the tangent is the limit as $h \to 0$, which appears to be about 6.8. Slope $\approx$ 6.8

15. $f(x) = e^{2x}$; $a = 0$

Technology formula for $f(x)$:

`Y1= e^(2x)` TI-83/84

`EXP(2*x)` Excel

To compute the average rate of change on the TI-83/84, use the following formulas:

$h = 1$: `(Y1(0+1)-Y1(0))/1`

$h = 0.01$: `(Y1(0+.01)-Y1(0))/.01`

$h = 0.001$: `(Y1(0+.001)-Y1(0))/.001`

Table:

h	Ave. Rate of Change
1	6.3891
0.01	2.0201
0.001	2.0020

The slope of the tangent is the limit as $h \to 0$, which appears to be 2.

Slope $\approx$ 2

16. $f(x) = \ln(2x)$; $a = 1$

Technology formula for $f(x)$:

`Y1= ln(2x)`

To compute the average rate of change on the TI-83/84, use the following formulas:

$h = 1$: `(Y1(1+1)-Y1(1))/1`

$h = 0.01$: `(Y1(1+.01)-Y1(1))/.01`

$h = 0.001$: `(Y1(1+.001)-Y1(1))/.001`

Table:

h	Ave. Rate of Change
1	0.6931
0.01	0.9950
0.001	0.9995

The slope of the tangent is the limit as $h \to 0$, which appears to be 1.

Slope $\approx$ 1

17. (i) P **(ii)** Q **(iii)** R **(iv)** S

18. (i) None **(ii)** R **(iii)** Q **(iv)** P

19. (i) Q **(ii)** None **(iii)** None **(iv)** None

20. (i) None **(ii)** R **(iii)** P and S **(iv)** Q

21. (a) (B): The graph starts on the tangent line and falls below it.

(b) (B): The graph starts above the tangent line and ends below it.

(c) (B): The graph is getting less steep.

(d) (A): The graph gets steeper until $x = 0$ and then gets less steep.

(e) (C): The value of $f(2)$ is the height of the graph at $x = 2$, which is about 2.5; the rate of change is the slope of the tangent line at that point, which is approximately 1.5.

22. (a) (B): The graph starts on the tangent line and falls below it.

(b) (C): The average rate of change of f over [0, 2] is 0, and the tangent is horizontal at $x = 1$.
(c) (A): The slope increases from a negative value at $x = -2$ to a positive value at $x = 0$.
(d) (A): The graph gets steeper until $x = 0$ and then gets less steep.
(e) (A): The value of $f(0)$ is the height of the graph at $x = 0$, which is 0; the rate of change is the slope of the tangent line at that point, which is approximately 1.5.

23. $f(x) = x^2 + x$

$$f'(x) = \lim_{h\to 0} \frac{f(x+h) - f(x)}{h}$$
$$= \lim_{h\to 0} \frac{(x+h)^2 + (x+h) - (x^2 + x)}{h}$$
$$= \lim_{h\to 0} \frac{x^2 + 2xh + h^2 + x + h - x^2 - x}{h}$$
$$= \lim_{h\to 0} \frac{2xh + h^2 + h}{h} = \lim_{h\to 0} (2x + h + 1)$$
$$= 2x + 1$$

24. $f(x) = 3x^2 - x + 1$

$$f'(x) = \lim_{h\to 0} \frac{f(x+h) - f(x)}{h}$$
$$= \lim_{h\to 0} \frac{3(x+h)^2-(x+h)+1 - (3x^2-x+1)}{h}$$
$$= \lim_{h\to 0} \frac{3x^2+6xh+3h^2-x-h+1 - (3x^2-x+1)}{h}$$
$$= \lim_{h\to 0} \frac{6xh+3h^2-h}{h}$$
$$= \lim_{h\to 0} \frac{h(6x+3h-1)}{h}$$
$$= \lim_{h\to 0} (6x+3h-1) = 6x - 1$$

25. $f(x) = 1 - \dfrac{2}{x}$

$$f'(x) = \lim_{h\to 0} \frac{f(x+h) - f(x)}{h}$$
$$= \lim_{h\to 0} \frac{1-2/(x+h) - (1-2/x)}{h}$$
$$= \lim_{h\to 0} \frac{-2/(x+h) + 2/x}{h}$$
$$= \lim_{h\to 0} \frac{-2/(x+h) + 2/x}{h}$$
$$= \lim_{h\to 0} \frac{-2x + 2(x+h)}{h(x+h)x}$$
$$= \lim_{h\to 0} \frac{2h}{h(x+h)x}$$
$$= \lim_{h\to 0} \frac{2}{(x+h)x} = 2/x^2$$

26. $f(x) = \dfrac{1}{x} + 1$

$$f'(x) = \lim_{h\to 0} \frac{f(x+h) - f(x)}{h}$$
$$= \lim_{h\to 0} \frac{1/(x+h) + 1 - (1/x + 1)}{h}$$
$$= \lim_{h\to 0} \frac{1/(x+h) - 1/x}{h}$$
$$= \lim_{h\to 0} \frac{x - (x + h)}{hx(x + h)}$$
$$= \lim_{h\to 0} \frac{-h}{hx(x + h)}$$
$$= \lim_{h\to 0} \frac{-1}{x(x + h)} = -\frac{1}{x^2}$$

27. $f'(x) = 50x^4 + 2x^3 - 1$

28. $f'(x) = (10x^{-5} + x^{-4}/2 - x^{-1} + 2)'$
$= -50x^{-6} - 4x^{-5}/2 + x^{-2}$
$= -50/x^6 - 2/x^5 + 1/x^2$

29. $f'(x) = (3x^3 + 3x^{1/3})' = 9x^2 + x^{-2/3}$

30. $f'(x) = (2x^{-2.1} - x^{0.1}/2)'$

$= -4.2x^{-3.1} - 0.1x^{-0.9}/2$

$= -4.2/x^{3.1} - 0.1/(2x^{0.9})$

31. $\frac{d}{dx}\left(x + \frac{1}{x^2}\right) = \frac{d}{dx}(x + x^{-2})$

$= 1 - 2x^{-3} = 1 - \frac{2}{x^3}$

32. $\frac{d}{dx}\left(2x - \frac{1}{x}\right) = \frac{d}{dx}(2x - x^{-1})$

$= 2 + x^{-2} = 2 + \frac{1}{x^2}$

33. $\frac{d}{dx}\left(\frac{4}{3x} - \frac{2}{x^{0.1}} + \frac{x^{1.1}}{3.2} - 4\right)$

$= \frac{d}{dx}\left(\frac{4}{3}x^{-1} - 2x^{-0.1} + \frac{1}{3.2}x^{1.1} - 4\right)$

$= -\frac{4}{3}x^{-2} + 0.2x^{-1.1} + \frac{1.1}{3.2}x^{0.1}$

$= -\frac{4}{3x^2} + \frac{0.2}{x^{1.1}} + \frac{1.1x^{0.1}}{3.2}$

34. $\frac{d}{dx}\left(\frac{4}{x} + \frac{x}{4} - |x|\right) = \frac{d}{dx}\left(4x^{-1} + \frac{1}{4}x - |x|\right)$

$= -4x^{-2} + 1/4 - |x|/x$

$= -4/x^2 + 1/4 - |x|/x$

The derivatives in 35–38 are the ones found in Exercises 27–30. The technology formulas are indicated below.

35. `50*x^4+2*x^3-1`

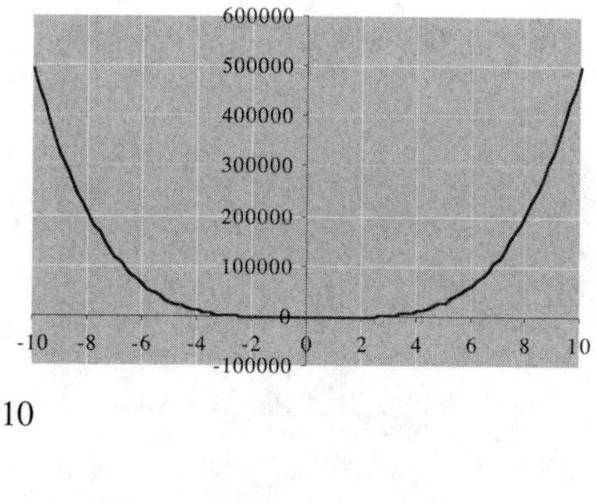

36. `-50/x^6-2/x^5+1/x^2`

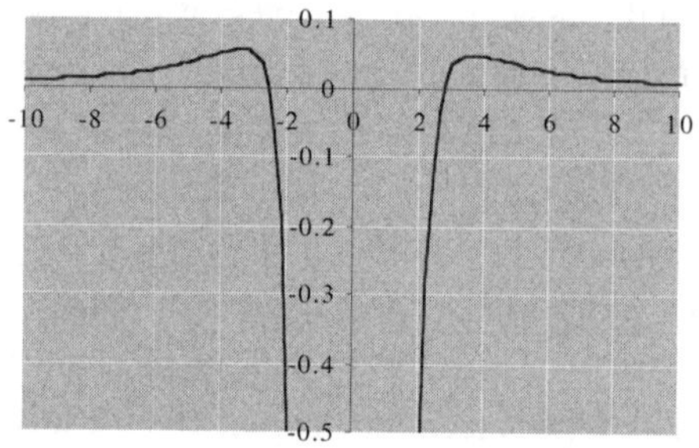

37. `9*x^2+1/(x^2)^(1/3)`

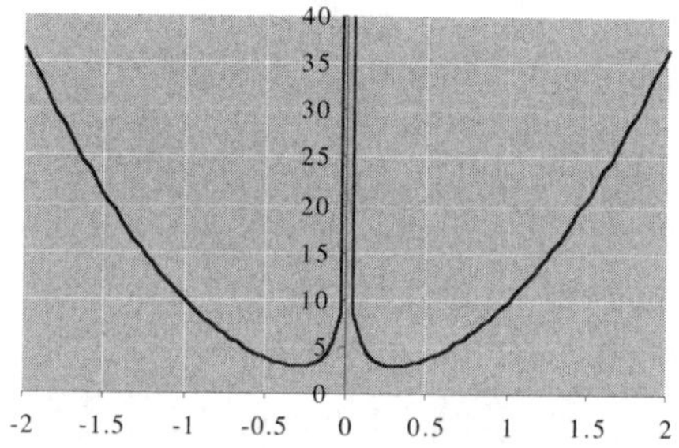

38. `-4.2/x^3.1-0.1/(2*x^0.9)`

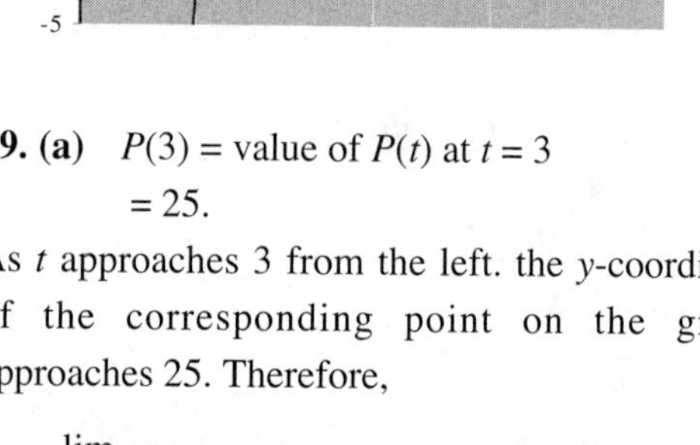

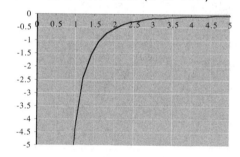

39. (a) $P(3)$ = value of $P(t)$ at $t = 3$

$= 25.$

As t approaches 3 from the left. the y-coordinate of the corresponding point on the graph approaches 25. Therefore,

$$\lim_{t\to 3^-} P(t) = 25$$

As t approaches 3 from the right. the y-coordinate of the corresponding point on the graph approaches 10. Therefore,

$$\lim_{t\to 3^+} P(t) = 10$$

Since the left and right-limits do not agree, $\lim_{t\to 3} P(t)$ does not exist.

Interpretation: $P(3) = 3$: O'Hagan purchased the stock at \$25. $\lim_{t\to 3^-} P(t) = 25$: The value of the stock had been approaching \$25 up the time he bought it. $\lim_{t\to 3^+} P(t) = 10$: The value of the stock dropped to \$10 immediately after he bought it.

(b) As t approaches 6 from either side, $P(t)$ approaches 5, which is also the value of $P(6)$. In other words,

$$\lim_{t\to 6} P(t) = 5 = P(6)$$

showing that P is continuous at $t = 6$. On the other hand, the graph comes to a sharp point at $t = 6$ so P is not differentiable at $t = 6$. Interpretation: the stock changed continuously but suddenly reversed direction (and started to go up) the instant O'Hagan sold it.

40. (a) As t approaches 2 from either side, $C(t)$ approaches 8000 (which is the same as $C(3)$; the function is continuous). Hence,

$$\lim_{t\to 2} C(t) = 8000$$

Close to 2 weeks after the start of the marketing campaign, the weekly cost was about \$8000.

As t goes to $+\infty$, the y-coordinate $C(t)$ appears to level off at around 10,000, so we estimate

$$\lim_{t\to +\infty} C(t) = 10{,}000$$

In the long term, the weekly cost of the marketing campaign will be around \$10,000.

(b) $C'(t)$ represents the slope of the tangent at time t. As t goes to $+\infty$, the graph levels off, so its slope approaches 0. so we estimate

$$\lim_{t\to +\infty} C'(t) = 0$$

To interpret the result, recall that $C'(t)$ measures the rate of change of $C(t)$. Since this is approaching 0, $C(t)$ is changing by smaller and smaller amounts. Thus, in the long term, the weekly cost of the marketing campaign will remain approximately constant.

41. (a) $(9000 - 6500)/5 = 500$ books per week

(b) [3, 4] (600 books per week), [4, 5] (700 books per week)

(c) [3, 5], when the average rate of increase was 650 books per week.

42. (a) Average rate of change $= \dfrac{C(6) - C(0)}{6 - 0} \approx \dfrac{10{,}000\text{–}2000}{6} = \dfrac{8000}{6} \approx \$1333$ per week

(b) $\dfrac{C(6) - C(2)}{6 - 2} \approx \dfrac{10{,}000\text{–}8000}{4} = \dfrac{2000}{4} = \$500$ per week

(c) Choice (B): The graph becomes less steep as t changes from 1 to 6, so the slope is deceasing. This means that the rate of change of cost is decreasing. Also, the cost itself is increasing.

43. (a) $w'(t) = -11.1t^2 + 149.2t + 135.5$, so $w'(1) \approx 274$ books per week

(b) $w'(7) = 636$ books per week

(c) It would not be realistic to use the function w through week 20: It begins to decrease after $t = 14$. Graph:

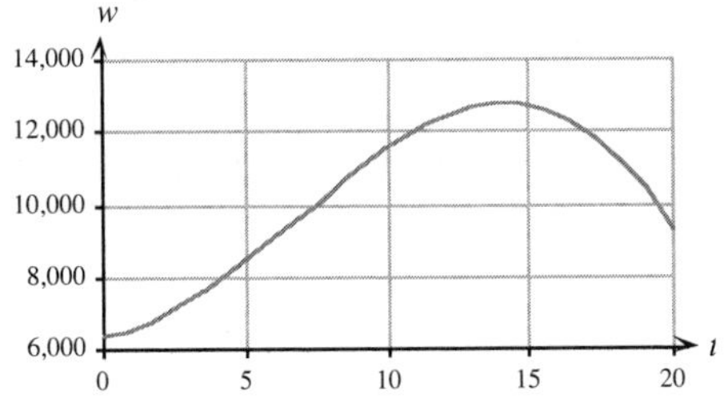

44. $6053 + \frac{4474}{1 + e^{-0.55(t-4.8)}}$

Technology formulas:

TI-83/84:

```
6053+4474/(1+e^(-0.55*(x-4.8)))
```

Excel:

```
=6053+4474/(1+EXP(-0.55*(x-4.8)))
```

(a) Since we don't yet know how to find $s'(6)$ algebraically, we estimate it:

h	Ave. Rate of Change
1	496.78652
0.1	547.825452
0.01	552.276972
0.001	552.713694
0.0001	552.75728

So, $s'(6) \approx 553$ books per week.

(b) We again estimate the rate of change at the point $a = 14$.

h	Ave. Rate of Change
1	11.8911425
0.1	15.0069534
0.01	15.3764323
0.001	15.4140372
0.0001	15.4178043

$s'(14) \approx 15$ books per week.

(c) Sales level off at 10,527 books per week, with a zero rate of change. Graph (using technology formula above):

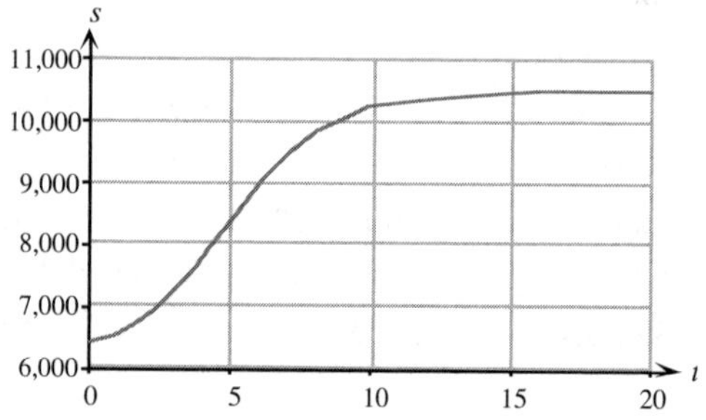

45. (a) $C'(x) = -0.00004x + 3.2$, so $C'(8000)$ = \$2.88 per book

(b) $\bar{C}(x) = -0.00002x + 3.2 + 5400/x$, $\bar{C}(8000)$ = \$3.715 per book

(c) $\bar{C}'(x) = -0.00002 - 5400/x^2$, so $\bar{C}'(8000) \approx$ −\$0.000104 per book, per additional book sold.

(d) At a sales level of 8000 books per week, the cost is increasing at a rate of \$2.88 per book (so that the 8001st book costs approximately \$2.88 to sell), and it costs an average of \$3.715 per book to sell the first 8000 books. Moreover, the average cost is decreasing at a rate of \$0.000104 per book, per additional book sold.

Chapter 4
4.1

The solutions to Exercises 1–12 show the calculation of the derivative using the product or quotient rule as appropriate.

1. Product rule: $f'(x) = (0)x + 3(1) = 3$

3. Product rule: $g'(x) = (1)x^2 + x(2x) = 3x^2$

5. Product rule: $h'(x) = (1)(x + 3) + x(1) = 2x + 3$

7. Product rule: $r'(x) = (0)x^{2.1} + 100(2.1x^{1.1}) = 210x^{1.1}$

9. Quotient rule: $s'(x) = \frac{(0)x - 2(1)}{x^2} = -\frac{2}{x^2}$

11. Quotient rule: $u'(x) = \frac{(2x)3 - x^2(0)}{3^2} = \frac{2x}{3}$

13. $\frac{dy}{dx} = 3(4x^2 - 1) + 3x(8x) = 36x^2 - 3$

15. $\frac{dy}{dx} = 3x^2(1 - x^2) + x^3(-2x) = 3x^2 - 5x^4$

17. $\frac{dy}{dx} = 2(2x + 3) + (2x + 3)(2) = 8x + 12$

19. $\frac{dy}{dx} = \sqrt{x} + \frac{x}{2\sqrt{x}} = \sqrt{x} + \frac{\sqrt{x}}{2} = \frac{3\sqrt{x}}{2}$

21. $\frac{dy}{dx} = (x^2 - 1) + (x + 1)(2x) = (x + 1)(3x - 1)$

23. $\frac{dy}{dx} = (x^{-0.5} + 4)(x - x^{-1}) + (2x^{0.5} + 4x - 5)\cdot(1 + x^{-2})$

25. $\frac{dy}{dx} = (4x - 4)(2x^2 - 4x + 1) + (2x^2 - 4x + 1)(4x - 4) = 8(2x^2 - 4x + 1)(x - 1)$

27. $\frac{dy}{dx} = \left(\frac{1}{3.2} - \frac{3.2}{x^2}\right)(x^2+1) + \left(\frac{x}{3.2} + \frac{3.2}{x}\right)(2x)$

29. $\frac{dy}{dx} = 2x(2x + 3)(7x + 2) + 2x^2(7x + 2) + 7x^2(2x + 3)$

31. $\frac{dy}{dx} = 5.3(1 - x^{2.1})(x^{-2.3} - 3.4) - 2.1x^{1.1}(5.3x - 1)(x^{-2.3} - 3.4) - 2.3x^{-3.3}(5.3x - 1)(1 - x^{2.1})$

33. $\frac{dy}{dx} = \frac{1}{2\sqrt{x}}\left(\sqrt{x} + \frac{1}{x^2}\right) + (\sqrt{x} + 1)\left(\frac{1}{2\sqrt{x}} - \frac{2}{x^3}\right)$

35. $\frac{dy}{dx} = \frac{2(3x - 1) - 3(2x + 4)}{(3x - 1)^2} = \frac{-14}{(3x - 1)^2}$

37. $\frac{dy}{dx} = \frac{(4x + 4)(3x - 1) - 3(2x^2 + 4x + 1)}{(3x - 1)^2} = \frac{6x^2 - 4x - 7}{(3x - 1)^2}$

39. $\frac{dy}{dx} = \frac{(2x - 4)(x^2 + x + 1) - (x^2 - 4x + 1)(2x + 1)}{(x^2 + x + 1)^2} = \frac{5x^2 - 5}{(x^2 + x + 1)^2}$

41. $\frac{dy}{dx} = \frac{(0.23x^{-0.77} - 5.7)(1 - x^{-2.9}) - 2.9x^{-3.9}(x^{0.23} - 5.7x)}{(1 - x^{-2.9})^2}$

43. $\dfrac{dy}{dx} = \dfrac{\frac{1}{2}x^{-1/2}(x^{1/2} - 1) - \frac{1}{2}x^{-1/2}(x^{1/2} + 1)}{(x^{1/2} - 1)^2} = \dfrac{-1}{\sqrt{x}\,(\sqrt{x} - 1)^2}$

45. $\dfrac{dy}{dx} = \dfrac{d}{dx}\left[\dfrac{\frac{1}{x^2}(x + 1)}{x(1 + x)}\right] = \dfrac{d}{dx}\left(\dfrac{1}{x^3}\right) = -\dfrac{3}{x^4}$ (sometimes it pays to simplify first)

47. $\dfrac{dy}{dx} = \dfrac{[(x + 1) + (x + 3)](3x - 1) - 3(x + 3)(x + 1)}{(3x - 1)^2} = \dfrac{3x^2 - 2x - 13}{(3x - 1)^2}$

49. $\dfrac{dy}{dx} = \dfrac{[(x + 1)(x + 2) + (x + 3)(x + 2) + (x + 3)(x + 1)](3x - 1) - 3(x + 3)(x + 1)(x + 2)}{(3x - 1)^2} =$
$\dfrac{6x^3 + 15x^2 - 12x - 29}{(3x - 1)^2}$

51. The calculation thought experiment tells us that the expression for y is a difference:

$\dfrac{d}{dx}[x^4 - (x^2+120)(4x-1)]$

$= \dfrac{d}{dx}[x^4] - \dfrac{d}{dx}[(x^2+120)(4x-1)]$

Now use the power rule for the first expression and the product rule for the second:

$= 4x^3 - [(2x)(4x-1) + (x^2+120)(4)]$

$= 4x^3 - 12x^2 + 2x - 480$

53. The calculation thought experiment tells us that the expression for y is a sum:

$\dfrac{d}{dx}\left[x+1 + 2\left(\dfrac{x}{x+1}\right)\right]$

$= \dfrac{d}{dx}[x] + \dfrac{d}{dx}[1] + 2\dfrac{d}{dx}\left(\dfrac{x}{x+1}\right)$

Now use the power rule for the expressions on the left and the quotient rule for the one on the right:

$= 1 + 0 + 2\dfrac{(1)(x+1) - x(1)}{(x+1)^2}$

$= 1 + \dfrac{2}{(x+1)^2}$

55. The calculation thought experiment tells us that the expression for y is a difference:

$\dfrac{d}{dx}\left[(x+1)(x-2) - 2\left(\dfrac{x}{x+1}\right)\right]$

$= \dfrac{d}{dx}[(x+1)(x-2)] - 2\dfrac{d}{dx}\left(\dfrac{x}{x+1}\right)$

Now use the product rule for the expressions on the left and the quotient rule for the one on the right:

$= (1)(x-2) + (x+1)(1) - 2\dfrac{(1)(x+1) - x(1)}{(x+1)^2}$

$= 2x - 1 - \dfrac{2}{(x+1)^2}$

57. $\dfrac{d}{dx}[(x^2 + x)(x^2 - x)]$

$= (2x + 1)(x^2 - x) + (x^2 + x)(2x - 1)$

$= 4x^3 - 2x$

59. $\dfrac{d}{dx}[(x^3 + 2x)(x^2 - x)]\Big|_{x=2}$

$= [(3x^2 + 2)(x^2 - x) + (x^3 + 2x)(2x - 1)]\Big|_{x=2}$

$= (5x^4 - 4x^3 + 6x^2 - 4x)\Big|_{x=2} = 64$

61. $\frac{d}{dt}[(t^2 - t^{0.5})(t^{0.5} + t^{-0.5})]\Big|_{t=1}$

$= [(2t - 0.5t^{-0.5})(t^{0.5} + t^{-0.5})$

$+ (t^2 - t^{0.5})(0.5t^{-0.5} - 0.5t^{-1.5})]\Big|_{t=1}$

$= (2.5t^{1.5} + 1.5t^{0.5} - 1)\Big|_{t=1} = 3$

63. $f'(x) = 2x(x^3 + x) + (x^2 + 1)(3x^2 + 1)$

$= 5x^4 + 6x^2 + 1,$

so $f'(1) = 12$ is the slope. The tangent line passes through $(1, f(1)) = (1, 4)$, so its equation is $y = 12x - 8$.

65. $f'(x) = \frac{(x + 2) - (x + 1)}{(x + 2)^2} = \frac{1}{(x + 2)^2}$, so $f'(0) = 1/4$. The tangent line passes through $(0, f(0)) = (0, 1/2)$, so its equation is $y = x/4 + 1/2$.

67. $f'(x) = \frac{2x(x) - (x^2 + 1)}{x^2} = \frac{x^2 - 1}{x^2}$, so $f'(-1) = 0$. The tangent line passes through $(-1, f(-1)) = (-1, -2)$, so its equation is $y = -2$.

69. Rate of change of monthly sales $= q'(t) = 2000 - 200t$.

When $t = 5$: $q'(5) = 2000 - 200(5) = 1000$ units/month

Therefore, sales are increasing at a rate of 1000 units per month).

Rate of change of price $= p'(t) = -2t$

When $t = 5$: $p'(5) = -2(5) = -\$10$/month.

Therefore, The price of a sound system is dropping at a rate of \$10 per month.

Revenue:

$R(t) = p(t)q(t)$

$= (1000-t^2)(2000t-100t^2)$

Rate of change of revenue:

$R'(t) = p'(t)q(t) + p(t)q'(t)$

$= (-2t)(2000t-100t^2)$

$+ (1000-t^2)(2000-200t)$

$R'(5) = [-2(5)][2000(5)-100(5)^2]$

$+ [1000-(5)^2][2000-200(5)]$

$= \$900{,}000$/month

Therefore, revenue is increasing at a rate of \$900,000 per month.

71. Revenue $R(t) = P(t)Q(t)$ million dollars.

In 2001, $t = 1$, so

$R(1) = P(1)Q(1)$

$= [5(1) + 25][0.082(1)^2 - 0.22(1) + 8.2]$

$= 241.86 \approx \$242$ million

To obtain the rate of change of daily revenue, we compute $R'(t)$.

By the product rule,

$R'(t) = P'(t)Q(t) + P(t)Q'(t)$

$= (5)(0.082t^2 - 0.22t + 8.2)$

$+ (5t + 25)(0.164t - 0.22)$

Therefore,

$R'(1) = (5)(0.082(1)^2 - 0.22(1) + 8.2)$

$+ (5(1) + 25)(0.164(1) - 0.22)$

$= 38.63 \approx \$39$ million per year.

73. Let $S(t)$ be the number of T-shirts sold per day. If $t = 0$ is now, we are told that $S(0) = 20$ and $S'(0) = -3$. Let $p(t)$ be the price of T-shirts. We are told that $p(0) = 7$ and $p'(0) = 1$. The revenue is then $R(t) = S(t)p(t)$, so $R'(0) = S'(0)p(0) + S(0)p'(0) = -3(7) + 20(1) = -1$. So, revenue is decreasing at a rate of \$1 per day.

75. The cost per passenger is $Q(t) = C(t)/P(t) = (10{,}000 + t^2)/(1000 + t^2)$. So, $Q'(t) = \frac{2t(1000 + t^2) - 2t(10{,}000 + t^2)}{(1000 + t^2)^2} = \frac{-18{,}000t}{(1000 + t^2)^2}$ and $Q'(6) \approx -0.10$. The cost per passenger is decreasing at a rate of \$0.10 per month.

77. $M'(x) = \frac{3000(3600x^{-2} - 1)}{(x + 3600x^{-1})^2}$, so $M'(10) \approx 0.7670$ mpg/mph. This means that, at a speed of

10 mph, the fuel economy is increasing at a rate of 0.7670 miles per gallon per one mph increase in speed. $M'(60) = 0$ mpg/mph. This means that, at a speed of 60 mph, the fuel economy is neither increasing nor decreasing with increasing speed. $M'(70) \approx -0.0540$. This means that, at 70 mph, the fuel economy is decreasing at a rate of 0.0540 miles per gallon per one mph increase in speed. 60 mph is the most fuel-efficient speed for the car. (In the next chapter we shall discuss how to locate largest values in general.)

79. Cost: $10{,}000t/3 + 80{,}000$
Personnel: $-12{,}500t + 1{,}500{,}000$ (t since 1995).
Rate of change at $t = 7$ is
$(10{,}000/3)(-12{,}500(7) + 1{,}500{,}000) + (10{,}000(7)/3 + 80{,}000)(-12{,}500)$
$= 3416{,}666{,}667 \approx 3{,}420{,}000{,}000.$
Total military personnel costs were increasing at a rate of about $3420 million per year in 2002.

81. $R'(p) = -\dfrac{5.625}{(1 + 0.125p)^2}$ so $R'(4) = -2.5$ thousand organisms per hour, per 1000 organisms. This means that the reproduction rate of organisms in a culture containing 4000 organisms is declining at a rate of 2500 organisms per hour, per 1000 additional organisms.

83. Let $P(t)$ be the number of eggs; then $P(t) = 30 - t$.
The total oxygen consumption is $P(t)C(t)$ and its rate of change is

$$P'(t)C(t) + P(t)C'(t)$$
$$= (-1)(-0.016t^4+1.1t^3-11t^2+3.6t) + (30-t)(-0.064t^3+3.3t^2-22t+3.6)$$

At $t = 25$ this is approximately $-1572 \approx -1600$. Thus, oxygen consumption is decreasing at a rate of about 1600 milliliters per day. This must be due to the fact that the number of eggs is decreasing, because $C'(25)$ is positive.

85. (a) $c(t) - m(t)$ = Combined market share – MSN market share, and therefore represents the combined market share of the other three providers (Comcast, Earthlink, and AOL).

$m(t)/c(t)$ = (MSN market share)/(Combined market share), MSN's market share as a fraction of the four providers considered.

(b) By the quotient rule

$$\frac{d}{dt}\left(\frac{m(t)}{c(t)}\right) = \frac{m'(t)c(t) - m(t)c'(t)}{[c(t)]^2}$$

$$= \frac{(-1.66t+3.8)(4.2t+36) - (-0.83t^2 + 3.8t + 6.8)(4.2)}{(4.2t+36)^2}$$

Instead of simplifying this, we can just evaluate directly at $t = 3$:

$$\frac{d}{dt}\left(\frac{m(t)}{c(t)}\right)\Bigg|_{t=3}$$

$$= \frac{[-1.66(3)+3.8][4.2(3)+36] - [-0.83(3)^2 + 3.8(3) + 6.8][4.2]}{[4.2(3)+36]^2}$$

≈ -0.043 per year, or -4.3 percentage points per year.

In June, 2003, ($t = 3$) MSN's market share as a fraction of the four providers considered was decreasing at a rate of about 0.043 (or 4.3 percentage points) per year.

87. The analysis is suspect, because it seems to be asserting that the annual increase in revenue, which we can think of as dR/dt, is the product of the annual increases, dp/dt in price, and dq/dt in sales. However, because $R = pq$, the product rule implies that dR/dt is not the product of dp/dt and dq/dt, but is instead

$$\frac{dR}{dt} = \frac{dp}{dt}\cdot q + p\cdot\frac{dq}{dt}.$$

89. Answers will vary; $q = -p + 1000$ is one example: $R(p) = pq = -p^2 + 1000p$, so $R'(p) = -2p + 1000$ and $R'(100) = 800 > 0$.

91. Mine; it is increasing twice as fast as yours. The rate of change of revenue is given by $R'(t) = p'(t)q(t)$ because $q'(t) = 0$ for both of us. Thus, in this case, $R'(t)$ does not depend on the selling price $p(t)$.

93. (A): If the marginal product was greater than the average product it would force the average product to increase. (See the formula for the rate of change of the average given in the solution to Exercise 82.)

4.2

1. $2(2x + 1)(2) = 4(2x + 1)$

3. $-(x - 1)^{-2}(1) = -(x - 1)^{-2}$

5. $-2(2 - x)^{-3}(-1) = 2(2 - x)^{-3}$

7. $0.5(2x + 1)^{-0.5}(2) = (2x + 1)^{-0.5}$

9. $-(4x - 1)^{-2}(4) = -4(4x - 1)^{-2}$

11. $-(3x - 1)^{-2}(3) = -3/(3x - 1)^2$

13. $4(x^2 + 2x)^3\frac{d}{dx}(x^2 + 2x) =$

$4(x^2 + 2x)^3(2x + 2)$

15. $-(2x^2 - 2)^{-2}\frac{d}{dx}(2x^2 - 2) = -4x(2x^2 - 2)^{-2}$

17. $-5(x^2 - 3x - 1)^{-6}\frac{d}{dx}(x^2 - 3x - 1) =$

$-5(2x - 3)(x^2 - 3x - 1)^{-6}$

19. $-3(x^2 + 1)^{-4}\frac{d}{dx}(x^2 + 1) = -6x/(x^2 + 1)^4$

21. $1.5(0.1x^2 - 4.2x + 9.5)^{0.5}\frac{d}{dx}(0.1x^2 - 4.2x + 9.5)$

$= 1.5(0.2x - 4.2)(0.1x^2 - 4.2x + 9.5)^{0.5}$

23. $4(s^2 - s^{0.5})^3\frac{d}{dx}(s^2 - s^{0.5}) =$

$4(2s - 0.5s^{-0.5})(s^2 - s^{0.5})^3$

25. $\frac{d}{dx}\sqrt{1 - x^2} = \frac{d}{dx}(1 - x^2)^{1/2}$

$= \frac{1}{2}(1 - x^2)^{-1/2}(-2x) = -\frac{x}{\sqrt{1 - x^2}}$

27. $\left(-\frac{1}{2}\right)2[(x + 1)(x^2 - 1)]^{-3/2}\frac{d}{dx}[(x + 1)(x^2 - 1)]$

$= -[(x + 1)(x^2 - 1)]^{-3/2}[(x^2 - 1) + (x + 1)(2x)]$

$= -[(x + 1)(x^2 - 1)]^{-3/2}(3x - 1)(x + 1)$

29. $2(3.1x - 2)^1(3.1) - (-2)(3.1x - 2)^{-3}(3.1)$

$= 6.2(3.1x - 2) + 6.2/(3.1x - 2)^3$

31. $2[(6.4x - 1)^2 + (5.4x - 2)^3]\frac{d}{dx}[(6.4x - 1)^2 + (5.4x - 2)^3]$

$= 2[(6.4x - 1)^2 + (5.4x - 2)^3][12.8(6.4x - 1) + 16.2(5.4x - 2)^2]$

33. $\frac{d}{dx}[(x^2 - 3x)^{-2}](1 - x^2)^{0.5} + (x^2 - 3x)^{-2}\frac{d}{dx}[(1 - x^2)^{0.5}]$

$= -2(x^2 - 3x)^{-3}(2x - 3)(1 - x^2)^{0.5} - x(x^2 - 3x)^{-2}(1 - x^2)^{-0.5}$

35. $2\left(\frac{2x + 4}{3x - 1}\right)\frac{d}{dx}\left(\frac{2x + 4}{3x - 1}\right)$

$= \left(\frac{4(x + 2)}{3x - 1}\right) \cdot \frac{2(3x - 1) - (2x + 4)(3)}{(3x - 1)^2} = \frac{-56(x + 2)}{(3x - 1)^3}$

37. $3\left(\frac{z}{1+z^2}\right)^2\frac{d}{dz}\left(\frac{z}{1+z^2}\right)$

$= \left(\frac{3z^2}{(1+z^2)^2}\right)\cdot\frac{(1+z^2)-z(2z)}{(1+z^2)^2}$

$= \frac{3z^2(1-z^2)}{(1+z^2)^4}$

39. $3[(1+2x)^4-(1-x)^2]^2\frac{d}{dx}[(1+2x)^4-(1-x)^2]$

$= 3[(1+2x)^4-(1-x)^2]^2[8(1+2x)^3+2(1-x)]$

41. $4.3[2+(x+1)^{-0.1}]^{3.3}\frac{d}{dx}[2+(x+1)^{-0.1}]$

$= -0.43(x+1)^{-1.1}[2+(x+1)^{-0.1}]^{3.3}$

43. $-(\sqrt{2x+1}-x^2)^{-2}\frac{d}{dx}[(2x+1)^{1/2}-x^2]$

$= -(\sqrt{2x+1}-x^2)^{-2}\left(\frac{1}{2}(2x+1)^{-1/2}(2)-2x\right)$

$= -\frac{\frac{1}{\sqrt{2x+1}}-2x}{(\sqrt{2x+1}-x^2)^2}$

45. $3\{1+[1+(1+2x)^3]^3\}^2\frac{d}{dx}\{1+[1+(1+2x)^3]^3\}$

$= 9\{1+[1+(1+2x)^3]^3\}^2[1+(1+2x)^3]^2\frac{d}{dx}[1+(1+2x)^3]$

$= 27\{1+[1+(1+2x)^3]^3\}^2[1+(1+2x)^3]^2(1+2x)^2(2) = 54(1+2x)^2[1+(1+2x)^3]^2\{1+[1+(1+2x)^3]^3\}^2$

47. $\frac{dy}{dt} = 100x^{99}\frac{dx}{dt} - 99x^{-2}\frac{dx}{dt}$

$= (100x^{99}-99x^{-2})\frac{dx}{dt}$

49. $\frac{ds}{dt} = \frac{d}{dt}(r^{-3}+r^{0.5})$

$= -3r^{-4}\frac{dr}{dt} + 0.5r^{-0.5}\frac{dr}{dt}$

$= (-3r^{-4}+0.5r^{-0.5})\frac{dr}{dt}$

51. $\frac{dV}{dt} = 4\pi r^2\frac{dr}{dt}$

53. $\frac{dy}{dt} = 3x^2\frac{dx}{dt} - x^{-2}\frac{dx}{dt}$,

so $\left.\frac{dy}{dt}\right|_{t=1} = 3(2)^2(-1) - 2^{-2}(-1) = -47/4$

55. $\frac{dx}{dy} = \frac{1}{dy/dx} = \frac{1}{3}$

57. $\frac{dy}{dx} = \frac{dy/dt}{dx/dt} = \frac{-5}{3}$

59. $\frac{dx}{dy} = \frac{1}{dy/dx}$

$= \left.\frac{1}{6x-2}\frac{dx}{dy}\right|_{x=1}$

$= \frac{1}{6(1)-2} = \frac{1}{4}$

61. To express y as a function of t, take the given equation for y:

$y = 35x^{-0.25}$

and substitute for x using $x = 7 + 0.2t$:

$y = 35(7+0.2t)^{-0.25}$

The rate of change of spending on food is given by dy/dt. By the chain rule,.

$\frac{dy}{dt} = 35(-0.25)(7+0.2t)^{-1.25}\frac{d}{dt}(7+0.2t)$

$= -8.75(7+0.2t)^{-1.25}(0.2)$

$= -1.75(7+0.2t)^{-1.25}$ percentage points per month.

(The units of dy/dt are units of y are unit of t; that is, percentage points per month.)

January 1 is represented by $t = 0$, and so November 1 is given by $t = 10$:

$$\left.\frac{dy}{dt}\right|_{t=10} = -1.75(7 + 0.2\times 10)^{-1.25}$$

≈ -0.11 percentage points per month.

63. $\frac{dP}{dq} = 5000 - 0.5q$ and $\frac{dq}{dn} = 30 + 0.02n$.

When $n = 10$,

$$q = 30(10) + 0.01(10)^2 = 301.$$

Hence, $\left.\frac{dP}{dn}\right|_{n=10} = \frac{dP}{dq}\frac{dq}{dn}$

$= (5000 - 0.5(301))(30 + 0.02(10))$

$= 146{,}454.9.$

At an employment level of 10 engineers, Paramount will increase its profit at a rate of \$146,454.90 per additional engineer hired.

65. We are told that

$\frac{dR}{dp} = \$40$ per \$1 increase in the price

$\frac{dp}{dq} = -\$0.75$ per additional ruby sold

Marginal revenue $= \frac{dR}{dq} = \frac{dR}{dp}\frac{dp}{dq} = (40)(-0.75) =$

$-\$30$ per additional ruby sold.

Interpretation: The revenue is decreasing at a rate of \$30 per additional ruby sold.

67. We are given $dx/dt = -3$ and we need to find dy/dt. From the linear relation we have

$$\frac{dy}{dt} = \frac{dy}{dx}\frac{dx}{dt}$$

$$= (1.5)(-2) = -3$$

Hence, $\frac{dy}{dt} = -3$ murders per 100,000 residents/yr each year.

69. $\frac{dM}{dt} = 2.48$ while $\frac{dB}{dt} = 15{,}700$. The chain rule gives $\frac{dM}{dt} = \frac{dM}{dB}\frac{dB}{dt}$, hence $\frac{dM}{dB} = \frac{dM}{dt}\Big/\frac{dB}{dt} =$ 2.48/15,700 = 0.000158 manatees per boat, or 15.8 manatees per 100,000 boats. Approximately 15.8 more manatees are killed each year for each additional 100,000 registered boats.

71. $\frac{dA}{dt} = 2\pi r\frac{dr}{dt} = 2\pi(3)(2) = 12\pi$ mi²/h

73. $\frac{dV}{dt} = 4\pi r^2\frac{dr}{dt} = 4\pi(10)^2(0.5) = 200\pi$ ft³/week;

$C = 1000V$, so

$$\frac{dC}{dt} = 1000\frac{dV}{dt} = \$200{,}000\pi/\text{week}$$

$\approx \$628{,}000$/week

75. **(a)** $q'(4) \approx (q(4 + 0.0001) - q(4 - 0.0001))/0.0002 \approx 333$ units per month

(b) $R = 800q$, so $dR/dq = \$800$/unit: The marginal revenue is the selling price.

(c) $\frac{dR}{dt} = \frac{dR}{dq}\frac{dq}{dt} \approx 800(333) \approx \$267{,}000$ per month.

77. Keeping r and p fixed,

$$\frac{dM}{dt} = 1.2y^{-0.4}r^{-0.3}p\frac{dy}{dt},$$

so $\frac{dM/dt}{M} = \frac{1.2y^{-0.4}r^{-0.3}p}{2y^{0.6}r^{-0.3}p}\frac{dy}{dt} = 0.6\frac{dy/dt}{y}$. We are given $\frac{dy/dt}{y} = 5\%$ per year, so

$$\frac{dM/dt}{M} = 0.6(5) = 3\% \text{ per year}$$

79. Keeping r fixed,

$$\frac{dM}{dt} = 1.2y^{-0.4}r^{-0.3}p\frac{dy}{dt} + 2y^{0.6}r^{-0.3}\frac{dp}{dt},$$

so $\frac{dM/dt}{M} = 0.6\frac{dy/dt}{y} + \frac{dp/dt}{p}$

$= 0.6(5) + 5 = 8\%$ per year.

81.The glob squared, times the derivative of the glob.

83. The derivative of a quantity cubed is three times the (original) quantity squared, times the derivative of the quantity. Thus, the correct answer is $3(3x^3-x)^2(9x^2-1)$.

85. Following the calculation thought experiment, pretend that you are evaluating the function at a specific value of x. If the last operation you would perform is addition or subtraction, look at each summand separately. If the last operation is multiplication, use the product rule first; if it is division, use the quotient rule first; if it is any other operation (such as raising a quantity to a power or taking a radical of a quantity), then use the chain rule first.

87. An example is

$$f(x) = \sqrt{x + \sqrt{x + \sqrt{x + \sqrt{x + \sqrt{x + 1}}}}}.$$

4.3

1. $\frac{d}{dx}[\ln(x-1)] = \frac{1}{x-1}\frac{d}{dx}(x-1)$

$= \frac{1}{x-1}\cdot 1 = \frac{1}{x-1}$

3. $1/(x \ln 2)$ (from the formula in the textbook)

5. $\frac{d}{dx}[\ln(x^2+3)] = \frac{1}{x^2+3}\frac{d}{dx}(x^2+3)$

$= \frac{1}{x^2+3}\cdot 2x = \frac{2x}{x^2+3}$

7. $\frac{d}{dx}e^{x+3} = e^{x+3}\frac{d}{dx}(x+3)$

$= e^{x+3}(1) = e^{x+3}$

9. $\frac{d}{dx}e^{-x} = e^{-x}\frac{d}{dx}(-x)$

$= e^{-x}(-1) = -e^{-x}$

11. $4^x \ln 4$ (from the formula in the textbook)

13. $\frac{d}{dx}2^u = 2^u \ln 2 \frac{du}{dx}$

$\frac{d}{dx}2^{x^2-1} = 2^{x^2-1}\ln 2 \frac{d}{dx}(x^2-1)$

$= 2^{x^2-1}\, 2x \ln 2$

15. $(1)\ln x + x\frac{d}{dx}\ln x$

$= \ln x + x\left(\frac{1}{x}\right) = 1 + \ln x$

17. $2x \ln x + (x^2+1)\left(\frac{1}{x}\right)$

$= 2x \ln x + \frac{x^2+1}{x}$

19. $5(x^2+1)^4(2x)\ln x + (x^2+1)^5\left(\frac{1}{x}\right)$

$= 10x(x^2+1)^4 \ln x + \frac{(x^2+1)^5}{x}$

21. $\frac{d}{dx}[\ln|3x-1|] = \frac{1}{3x-1}\frac{d}{dx}(3x-1)$

$= \frac{1}{3x-1}\cdot 3 = \frac{3}{3x-1}$

23. $\frac{d}{dx}[\ln|2x^2+1|] = \frac{1}{2x^2+1}\frac{d}{dx}(2x^2+1)$

$= \frac{1}{2x^2+1}\cdot 4x = \frac{4x}{2x^2+1}$

25. $\frac{d}{dx}[\ln|x^2-2.1x^{0.3}|]$

$= \frac{1}{x^2-2.1x^{0.3}}\frac{d}{dx}(x^2-2.1x^{0.3})$

$= \frac{1}{x^2-2.1x^{0.3}}\cdot(2x-0.63x^{-0.7})$

$= \frac{2x-0.63x^{-0.7}}{x^2-2.1x^{0.3}}$

27. $\frac{d}{dx}[\ln(-2x+1) + \ln(x+1)]$

$= \frac{-2}{-2x+1} + \frac{1}{x+1}$

29. $\frac{d}{dx}[\ln(3x+1) - \ln(4x-2)]$

$= \frac{3}{3x+1} - \frac{4}{4x-2}$

31. $\frac{d}{dx}[\ln|x+1| + \ln|x-3| - \ln|-2x-9|]$

$= \frac{1}{x+1} + \frac{1}{x-3} - \frac{-2}{-2x-9}$

$= \frac{1}{x+1} + \frac{1}{x-3} - \frac{2}{2x+9}$

33. $\frac{d}{dx}[1.3\ln(4x-2)] = \frac{5.2}{4x-2}$

35. $\frac{d}{dx}[\ln|x+1|^2 - \ln|3x-4|^3 - \ln|x-9|]$

$= \frac{d}{dx}[2\ln|x+1| - 3\ln|3x-4| - \ln|x-9|] =$

$\frac{2}{x+1} - \frac{9}{3x-4} - \frac{1}{x-9}$

37. $\frac{d}{dx}\log_2(x+1) = \frac{1}{(x+1)\ln 2}\frac{d}{dx}(x+1)$

$= \frac{1}{(x+1)\ln 2}$

39. $\frac{d}{dt}\log_3(t+1/t) = \frac{1}{(t+1/t)\ln 3}\frac{d}{dt}((t+1/t))$

$= \frac{1}{(t+1/t)\ln 3}(1-1/t^2)$

$= \frac{1-1/t^2}{(t+1/t)\ln 3}$

41. $2(\ln|x|)\left(\frac{1}{x}\right) = \frac{2\ln|x|}{x}$

43. $\frac{d}{dx}\{2\ln x - [\ln(x-1)]^2\}$

$= \frac{2}{x} - 2[\ln(x-1)]\frac{1}{x-1}$

$= \frac{2}{x} - \frac{2\ln(x-1)}{x-1}$

45. $e^x + xe^x = e^x(1+x)$

47. $\frac{1}{x+1} + 9x^2e^x + 3x^3e^x$

$= \frac{1}{x+1} + 3e^x(x^3 + 3x^2)$

49. $e^x\ln|x| + e^x\left(\frac{1}{x}\right) = e^x(\ln|x| + 1/x)$

51. $2e^{2x+1}$

53. $(2x-1)e^{x^2-x+1}$

55. $2xe^{2x-1} + x^2(2e^{2x-1}) = 2xe^{2x-1}(1+x)$

57. $2e^{2x-1}\frac{d}{dx}e^{2x-1} = 2e^{2x-1}(2e^{2x-1}) = 4(e^{2x-1})^2$

OR

$\frac{d}{dx}[(e^{2x-1})^2] = \frac{d}{dx}(e^{4x-2}) = 4e^{4x-2} = 4(e^{2x-1})^2$

59. $t(x) = 3^{2x-4}$

$t'(x) = 3^{2x-4}\ln 3\frac{d}{dx}[2x-4]$

$= 2\cdot 3^{2x-4}\ln 3$

61. $v(x) = 3^{2x+1} + e^{3x+1}$

$v'(x) = 3^{2x+1}\ln 3(2) + e^{3x+1}(3)$

$= 2\cdot 3^{2x+1}\ln 3 + 3e^{3x+1}$

63. $u(x) = \frac{3^{x^2}}{x^2+1}$

$u'(x) = \frac{3^{x^2}\ln 3\,(2x)(x^2+1) - 3^{x^2}(2x)}{(x^2+1)^2}$

$= \frac{2x3^{x^2}[(x^2+1)\ln 3 - 1]}{(x^2+1)^2}$

65. $\frac{(e^x - e^{-x})(e^x - e^{-x}) - (e^x + e^{-x})(e^x + e^{-x})}{(e^x - e^{-x})^2}$

$= \frac{e^{2x} - 2 + e^{-2x} - (e^{2x} + 2 + e^{-2x})}{(e^x - e^{-x})^2}$

$= \frac{-4}{(e^x - e^{-x})^2}$

67. $\frac{d}{dx}[e^{(3x-1)+(x-2)+x}] = \frac{d}{dx}e^{5x-3} = 5e^{5x-3}$

69. $\frac{d}{dx}(x\ln x)^{-1} = -(x\ln x)^{-2}[\ln x + x(1/x)]$

$= -\frac{\ln x + 1}{(x\ln x)^2}$

71. Note that $\ln(e^x) = x$, so $f(x) = x^2 - 2\ln(e^x) = x^2 - 2x$, so $f'(x) = 2x - 2 = 2(x-1)$

73. $\frac{1}{\ln x}\frac{d}{dx}\ln x = \frac{1}{x\ln x}$

75. $\frac{d}{dx}\ln(\ln x)^{1/2} = \frac{d}{dx}\left(\frac{1}{2}\ln(\ln x)\right) = \frac{1}{2}\left(\frac{1}{\ln x}\right)\left(\frac{1}{x}\right)$

$= \frac{1}{2x\ln x}$

77. $\frac{dy}{dx} = e^x \log_2 x + \frac{e^x}{x \ln 2} = \frac{e}{\ln 2}$ when $x = 1$.

Therefore the slope of the desired line is $e/\ln 2$, and a point on the line is (1, 0).

So, the equation of the line is $y = (e/\ln 2)(x - 1) \approx 3.92(x - 1)$.

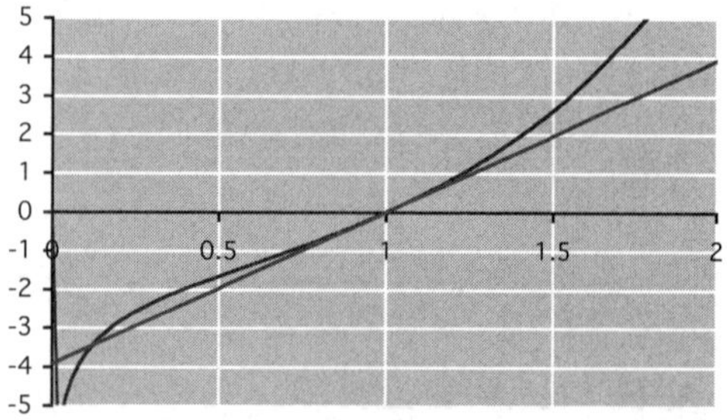

79. $\frac{dy}{dx} = \frac{d}{dx}\left(\frac{1}{2}\ln(2x + 1)\right) = \frac{1}{2x + 1} = 1$ when $x = 0$. The tangent line has slope 1 and passes through (0, 0), so its equation is $y = x$.

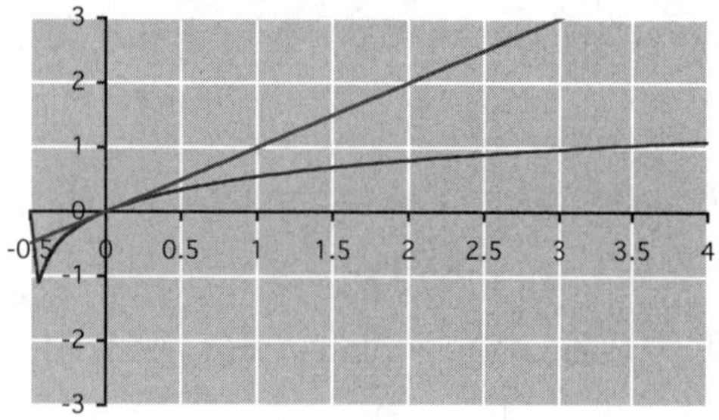

81. $\frac{dy}{dx} = 2xe^{x^2} = 2e$ when $x = 1$. The line at right angles to the graph has slope $-1/(2e)$ and passes through $(1, e)$, so its equation is

$y = -[1/(2e)](x - 1) + e \approx -0.1839x + 2.9022$.

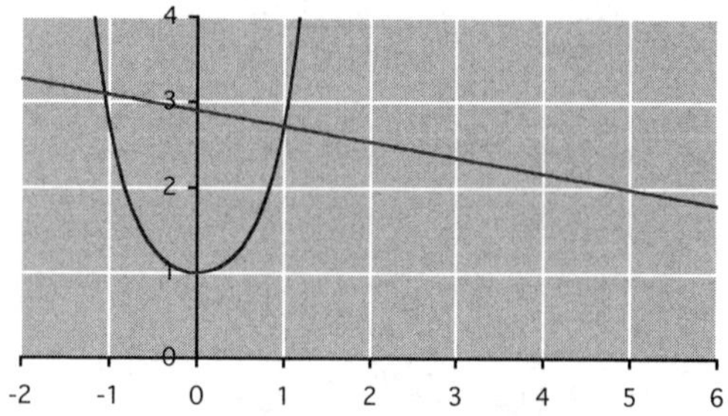

83. In 2003, $t = 9$, and so the price of an apartment was

$p(3) = 0.33e^{0.16(9)} = 0.33e^{1.44} \approx \$1.4$ million dollars

For the rate of change, we compute the derivative:

$p'(t) = \frac{d}{dt}[0.33e^{0.16t}]$

$= (0.33)(0.16)e^{0.16t} = 0.0528e^{0.16t}$

In 2003,

$p'(9) = 0.0528e^{0.16(9)} = 0.0528e^{1.44}$

≈ 0.22 million dollars/year,

or \$220,000 per year. Thus, in 2003 the average price of a two-bedroom apartment in downtown New York City was increasing at a rate of about \$220,000 per year.

85. From the continuous compounding formula, the value of the balance at time t years is $A(t) = 10{,}000e^{0.04t}$. Its derivative is $A'(t) = 400e^{0.04t}$, so, after 3 years, the balance is growing at the rate of $A'(3) = \$451.00$ per year.

87. From the compound interest formula, the value of the balance at time t years is $A(t) = 10{,}000(1 + 0.04/2)^{2t} = 10{,}000(1.02)^{2t}$. Its derivative is $A'(t) = [20{,}000 \ln(1.02)]\,(1.02)^{2t}$, so, after 3 years, the balance is growing at the rate of $A'(3) = \$446.02$ per year.

89. $A(t) = \dfrac{7.0}{1 + 5.4e^{-0.18t}}$

$= 7.0(1 + 5.4e^{-0.18t})^{-1}$

$A'(t) = -7.0(1 + 5.4e^{-0.18t})^{-2}(5.4e^{-0.18t})(-0.18)$

$= \dfrac{6.804e^{-0.18t}}{(1 + 5.4e^{-0.18t})^2}$

$A'(7) = \dfrac{6.804e^{-0.18(7)}}{(1 + 5.4e^{-0.18(7)})^2} \approx 0.30$ thousand articles pear year, or 300 articles per year.

91. $P(t) = \dfrac{150}{1 + 15{,}000e^{-0.35t}}$

$= 150(1 + 15{,}000e^{-0.35t})^{-1}$.

$P'(t) = -150(1 + 15{,}000e^{-0.35t})^{-2}[15{,}000(-0.35)e^{-0.35t}]$

$= 787{,}500e^{-0.35t}/(1 + 15{,}000e^{-0.35t})^2$.

$P'(20) \approx 3.3$ million cases/week, or 330,000 cases/week

$P'(30) \approx 11.0$ million cases/week, or 11,000,000 cases/week

$P'(40) \approx 0.64$ million cases/week, or 640,000 cases/week

93. $A(t) = \dfrac{7.0}{1 + 5.4(1.2)^{-t}} = 7.0(1 + 5.4(1.2)^{-t})^{-1}$

$A'(t) = -7.0(1 + 5.4(1.2)^{-t})^{-2}5.4(1.2)^{-t}(-1)\ln 1.2$

$= \dfrac{6.8912(1.2)^{-t}}{(1 + 5.4(1.2)^{-t})^2}$

$A'(7) = \dfrac{6.8915.4(1.2)^{-7}}{(1 + 5.4(1.2)^{-7})^2} \approx 0.31$ thousand articles pear year, or 310 articles per year.

95. The population t years after 1995 was given by $P(t) = 4{,}000{,}000(2^{t/10})$. Its derivative is $P'(t) = [400{,}000 \ln 2]\ 2^{t/10}$, so, at the start of 1995, the population was growing at the rate of $P'(0) \approx 277{,}000$ people/year.

97. The amount of Plutonium-239 left after t years is $P(t) = 10(0.5)^{t/24{,}400}$. Its derivative is $P'(t) = [(10/24{,}400) \ln 0.5]\ (0.5)^{t/24{,}400}$, so, after 100 years, the rate of change is $P'(100) \approx -0.000283$ g/year. That is, the Plutonium is decaying at the rate of 0.000283 g/year.

99. (a) (A): The data would best be modeled by a function that approaches a value in the high 400s exponentially. The function in (A) is the only one that does that: the one in (B) decays to 0, the one in (C) increases without bound, and the one in (D) decreases without bound.

(b) $S'(x) = 136(0.0000264)e^{-0.0000264x}$, so $S'(45{,}000) \approx 0.001$. At an income level of \$45,000, the average verbal SAT increases by approximately $1000(0.001) = 1$ point for each \$1000 increase in income.

(c) $S'(x)$ decreases with increasing x, so that as parental income increases, the effect on SAT scores decreases.

101. (a) $a'(x) = 120(0.172)^x \ln(0.172)$

$a'(2) = 120(0.172)^2 \ln(0.172)$

≈ -6.25 years/child

(We rounded to 3 significant digits because the given coefficients are only given to 3 digits.) The answer tells us that, when the fertility rate is 2 children per woman, the average age of a population is dropping at a rate of 6.26 years per one-child increase in the fertility rate.

(b) From part (a), the average age of a population is dropping at a rate of 6.26 years per one-child increase in the fertility rate. In other words:

Given: 1 child per woman increase → 6.26 year drop in average age

Want: x child per woman increase → 1 year drop in average age

Solution: $x = 1/6.25 = 0.160$ children per woman.

103. (a) $W'(t) = -1500[1 + 0.77(1.16)^{-t}]^{-2} \cdot[0.77(-1)\ln(1.16)\ (1.16)^{-t}] \approx \dfrac{171.425(1.16)^{-t}}{(1 + 0.77(1.16)^{-t})^2}$; $W'(6) \approx 41$ to two significant digits. The constants in the model are specified to two and three significant digits, so we cannot expect the answer to be accurate to more than two digits. In other words, all digits from the third on are probably meaningless. The answer tells us that in 1996, the number of authorized wiretaps was increasing at a rate of approximately 41 wiretaps per year.

(b) (A): As seen in the following graph, the derivative is positive (W is increasing…) but getting smaller (at a decreasing rate).

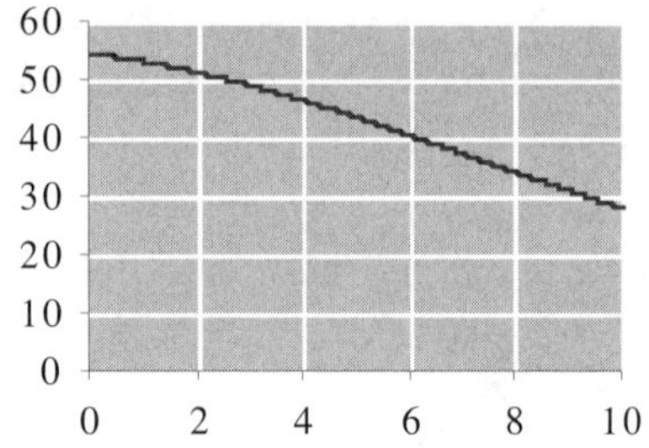

105. (a)

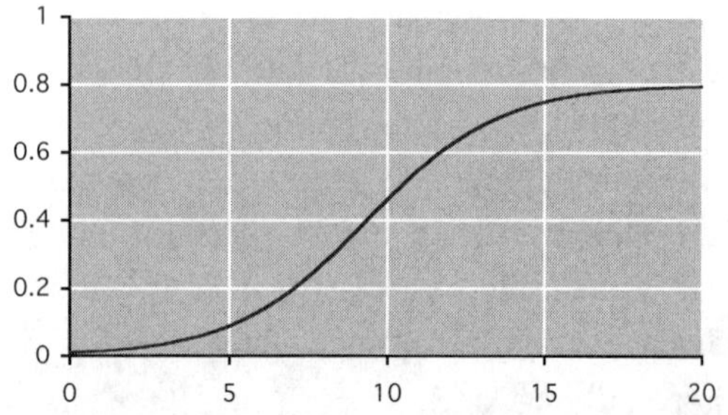

$p'(10) \approx 0.09$, so the percentage of firms using numeric control is increasing at a rate of 9 percentage points per year after 10 years.

(b) $\lim_{t\to+\infty} p(t) = 0.80$. Thus, in the long run, 80% of all firms will be using numeric control.

(c) $p'(t) = 0.3816e^{4.46\ -0.477t}/(1 + e^{4.46\ -0.477t})^2$. $p'(10) = 0.0931$. Graph:

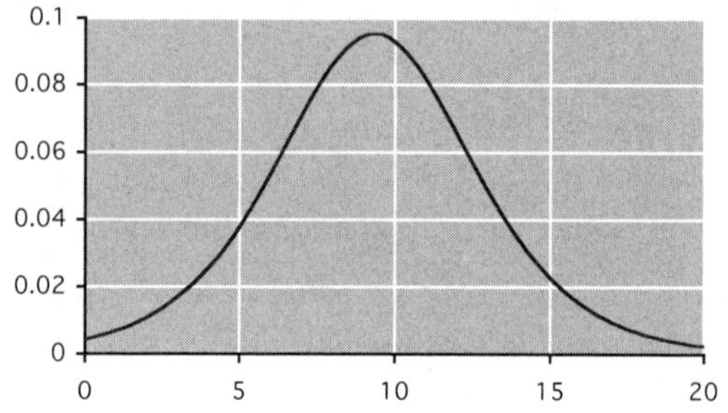

(d) $\lim_{t\to+\infty} p'(t) = 0$. Thus, in the long run, the percentage of firms using numeric control will stop increasing.

107. $R(t) = 350e^{-0.1t}(39t + 68)$ million dollars, so $R(2) \approx \$42{,}000$ million = \$42 billion.

$R'(t) = -35e^{-0.1t}(39t + 68) + 350(39)e^{-0.1t}$

$= (11{,}270 - 1365t)e^{-0.1t}$,

so $R'(2) \approx \$7000$ million per year = \$7 billion per year.

109. e raised to the glob, times the derivative of the glob.

111. 2 raised to the glob, times the derivative of the glob, times the natural logarithm of 2.

113. The power rule does not apply when the exponent is not constant. The derivative of 3 raised to a quantity is 3 raised to the quantity, times the derivative of the quantity, times ln 3. Thus, the correct answer is $3^{2x}\,2\ln 3$

115. No. If $N(t)$ is exponential, so is its derivative.

117. If $f(x) = e^{kx}$, then the fractional rate of change is $\dfrac{f'(x)}{f(x)} = \dfrac{ke^{kx}}{e^{kx}} = k$, the fractional growth rate.

119. If $A(t)$ is growing exponentially, then $A(t) = A_0e^{kt}$ for constants A_0 and k. Its percentage rate of change is then

$$\frac{A'(t)}{A(t)} = \frac{kA_0e^{kt}}{A_0e^{kt}} = k, \text{ a constant.}$$

4.4

1. Implicit differentiation:

$$2 + 3\frac{dy}{dx} = 0$$

$$\frac{dy}{dx} = -2/3$$

Solving for y first and then taking the derivative:

$$y = (7 - 2x)/3$$

$$\frac{dy}{dx} = -2/3$$

3. Implicit differentiation:

$$2x - 2\frac{dy}{dx} = 0$$

$$\frac{dy}{dx} = x$$

Solving for y first and then taking the derivative:

$$y = (x^2 - 6)/2$$

$$\frac{dy}{dx} = x$$

5. Implicit differentiation:

$$2 + 3\frac{dy}{dx} = y + x\frac{dy}{dx}$$

$$(3 - x)\frac{dy}{dx} = y - 2$$

$$\frac{dy}{dx} = (y - 2)/(3 - x)$$

Solving for y first and then taking the derivative:

$$y = -2x/(3 - x)$$

$$\frac{dy}{dx} = \frac{-2(3 - x) - 2x}{(3 - x)^2}$$

$$= \frac{-2}{3 - x} - \frac{2x}{(3 - x)^2}$$

$$= \frac{-2}{3 - x} + \frac{y}{3 - x} = \frac{y - 2}{3 - x}$$

7. Implicit differentiation:

$$e^x y + e^x \frac{dy}{dx} = 0$$

$$\frac{dy}{dx} = -y$$

Solving for y first and then taking the derivative:

$$y = 1/e^x = e^{-x},$$

$$\frac{dy}{dx} = -e^{-x} = -y$$

9. Implicit differentiation:

$$\frac{dy}{dx}\ln x + \frac{y}{x} + \frac{dy}{dx} = 0$$

$$(1 + \ln x)\frac{dy}{dx} = -\frac{y}{x}$$

$$\frac{dy}{dx} = -\frac{y}{x(1 + \ln x)}$$

Solving for y first and then taking the derivative:

$$y = 2/(1 + \ln x),$$

$$\frac{dy}{dx} = -2(1 + \ln x)^{-2}\left(\frac{1}{x}\right)$$

$$= -\frac{2}{1 + \ln x} \cdot \frac{1}{x(1 + \ln x)} = -\frac{y}{x(1 + \ln x)}$$

11. $2x + 2y\frac{dy}{dx} = 0$, $\frac{dy}{dx} = -x/y$

13. $2xy + x^2\frac{dy}{dx} - 2y\frac{dy}{dx} = 0$

$$(x^2 - 2y)\frac{dy}{dx} = -2xy$$

$$\frac{dy}{dx} = -2xy/(x^2 - 2y)$$

15. $3y + 3x\frac{dy}{dx} - \frac{1}{3}\frac{dy}{dx} = -\frac{2}{x^2}$

$$(x^3 - x^2)\frac{dy}{dx} = -(6 + 9x^2y$$

$$\frac{dy}{dx} = -(6 + 9x^2y)/(9x^3 - x^2)$$

17. $2x\frac{dx}{dy} - 6y = 0$

$$\frac{dx}{dy} = 3y/x$$

19. $2p\frac{dp}{dq} - q\frac{dp}{dq} - p = 10pq^2\frac{dp}{dq} + 10p^2q$

$$(2p - q - 10pq^2)\frac{dp}{dq} = p + 10p^2q$$

$$\frac{dp}{dq} = (p + 10p^2q)/(2p - q - 10pq^2)$$

21. $e^y + xe^y \frac{dy}{dx} - e^x \frac{dy}{dx} - ye^x = 0$

$(xe^y - e^x) \frac{dy}{dx} = ye^x - e^y$

$\frac{dy}{dx} = (ye^x - e^y)/(xe^y - e^x)$

23. $e^{st}\left(t \frac{ds}{dt} + s\right) = 2s \frac{ds}{dt}$

$(2s - te^{st}) \frac{ds}{dt} = se^{st}$

$\frac{ds}{dt} = se^{st}/(2s - te^{st})$

25. $\frac{e^x y^2 - 2ye^x(dy/dx)}{y^4} = e^y \frac{dy}{dx}$

$(2e^x + y^3e^y) \frac{dy}{dx} = ye^x$

$\frac{dy}{dx} = ye^x/(2e^x + y^3e^y)$

27. $\frac{2y - 1}{y^2 - y} \frac{dy}{dx} + 1 = \frac{dy}{dx}$

$[(2y - 1) - (y^2 - y)] \frac{dy}{dx} = -(y^2 - y)$

$\frac{dy}{dx} = (y - y^2)/(-1 + 3y - y^2)$

29. $\frac{y + x\frac{dy}{dx} + 2y\frac{dy}{dx}}{xy + y^2} = e^y \frac{dy}{dx}$

$[x + 2y - e^y(xy + y^2)] \frac{dy}{dx} = -y$

$\frac{dy}{dx} = -y/(x + 2y - xye^y - y^2e^y)$

31. (a) $4x^2 + 2y^2 = 12$

$8x + 4y \frac{dy}{dx} = 0$

$\frac{dy}{dx} = -2x/y$

$\left.\frac{dy}{dx}\right|_{(1,-2)} = -2(1)/(-2) = 1$

(b) Point: (1, –2) Slope: 1

Intercept: $b = y_1 - mx_1 = -2 - (1)(1) = -3$

Equation of tangent line:

$y = x - 3$

33. (a) $2x^2 - y^2 = xy$

$4x - 2y \frac{dy}{dx} = y + x\frac{dy}{dx}$

$\frac{dy}{dx} = \frac{4x - y}{x + 2y}$

$\left.\frac{dy}{dx}\right|_{(-1,2)} = \frac{4(-1) - 2}{-1 + 2(2)} = -2$

(b) Point: (−1, 2) Slope: −2

Intercept: $b = y_1 - mx_1 = 2 - (-2)(-1) = 0$

Equation of tangent line:

$y = -2x$

35. (a) $x^2y - y^2 + x = 1$

$2xy + x^2\frac{dy}{dx} - 2y\frac{dy}{dx} + 1 = 0$

$\frac{dy}{dx}(x^2 - 2y) = -(1 + 2xy)$

$\frac{dy}{dx} = -\frac{1 + 2xy}{x^2 - 2y}$

$\left.\frac{dy}{dx}\right|_{(1,0)} = -\frac{1 + 2(1)(0)}{1^2 - 2(0)} = -1$

(b) Point: (1, 0) Slope: –1

Intercept: $b = y_1 - mx_1 = 0 - (-1)(1) = 1$

Equation of tangent line:

$y = -x + 1$

37. (a) $xy - 2000 = y$

$y + x\frac{dy}{dx} = \frac{dy}{dx}$

$y = \frac{dy}{dx}(1-x)$

$\frac{dy}{dx} = \frac{y}{1-x}$

When $x = 2$, the corresponding value of y is obtained from the original equation $xy - 2000 = y$:

$xy - 2000 = y$

$2y - 2000 = y$

$y = 2000$

We now evaluate the derivative:

$$\frac{dy}{dx} = \frac{y}{1-x} = \frac{2000}{1-2} = -2000$$

(b) Point: (2, 2000) Slope: –2000

Intercept: $b = y_1 - mx_1$

$= 2000 - (-2000)(2) = 6000$

Equation of tangent line:

$y = -2000x + 6000$

39. (a) $\ln(x+y) - x = 3x^2$

$$\frac{1 + dy/dx}{x + y} - 1 = 6x$$

$$\frac{dy}{dx} = (6x + 1)(x + y) - 1$$

When $x = 0$, the corresponding value of y is obtained from the original equation $\ln(x+y) - x = 3x^2$ by substituting:

$\ln y = 0,\ y = e^0 = 1.$

We can now evaluate the derivative:

$$\frac{dy}{dx} = (0 + 1)(0 + 1) - 1 = 0$$

(b) Point: (0, 1) Slope: 0

Intercept: $b = y_1 - mx_1 = 1$

Equation of tangent line:

$y = 0x + 1$

$y = 1$

41. (a) $e^{xy} - x = 4x$

$$e^{xy}\left(y + x\frac{dy}{dx}\right) - 1 = 4$$

$$\frac{dy}{dx} = \frac{5 - ye^{xy}}{xe^{xy}}$$

When $x = 3$, the corresponding value of y is obtained from the original equation $e^{xy} - x = 4x$ by substituting:

$e^{3y} - 3 = 12,$

$e^{3y} = 15$

$3y = \ln 15$

$y = \frac{1}{3}\ln 15 \approx 0.902683,$

So $\frac{dy}{dx} \approx \frac{5 - (0.902683)(15)}{3(15)} \approx -0.1898$

(using the fact that $e^{3y} = 15$)

(b) Point: (3, 0.9027) Slope: −0.1898

Intercept: $b \approx 0.9027 - 3(-0.1898) = 1{,}4721$

Equation of tangent line:

$y = -0.1898x + 1{,}4721$

43. $y = \frac{2x+1}{4x-2}$

$$\ln y = \ln\left(\frac{2x+1}{4x-2}\right) = \ln(2x+1) - \ln(4x-2)$$

Take d/dx of both sides:

$$\frac{1}{y}\frac{dy}{dx} = \frac{2}{2x+1} - \frac{4}{4x\text{-}2}$$

$$\frac{dy}{dx} = y\left[\frac{2}{2x+1} - \frac{4}{4x\text{-}2}\right]$$

$$= \frac{2x+1}{4x-2}\left[\frac{2}{2x+1} - \frac{4}{4x-2}\right]$$

45. $y = \frac{(3x+1)^2}{4x(2x-1)^3}$

$$\ln y = \ln\left[\frac{(3x+1)^2}{4x(2x-1)^3}\right]$$

$= \ln((3x+1)^2 - \ln(4x) - \ln(2x-1)^3$

$= 2\ln(3x+1)) - \ln(4x) - 3\ln(2x-1)$

Take d/dx of both sides:

$$\frac{1}{y}\frac{dy}{dx} = \frac{6}{3x+1} - \frac{1}{x} - \frac{6}{2x-1}$$

$$\frac{dy}{dx} = y\left[\frac{6}{3x+1} - \frac{1}{x} - \frac{6}{2x-1}\right]$$

$$= \frac{(3x+1)^2}{4x(2x-1)^3}\left[\frac{6}{3x+1} - \frac{1}{x} - \frac{6}{2x-1}\right]$$

47. $y = (8x-1)^{1/3}(x-1)$

$\ln y = \ln[(8x-1)^{1/3}(x-1)]$

$= \ln(8x-1)^{1/3} + \ln(x-1)$

$= \frac{1}{3}\ln(8x-1) + \ln(x-1)$

$$\frac{1}{y}\frac{dy}{dx} = \frac{8}{3(8x-1)} + \frac{1}{x-1}$$

$$\frac{dy}{dx} = y\left[\frac{8}{3(8x-1)} + \frac{1}{x-1}\right]$$
$$= (8x-1)^{1/3}(x-1)\left[\frac{8}{3(8x-1)} + \frac{1}{x-1}\right]$$

49. $\ln y = \ln[(x^3 + x)\sqrt{x^3 + 2}]$

$= \ln(x^3 + x) + \frac{1}{2}\ln(x^3 + 2)$

$$\frac{1}{y}\frac{dy}{dx} = \frac{3x^2 + 1}{x^3 + x} + \frac{1}{2}\frac{3x^2}{x^3 + 2}$$
$$\frac{dy}{dx} = (x^3 + x)\sqrt{x^3 + 2}\left(\frac{3x^2 + 1}{x^3 + x} + \frac{1}{2}\frac{3x^2}{x^3 + 2}\right)$$

51. $\ln y = \ln(x^x) = x \ln x$

$$\frac{1}{y}\frac{dy}{dx} = \ln x + x\left(\frac{1}{x}\right) = 1 + \ln x$$
$$\frac{dy}{dx} = x^x(1 + \ln x)$$

53. $P = x^{0.6}y^{0.4}$

Taking d/dx of both sides gives

$$0 = 0.6x^{-0.4}y^{0.4} + 0.4x^{0.6}y^{-0.6}\frac{dy}{dx}$$
$$-0.6x^{-0.4}y^{0.4} + 0.4x^{0.6}y^{-0.6}\frac{dy}{dx}$$
$$\frac{dy}{dx} = -\frac{0.6x^{-0.4}y^{0.4}}{0.4x^{0.6}y^{-0.6}} = -\frac{3y}{2x}$$
$$\left.\frac{dy}{dx}\right|_{x=100,\ y=200{,}000} = -\frac{3(200{,}000)}{2(100)}$$
$= -\$3000$ per worker

The monthly budget to maintain production at the fixed level P is decreasing by approximately \$3000 per additional worker at an employment level of 100 workers and a monthly operating budget of \$200,000. In other words, increasing the workforce by one worker will result in a saving of approximately \$3000 per month.

55. $xy - 2000 = y$

Taking d/dx of both sides gives

$$y + x\frac{dy}{dx} = \frac{dy}{dx}$$
$$y = \frac{dy}{dx}(1-x)$$
$$\frac{dy}{dx} = \frac{y}{1-x}$$

When $x = 5$, the corresponding value of y is obtained from the original equation $xy - 2000 = y$:

$xy - 2000 = y$

$5y - 2000 = y$

$4y = 2000$

$y = 500$

We now evaluate the derivative:

$\frac{dy}{dx} = \frac{y}{1-x} = \frac{500}{1-5} = -125$ T-shirts per dollar. Thus, when the price is set at \$5, the demand is dropping by 125 T-shirts per \$1 increase in price.

57. Set $C = 200{,}000$ and differentiate the equation with respect to e:

$$0 = 100k\frac{dk}{de} + 120e$$
$$\frac{dk}{de} = -\frac{120e}{100k} = -\frac{6e}{5k}.$$

When $e = 15$

$200{,}000 = 15{,}000 + 50k^2 + 60(15)^2$

$= 50k^2 + 28{,}500,$

$50k^2 = 171{,}500,$

$k \approx 58.57$, so $\left.\frac{dk}{de}\right|_{e=15} = -\frac{6(15)}{5(58.57)}$

≈ -0.307 carpenters per electrician.

This means that, for a \$200,000 house whose construction employs 15 electricians, adding one more electrician would cost as much as approximately 0.307 additional carpenters. In other words, one electrician is worth approximately 0.307 carpenters.

59. (a) Set $x = 3.0$ and $g = 80$ and solve for t:

$80 = 12t - 0.2t^2 - 90,\ 0.2t^2 - 12t + 170 = 0,$

$t \approx 22.93$ hours by the quadratic formula. (The other root is rejected because it is larger than 30.)

(b) Set $g = 80$ and differentiate the equation with respect to x:

$$0 = 4x\frac{dt}{dx} + 4t - 0.4t\frac{dt}{dx} - 20x,$$

$$\frac{dt}{dx} = \frac{4t - 20x}{0.4t - 4x} = \frac{t - 5x}{0.1t - x},$$

so $\left.\frac{dt}{dx}\right|_{x=3.0} \approx \frac{22.93 - 5(3.0)}{0.1(22.93) - 3.0} \approx -11.2$ hours per grade point. This means that, for a 3.0 student who scores 80 on the examination, 1 grade point is worth approximately 11.2 hours.

61. $0 = 1.2y^{-0.4}r^{-0.3}p - 0.6y^{0.6}r^{-1.3}p\frac{dr}{dy}$

$\frac{dr}{dy} = 2\frac{r}{y}$, so $\frac{dr}{dt} = \frac{dr}{dy}\frac{dy}{dt} = 2\frac{r}{y}\frac{dy}{dt}$

by the chain rule.

63. x, y, y, x

65. Let $y = f(x)g(x)$. Then

$$\ln y = \ln f(x) + \ln g(x),$$

and $\frac{1}{y}\frac{dy}{dx} = \frac{f'(x)}{f(x)} + \frac{g'(x)}{g(x)}$,

so $\frac{dy}{dx} = y\left(\frac{f'(x)}{f(x)} + \frac{g'(x)}{g(x)}\right)$

$$= f(x)g(x)\left(\frac{f'(x)}{f(x)} + \frac{g'(x)}{g(x)}\right)$$

$$= f'(x)g(x) + f(x)g'(x)$$

67. Writing $y = f(x)$ specifies y as an explicit function of x. This can be regarded as an equation giving y as an *implicit* function of x. The procedure of finding dy/dx by implicit differentiation is then the same as finding the derivative of y as an explicit function of x: we take d/dx of both sides.

69. Differentiate both sides of the equation $y = f(x)$ with respect to y to get

$$1 = f'(x)\cdot\frac{dx}{dy},$$

giving

$$\frac{dx}{dy} = \frac{1}{f'(x)} = \frac{1}{dy/dx}$$

Chapter 4 Review Exercises

1. $f(x) = e^x(x^2 - 1)$

Product rule:

$$f'(x) = e^x(x^2 - 1) + e^x(2x)$$
$$= e^x(x^2 + 2x - 1)$$

2. $f(x) = \dfrac{x^2 + 1}{x^2 - 1}$

Quotient rule:

$$f'(x) = \frac{2x(x^2 - 1) - (x^2 + 1)(2x)}{(x^2 - 1)^2}$$
$$= \frac{-4x}{(x^2 - 1)^2}$$

3. $f(x) = (x^2 - 1)^{10}$

Chain rule:

$$f'(x) = 10(x^2 - 1)^9(2x)$$
$$= 20x(x^2 - 1)^9$$

4. $f(x) = \dfrac{1}{(x^2 - 1)^{10}} = (x^2 - 1)^{-10}$

Gneralized power ulre:

$$f'(x) = -10(x^2 - 1)^{-11}(2x)$$
$$= \frac{-20x}{(x^2 - 1)^{11}}$$

5. $f(x) = e^x(x^2 + 1)^{10}$

Product rule:

$$f'(x) = e^x(x^2 + 1)^{10} + e^x \cdot 10(x^2 + 1)^9(2x)$$
$$= e^x(x^2 + 1)^9(x^2 + 20x + 1)$$

6. $f(x) = \left[\dfrac{x - 1}{3x + 1}\right]^3$

Chain rule:

$$f'(x) = 3\left[\frac{x - 1}{3x + 1}\right]^2 \frac{(3x + 1) - 3(x - 1)}{(3x + 1)^2}$$
$$= \frac{12(x - 1)^2}{(3x + 1)^4}$$

7. $f(x) = \dfrac{3^x}{x-1}$

Quotient rule:

$$f'(x) = \frac{3^x \ln 3\ (x-1) - 3^x(1)}{(x-1)^2}$$
$$= \frac{3^x[(x-1)\ln 3 - 1]}{(x-1)^2}$$

8. $f(x) = 4^{-x}(x+1)$

Product rule:

$$f'(x) = (4^{-x})\ln 4(-1)(x+1) + 4^{-x}(1)$$
$$= 4^{-x}[-(x+1)\ln 4 + 1]$$

9. $f(x) = e^{x^2-1}$

Chain rule:

$$f'(x) = 2xe^{x^2-1}$$

10. $f(x) = (x^2 + 1)e^{x^2-1}$

Product rule:

$$f'(x) = 2xe^{x^2-1} + (x^2 + 1)(2x)e^{x^2-1}$$
$$= 2x(x^2 + 2)e^{x^2-1}$$

11. $\ln(x^2 - 1)$

Chain rule:

$$f'(x) = \frac{2x}{x^2 - 1}$$

12. $f(x) = \dfrac{\ln(x^2 - 1)}{x^2 - 1}$

Quotient rule:

$$f'(x) = \frac{\dfrac{2x}{x^2 - 1}(x^2 - 1) - 2x\ln(x^2 - 1)}{(x^2 - 1)^2}$$
$$= \frac{2x - 2x\ln(x^2 - 1)}{(x^2 - 1)^2}$$

13. $y = x - e^{2x-1}$

$$y' = 1 - 2e^{2x-1}$$

Set $y' = 0$:

$$1 - 2e^{2x-1} = 0$$

$2x - 1 = \ln(1/2) = -\ln 2$

$x = (1 - \ln 2)/2$

14. $y = e^{x^2}$

$y' = 2xe^{x^2}$

Set $y' = 0$:

$2xe^{x^2} = 0$ when $x = 0$

15. $y = \dfrac{x}{x+1}$

$$y' = \frac{(x + 1) - x}{(x + 1)^2} = \frac{1}{(x + 1)^2}$$

This is never 0, so there are no points at which the tangent line is horizontal.

16. $y = \sqrt{x}(x-1)$

$y' = \frac{3}{2}x^{1/2} - \frac{1}{2}x^{-1/2} = 0$

$3x^{1/2} = x^{-1/2}$

$3x = 1$, so $x = 1/3$

17. $x^2 - y^2 = x$

Take d/dx of both sides:

$$2x - 2y\frac{dy}{dx} = 1$$

Solve for $\dfrac{dy}{dx}$:

$$\frac{dy}{dx} = \frac{2x - 1}{2y}$$

18. $2xy + y^2 = y$

Take d/dx of both sides:

$$2y + 2x\frac{dy}{dx} + 2y\frac{dy}{dx} = \frac{dy}{dx}$$

Solve for $\dfrac{dy}{dx}$:

$$\frac{dy}{dx} = -\frac{2y}{2(x + y) - 1}$$

19. $e^{xy} + xy = 1$

Take d/dx of both sides:

$$\left(y + x\frac{dy}{dx}\right)e^{xy} + y + x\frac{dy}{dx} = 0,$$

Solve for $\dfrac{dy}{dx}$:

$$\frac{dy}{dx} = \frac{-y(e^{xy} + 1)}{x(e^{xy} + 1)} = -\frac{y}{x}$$

20. $\ln\left(\dfrac{y}{x}\right) = y$

Take d/dx of both sides:

$$\left(\frac{1}{y/x}\right)\frac{(dy/dx)\,x - y}{x^2} = \frac{dy}{dx}$$

$$x\frac{dy}{dx} - y = xy\frac{dy}{dx},$$

Solve for $\dfrac{dy}{dx}$:

$$\frac{dy}{dx} = \frac{y}{x(1 - y)}$$

21. $y = \dfrac{(2x-1)^4(3x+4)}{(x+1)(3x-1)^3}$

We use logarithmic differentiation:

$$\ln y = \ln\left[\frac{(2x-1)^4(3x+4)}{(x+1)(3x-1)^3}\right]$$

$$= 4\ln(2x-1) + \ln(3x+4) - \ln(x+1) - 3\ln(3x-1)$$

$$\frac{1}{y}\frac{dy}{dx} = \frac{8}{2x-1} + \frac{3}{3x+4} - \frac{1}{x+1} - \frac{9}{3x-1}$$

$$\frac{dy}{dx} = y\left[\frac{8}{2x-1} + \frac{3}{3x+4} - \frac{1}{x+1} - \frac{9}{3x-1}\right]$$

$$= \frac{(2x-1)^4(3x+4)}{(x+1)(3x-1)^3}\left[\frac{8}{2x-1} + \frac{3}{3x+4} - \frac{1}{x+1} - \frac{9}{3x-1}\right]$$

22. $y = x^{x-1}3^x$

We use logarithmic differentiation:

$\ln y = (x-1)\ln x + x\ln 3$

$$\frac{1}{y}\frac{dy}{dx} = (1)\ln x + (x-1)/x + \ln 3$$

$$\frac{dy}{dx} = y[\ln x + (x-1)/x + \ln 3]$$

$$= x^{x-1}3^x[\ln x + (x-1)/x + \ln 3]$$

23. $xy - y^2 = x^2 - 3$

$$y + x\frac{dy}{dx} - 2y\frac{dy}{dx} = 2x$$

$$\frac{dy}{dx}[x-2y] = 2x-y$$

$$\frac{dy}{dx} = \frac{2x-y}{x-2y}$$

$$\left.\frac{dy}{dx}\right|_{(-1,1)} = \frac{2(-1)-1}{-1-2(1)} = 1$$

Tangent line:

Point: $(-1, 1)$ Slope: 1

Intercept: $b = y_1 - mx_1 = 1 - (1)(-1) = 2$

Equation of tangent line:

$y = x + 2$

24. $\ln(xy) + y^2 = 1$

$$\frac{1}{xy}\left[y + x\frac{dy}{dx}\right] + 2y\frac{dy}{dx} = 0$$

$$\frac{dy}{dx}[1/y + 2y] = -1/x$$

$$\frac{dy}{dx} = \frac{-1}{x(1/y+2y)}$$

$$\left.\frac{dy}{dx}\right|_{(-1,-1)} = \frac{-1}{(-1)(1/(-1)+2(-1))} = -1/3$$

Tangent line:

Point: $(-1, -1)$ Slope: $-1/3$

Intercept: $b = y_1 - mx_1 = -1 - (-1/3)(-1) = -4/3$

Equation of tangent line:

$y = -x/3-4/3$

25. Let $q(t)$ be the number of books sold per week and let $p(t)$ be the price per book. We are given that $q(0) = 1000$ and $q'(0) = 200$ (taking $t = 0$ as now), and that $p(0) = 20$ and $p'(0) = -1$. Since revenue is $R(t) = p(t)q(t)$, we have

$R'(0) = p'(0)q(0) + p(0)q'(0)$

$= (-1)(1000) + 20(200)$

$=$ \$3000 per week (rising).

26. Let $q(t)$ be the number of books sold per week and let $p(t)$ be the price per book. We are given that $q(0) = 1000$, $p(0) = 20$ and $p'(0) = -1$ (taking $t = 0$ as now). We desire $R'(0) = 5000$ and need to compute $q'(0)$. Since $R(t) = p(t)q(t)$, we have

$(-1)(1000) + 20q'(0) = 5000$

$q'(0) = 300$ books per week

27. $R = pq$ gives $R' = p'q + pq'$. Thus, $R'/R = R'/(pq) = (p'q + pq')/pq = p'/p + q'/q$.

28. Let P be the stock price and E the earnings. We are given $P = 100$, $P' = 50$, $E = 1$, and $E' = 0.10$.

$$\frac{d}{dt}\left(\frac{P}{E}\right) = \frac{P'E - PE'}{E^2}$$

$$= \frac{50(1) - 100(0.10)}{1^2} = 40,$$

so the P/E ratio was rising at a rate of 40 units per year.

29. Let P be the stock price and E the earnings. We are given $P = 100$, $E = 1$, $E' = 0.10$, and $\frac{d}{dt}\left(\frac{P}{E}\right) = 100$.

$$\frac{d}{dt}\left(\frac{P}{E}\right) = \frac{P'E - PE'}{E^2}$$

$$100 = \frac{P' - 100(0.10)}{1^2}$$

Solve for p' to get

$P' =$ \$110 per year.

30. $Q = P/E$ gives

$Q' = (P'E - PE')/E^2$.

Thus,

$Q'/Q = Q'/(P/E)$

$= (P'E - PE')/PE = P'/P - E'/E$.

31. The rate of increase of weekly sales is

$$s'(t) = \frac{2460.7\, e^{-0.55(t-4.8)}}{(1 + e^{-0.55(t-4.8)})^2}$$

$t = 6$:

$s'(6) \approx 553$ books per week

32. The rate of increase of weekly sales is

$$s'(t) = \frac{2460.7\ e^{-0.55(t-4.8)}}{(1 + e^{-0.55(t-4.8)})^2}$$

$t = 15$:

$s'(15) \approx 15$ books per week.

33. If $H(t)$ is the daily number of hits t weeks into the year, then

$$H(t) = 1000(1.05)^t.$$

So,

$$H'(t) = 1000 \ln(1.05)\ (1.05)^t$$

and

$$H'(52) = 1000 \cdot \ln(1.05)\ (1.05)^{52}$$

≈ 616.8 hits per day per week.

34. (a) $100q + 100p \dfrac{dq}{dp} + 2q \dfrac{dq}{dp} = 0$

$$\frac{dp}{dq} = \frac{-100q}{100p + 2q}$$

When $p = 40$ and $q = 1000$

$$\frac{dp}{dq} = \frac{-100(1000)}{100(40) + 2(1000)}$$

≈ -16.67 copies per \$1.

The demand for the gift edition of *Lord of the Rings* is dropping at a rate of about 16.67 copies per \$1 increase in the price.

(b) Since $R = pq$,

$$\frac{dR}{dp} = q + p\frac{dq}{dp} \approx 1000 + 40(-16.67)$$

$\approx$ \$333 per dollar increase in price. The derivative is positive, so the price should be raised.

Chapter 5

5.1

1. Absolute min.: $(-3, -1)$, relative max: $(-1, 1)$, relative min: $(1, 0)$, absolute max: $(3, 2)$

3. Absolute min: $(3, -1)$ and $(-3, -1)$, absolute max: $(1, 2)$

5. Absolute min: $(-3, 0)$ and $(1, 0)$, absolute max: $(-1, 2)$ and $(3, 2)$

7. Relative min: $(-1, 1)$

9. Absolute min: $(-3, -1)$, relative max: $(-2, 2)$, relative min: $(1, 0)$, absolute max: $(3, 3)$

11. Relative max: $(-3, 0)$, absolute min: $(-2, -1)$, stationary non-extreme point: $(1, 1)$.

13. $f(x) = x^2 - 4x + 1$ with domain $[0, 3]$

$f'(x) = 2x - 4$

$f'(x) = 0$ when

$2x - 4 = 0$

$x = 2$

$f'(x)$ is defined for all x in $[0, 3]$. Thus, we have the stationary point at $x = 2$ and the endpoints $x = 0$ and $x = 3$:

x	0	2	3
$f(x)$	1	-3	-2

The graph must decrease from $x = 0$ until $x = 2$, then increase again until $x = 3$.

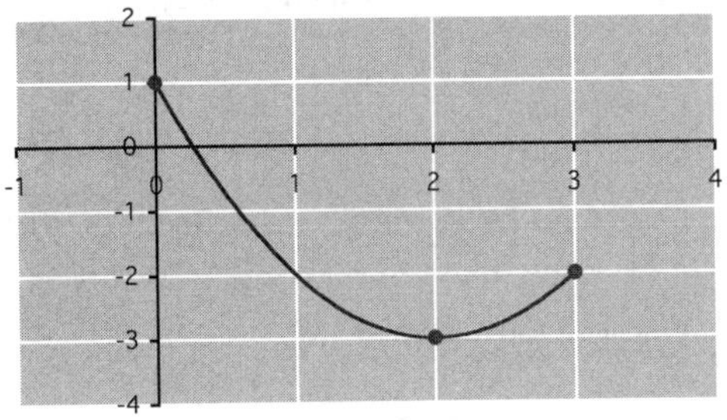

This gives an absolute max at $(0, 1)$, an absolute min at $(2, -3)$, and a relative max at $(3, -2)$.

15. $g(x) = x^3 - 12x$ with domain $[-4, 4]$

$g'(x) = 3x^2 - 12$

$g'(x) = 0$ when

$3x^2 - 12 = 0$

$x = \pm 2$

$g'(x)$ is defined for all x in $[-4, 4]$. Thus, we have stationary points at $x = \pm 2$ and the endpoints $x = \pm 4$:

x	-4	-2	2	4
$g(x)$	-16	16	-16	16

The graph increases from $x = -4$ until $x = -2$, then decreases until $x = 2$, then increases until $x = 4$.

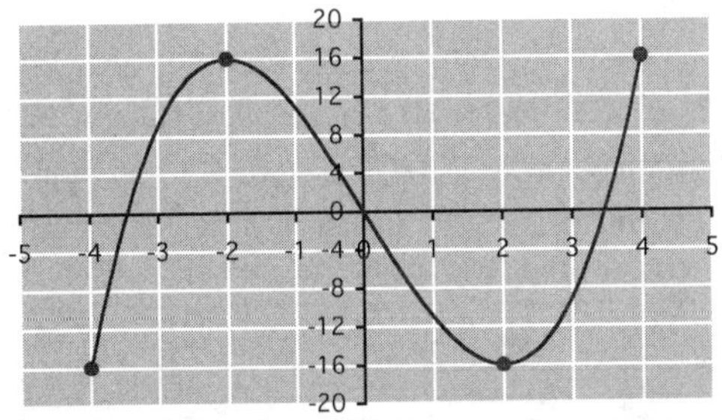

This gives an absolute min at $(-4, -16)$, an absolute max at $(-2, 16)$, an absolute min at $(2, -16)$, and an absolute max at $(4, 16)$.

17. $f(t) = t^3 + t$ with domain $[-2, 2]$

$f'(t) = 3t^2 + 1$

$f'(t) = 0$ when

$3t^2 + 1 = 0,$

which has no solution; $f'(t)$ is defined for all t in $[-2, 2]$. Thus, there are no critical points in the domain, just the endpoints ± 2:

t	-2	2
$f(t)$	-10	10

The graph increases from $x = -2$ until $x = 2$.

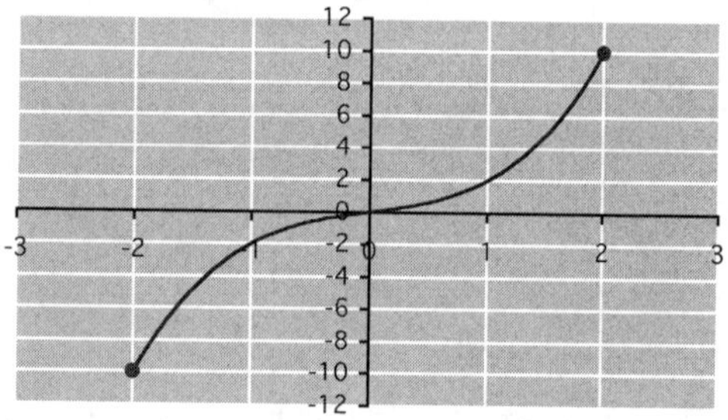

This gives an absolute min at $(-2, -10)$ and an absolute max at $(2, 10)$.

19. $h(t) = 2t^3 + 3t^2$ with domain $[-2, +\infty)$

$h'(t) = 6t^2 + 6t$

$h'(t) = 0$ when

$6t^2 + 6t = 0$

$t = 0$ or $t = -1$

$h'(t)$ is defined for all t in $[-2, \infty)$. Thus, we have stationary points at $t = 0$ and $t = -1$ and the endpoint at $t = -2$. In addition to these we test one point to the right of $t = 0$:

t	-2	-1	0	1
$h(t)$	-4	1	0	5

The graph increases from $x = -2$ until $x = -1$, then decreases until $x = 0$, then increases from that point on. (Remember that $x = 1$ is just a test point to see whether the graph is increasing or decreasing to the right of $x = 0$.)

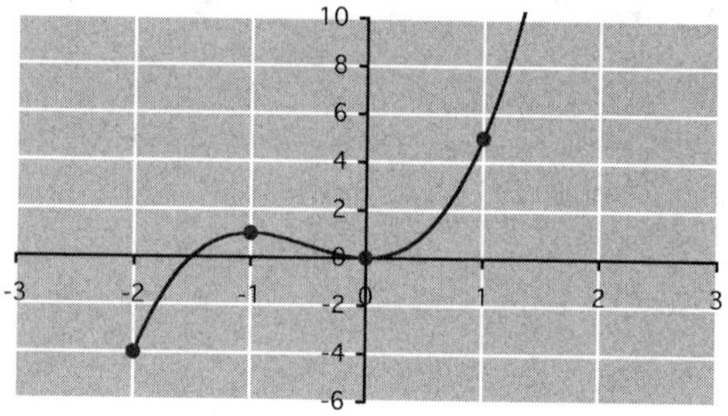

This gives an absolute min at $(-2, -4)$, a relative max at $(-1, 1)$, and relative min at $(0, 0)$.

21. $f(x) = x^4 - 4x^3$ with domain $[-1, +\infty)$

$f'(x) = 4x^3 - 12x^2$

$f'(x) = 0$ when

$4x^3 - 12x^2 = 0$

$4x^2(x - 3) = 0$

$x = 0$ or $x = 3$

$f'(x)$ is defined for all x in $[-1, \infty)$. Thus, we have stationary points at $x = 0$ and $x = 3$ and the endpoint at $x = -1$. In addition to these we test one point to the right of $x = 3$:

x	-1	0	3	4
$f(x)$	5	0	-27	0

The graph decreases from $x = -1$ until $x = 0$, continues to decrease until $x = 3$, then increases from that point on.

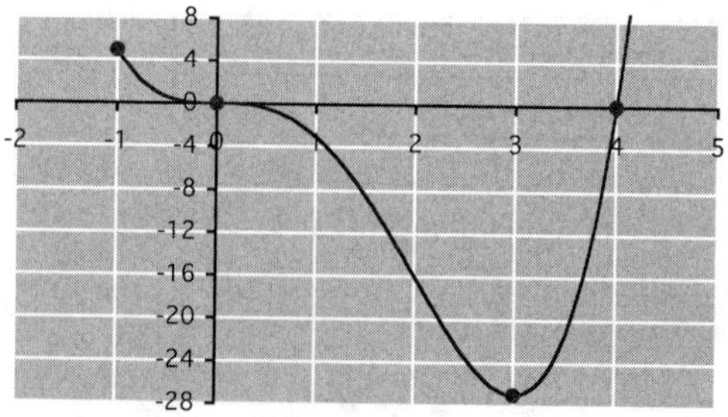

This gives a relative max at $(-1, 5)$ and an absolute min at $(3, -27)$.

23. $g(t) = \frac{1}{4}t^4 - \frac{2}{3}t^3 + \frac{1}{2}t^2$ with domain $(-\infty, +\infty)$

$g'(t) = t^3 - 2t^2 + t$

$g'(t) = 0$ when

$t^3 - 2t^2 + t = 0$

$t(t-1)^2 = 0$

$t = 0$ or $t = 1$

$g'(t)$ is defined for all t. Thus, we have stationary points at $t = 0$ and $t = 1$. Since we have no endpoints, we test a point to the left of $t = 0$ and a point to the right of $t = 1$:

t	-1	0	1	2
$g(t)$	17/12	0	1/12	2/3

The graph decreases until $t = 0$, increases until $t = 1$, then continues to increase to the right of $t = 1$.

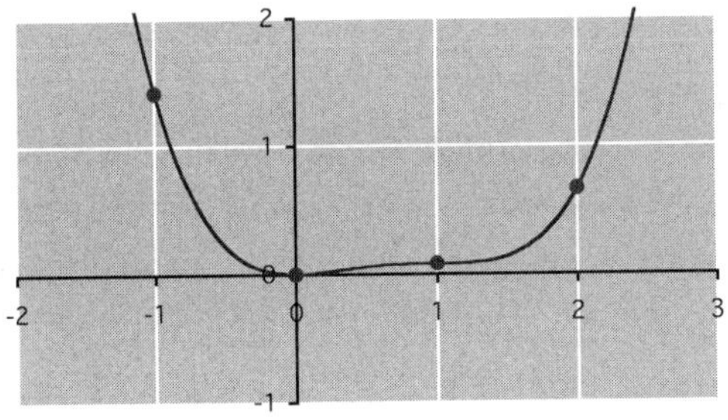

This gives an absolute min at (0, 0) and no other extrema.

25. $h(x) = (x-1)^{2/3}$ with domain [0, 2]

Endpoints: 0, 2

$$h'(x) = \frac{2}{3}(x-1)^{-1/3} = \frac{2}{3(x-1)^{1/3}}$$

Stationary points: $h'(x) = 0$ when

$$\frac{2}{3(x-1)^{1/3}} = 0,$$

which is impossible, so there are no stationary points.

Singular points: $h'(x)$ is undefined when $x = 1$, so we have a singular point at $x = 1$.

x	0	1	2
$h(x)$	1	0	1

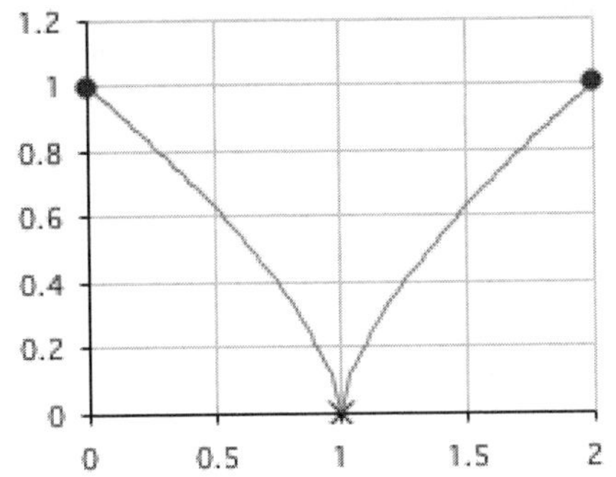

This gives absolute maxima at (0, 1) and (2, 1) and an absolute min at (1, 0).

27. $k(x) = \frac{2x}{3} + (x+1)^{2/3}$ with domain $(-\infty, 0]$

Endpoint: 0

$$k'(x) = \frac{2}{3} + \frac{2}{3}(x+1)^{-1/3}$$
$$= \frac{2}{3} + \frac{2}{3(x+1)^{1/3}}$$

Stationary points: $k'(x) = 0$ when

$$\frac{2}{3} + \frac{2}{3(x+1)^{1/3}} = 0$$
$$\frac{2}{3} = -\frac{2}{3(x+1)^{1/3}}$$

cross-multiply:

$6(x+1)^{1/3} = -6$

$(x+1)^{1/3} = -1$

$(x+1) = (-1)^3 = -1$

$x = -2,$

so we have a stationary point at $x = -2$.

Singular points: $k'(x)$ is undefined when $x = -1$, so we have a singular point at $x = -1$.

Thus we have an endpoint at $x = 0$, a stationary point at $x = -2$ and a singular point at $x = -1$. In addition we use a test point to the left of $x = -2$

x	-3	-2	-1	0
$f(x)$	-0.41	$-1/3$	$-2/3$	1

Notice that the singular point $(-1, -2/3)$ is not an absolute minimum because the graph eventually gets lower on the left.

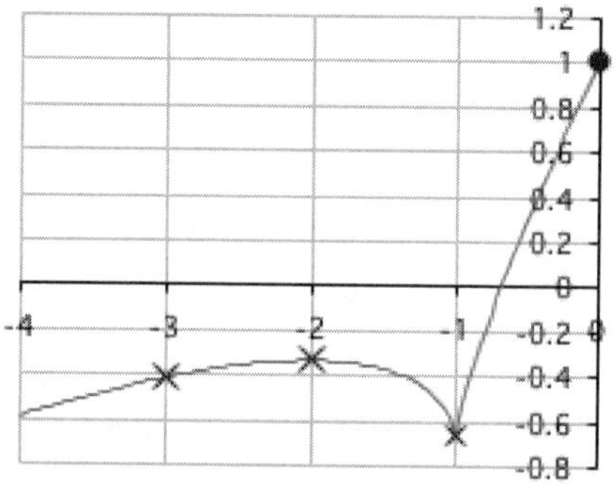

From the graph we see that we have a relative maximum at $(-2, -1/3)$, a relative minimum at $(-1, -2/3)$ and an absolute maximum at $(0, 1)$.

29. $f(t) = \dfrac{t^2 + 1}{t^2 - 1}$; $-2 \le t \le 2$, $t \ne \pm 1$

$$f'(t) = \frac{2t(t^2 - 1) - 2t(t^2 + 1)}{(t^2 - 1)^2}$$

$$= \frac{-4t}{(t^2 - 1)^2}$$

$f'(t) = 0$ when $t = 0$; $f'(t)$ is not defined when $t = \pm 1$, but these points are not in the domain of f. Thus, we have a stationary point at $t = 0$ and the endpoints at $t = \pm 2$. We also test points on either side of $t = \pm 1$ to see how f behaves near these points where it goes to $\pm\infty$ (that is, it has vertical asymptotes there, because the denominator goes to 0):

t	-2	$-3/2$	$-1/2$	0
$f(t)$	5/3	13/5	$-5/3$	-1
t	1/2	3/2	2	
$f(t)$	$-5/3$	13/5	5/3	

The graph increases from $t = -2$, approaching the vertical asymptote at $t = -1$; on the other side of the asymptote it increases until $t = 0$ then decreases as it approaches another vertical asymptote at $t = 1$; on the other side of the second asymptote it decreases until $t = 2$.

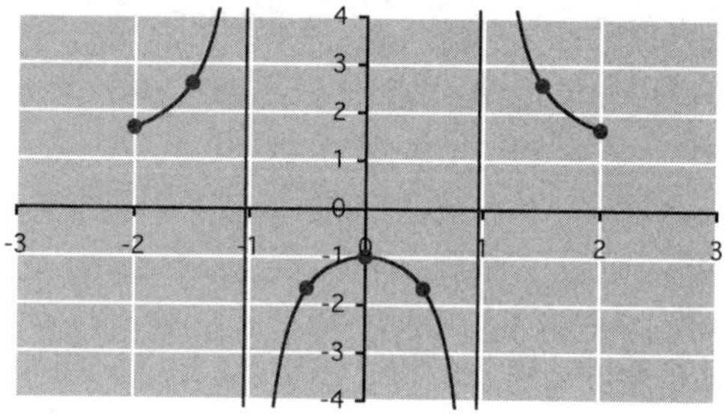

This gives a relative min at $(-2, 5/3)$, a relative max at $(0, -1)$, and a relative min at $(2, 5/3)$.

31. $f(x) = \sqrt{x}(x-1)$; $x \ge 0$

$$f'(x) = \frac{3}{2}x^{1/2} - \frac{1}{2}x^{-1/2}$$

$f'(x) = 0$ when

$$\frac{3}{2}x^{1/2} - \frac{1}{2}x^{-1/2} = 0$$

$$3x - 1 = 0$$

$$x = 1/3$$

$f'(x)$ is not defined at $x = 0$. Thus, we have a stationary point at $x = 1/3$ and a singular point at $x = 0$, which is also an endpoint. In addition, we test a point to the right of $x = 1/3$:

x	0	1/3	1
$f(x)$	0	$-2\sqrt{3}/9$	0

The graph decreases from $t = 0$ until $t = 1/3$, then increases from that point on.

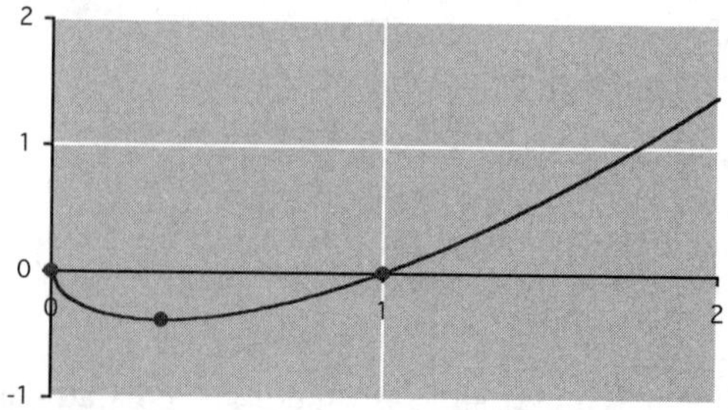

This gives a relative max at $(0, 0)$ and an absolute min at $(1/3, -2\sqrt{3}/9)$.

33. $g(x) = x^2 - 4\sqrt{x}$

Since the domain of g is not specified, we take it to be the set of all x for which g is defined, which is $[0, \infty)$.

$g'(x) = 2x - 2x^{-1/2}$

$g'(x) = 0$ when

$2x - 2x^{-1/2} = 0$

$x = x^{-1/2}$

$x^{3/2} = 1$

$x = 1$

$g'(x)$ is not defined when $x = 0$, which is also an endpoint. Thus, we have a singular endpoint at $x = 0$ and a stationary point at $x = 1$. We test one point to the right of $x = 1$:

x	0	1	2
$g(x)$	0	−3	−1.7

The graph decreases from $x = 0$ to $x = 1$, then increases from that point on.

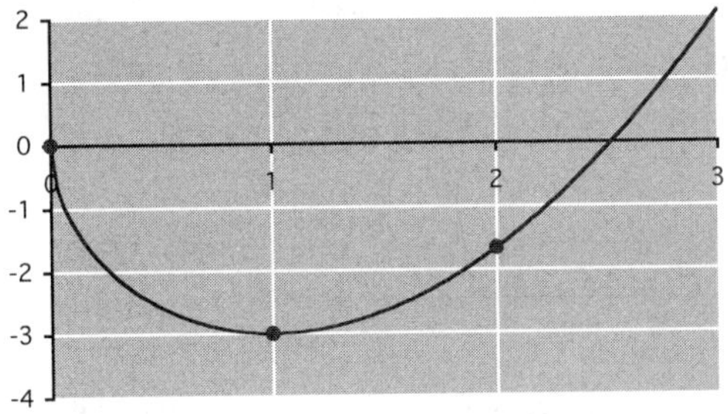

Thus, there is a relative max at (0, 0) and an absolute min at (1, −3).

35. $g(x) = \dfrac{x^3}{x^2 + 3}$

The domain of g is $(-\infty, \infty)$.

$$g'(x) = \frac{3x^2(x^2 + 3) - x^3(2x)}{(x^2 + 3)^2}$$

$$= \frac{x^4 + 9x^2}{(x^2 + 3)^2}$$

$g'(x) = 0$ when

$x^4 + 9x^2 = 0$

$x^2(x^2 + 9) = 0$

$x = 0$

$g'(x)$ is always defined. Thus, we have only one stationary point, at $x = 0$. In addition, we test a point on either side:

x	−1	0	1
$g(x)$	−1/4	0	1/4

The graph is always increasing.

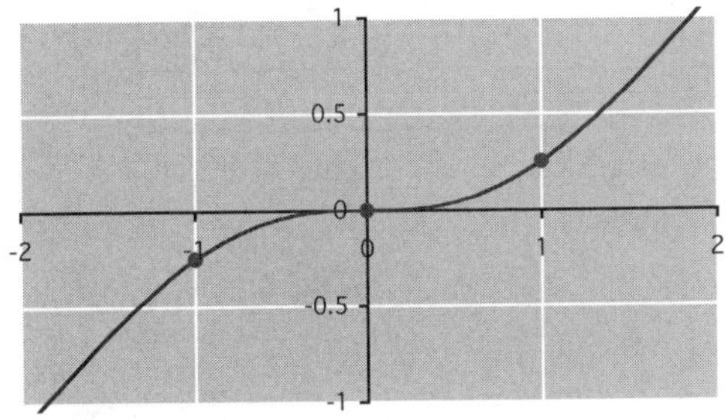

Thus, there are no relative extrema.

37. $f(x) = x - \ln x$ with domain $(0, +\infty)$

$f'(x) = 1 - 1/x$

$f'(x) = 0$ when

$1 - 1/x = 0$

$x = 1$

$f'(x)$ is defined for all x in the domain of f. We test the one stationary point at $x = 1$ and a point on either side:

x	1/2	1	2
$g(x)$	1.19	1	1.31

The graph decreases until $x = 1$ and then increases from that point on.

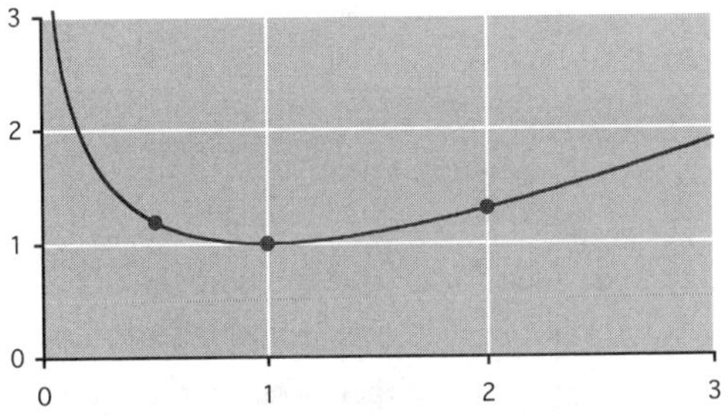

Thus, there is an absolute min at (1, 1).

39. $g(t) = e^t - t$ with domain $[-1, 1]$

$g'(t) = e^t - 1$

$g'(t) = 0$ when

$e^t - 1 = 0$

$e^t = 1$

$t = \ln 1 = 0$

$g'(t)$ is always defined. We need to test only the stationary point and the endpoints:

t	-1	0	1
$g(t)$	1.37	1	1.72

The graph decreases from $t = -1$ to $t = 0$ and then increases to $t = 1$.

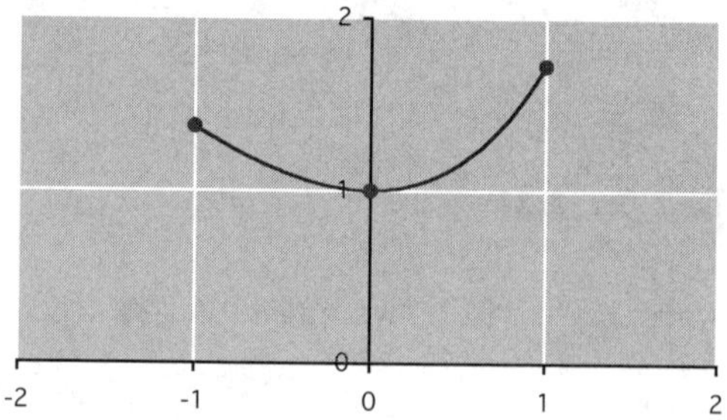

Thus, there is a relative max at $(-1, e^{-1} + 1)$, an absolute min at $(0, 1)$, and an absolute max at $(1, e - 1)$.

41. $f(x) = \dfrac{2x^2-24}{x+4}$

The domain of f is all $x \neq 4$.

$$f'(x) = \frac{4x(x+4) - (2x^2 - 24)}{(x+4)^2}$$

$$= \frac{2x^2 + 16x + 24}{(x+4)^2}$$

$f'(x) = 0$ when

$2x^2 + 16x + 24 = 0$

$2(x + 2)\cdot(x + 6) = 0$

$x = -2$ or -6

$f'(x)$ is defined for all x in the domain of f. Thus, we have stationary points at $x = -2$ and $x = -6$. We test points on either side of the stationary points and between them and the asymptote at $x = -4$:

x	-7	-6	-5
$f(x)$	-24.7	-24	-26
x	-3	-2	-1
$f(x)$	-6	-8	-7.3

The graph increases to $x = -6$ then decreases approaching the asymptote at $x = -4$. On the other side of the asymptote the graph decreases to $x = -2$ and then increases again.

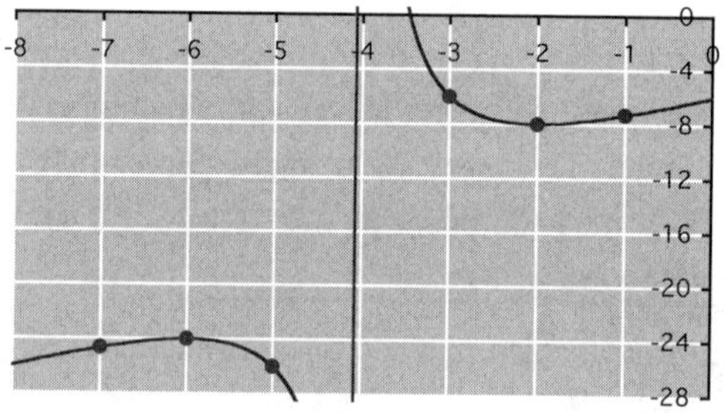

Thus, there is a relative max at $(-6, -24)$ and a relative min at $(-2, -8)$.

43. $f(x) = xe^{1-x^2}$

$f'(x) = e^{1-x^2} + x(-2x)e^{1-x^2}$

$= (1 - 2x^2)e^{1-x^2}$

$f'(x) = 0$ when

$1 - 2x^2 = 0$

$x = \pm 1/\sqrt{2}$

$f'(x)$ is always defined. We test the two stationary points and a point on either side:

x	-1	$-1/\sqrt{2}$	$1/\sqrt{2}$	1
$f(x)$	-1	-1.17	1.17	1

The graph decreases until $x = -1/\sqrt{2}$, increases until $x = 1/\sqrt{2}$, then decreases again.

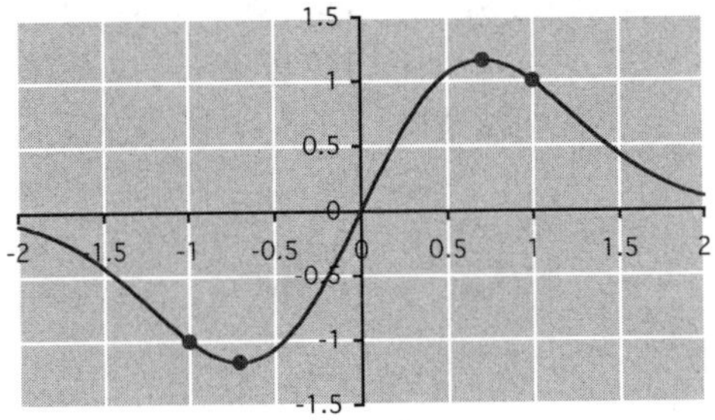

Thus, there is an absolute max at $(1/\sqrt{2}, \sqrt{e/2})$ and an absolute min at $(-1/\sqrt{2}, -\sqrt{e/2})$.

45. $y = x^2 + \dfrac{1}{x-2}$

$$y' = 2x - \frac{1}{(x-2)^2}.$$

The graphs of f and of f' look like this:

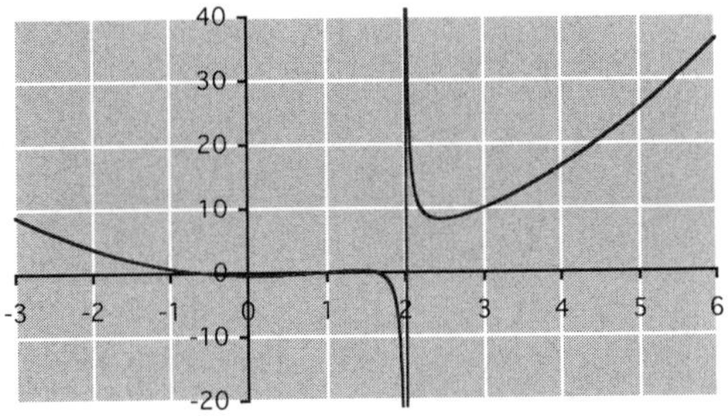

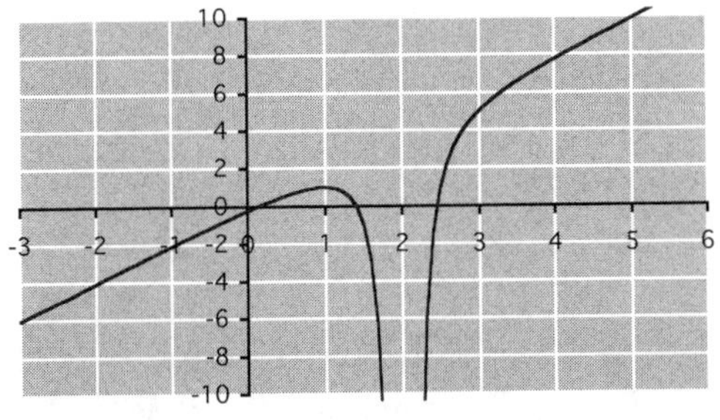

Looking closely at the graph of f' we can see three places where it crosses the x axis. Zooming in we can locate these points approximately as $x \approx 0.15$, $x \approx 1.40$, and $x \approx 2.45$. Thus, we have relative minima at (0.15, −0.52) and (2.45, 8.22) and a relative maximum at (1.40, 0.29).

47. $f(x) = (x-5)^2(x+4)(x-2)$ with domain $[-5, 6]$

$$f'(x) = 2(x-5)(x+4)(x-2) + (x-5)^2 \cdot (x-2) + (x-5)^2(x+4)$$
$$= (x-5)[2(x+4)\cdot(x-2) + (x-5)(x-2) + (x-5)(x+4)]$$
$$= (x-5)(4x^2 - 4x - 26).$$

The graphs of f and f' look like this:

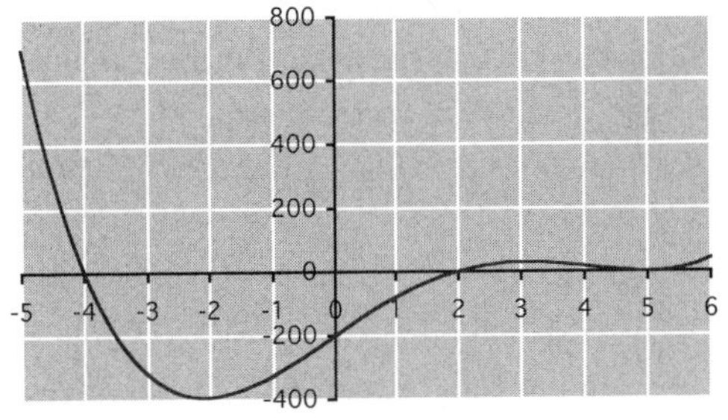

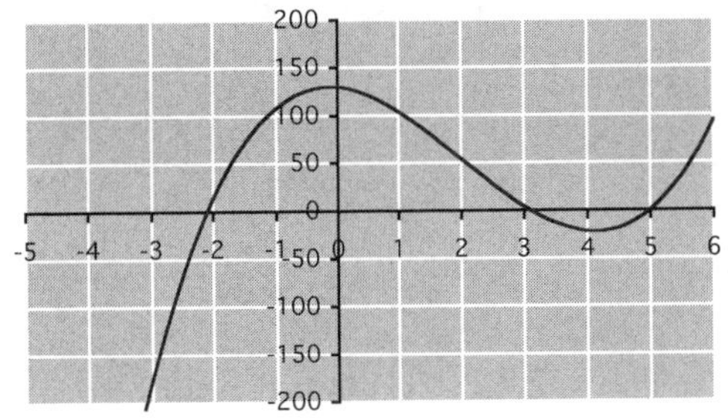

The graph of f' crosses the x axis in three places, hence f has three stationary points. Zooming in we can approximate these as $x \approx -2.10$, $x \approx 3.10$, and $x = 5$. Substituting these and the endpoints −5 and 6 into f, we find that we have an absolute maximum at (−5, 700), relative maxima at (3.10, 28.19) and (6, 40), an absolute minimum at (−2.10, −392.69), and a relative minimum at (5, 0).

49. The derivative is zero at $x = -1$. To the left of $x = -1$ the derivative is negative, indicating that the graph of f is decreasing. To the right of $x = -1$ the derivative is positive, indicating that the graph of f is increasing. Thus, f has a stationary minimum at $x = -1$.

51. Stationary minima at $x = -2$ and $x = 2$, stationary maximum at $x = 0$ (see the solution to #49).

53. Singular minimum at $x = 0$. At $x = 1$, the derivative of also zero, but to both the left and right of $x = 0$ the derivative is positive, indicating that the graph of f is increasing on both sides of $x = 0$. Thus, f has a stationary non-extreme point at $x = 1$

55. Stationary minimum at $x = -2$, singular non-extreme point at $x = -1$, singular non-extreme point at $x = 1$, stationary maximum at $x = 2$

57.

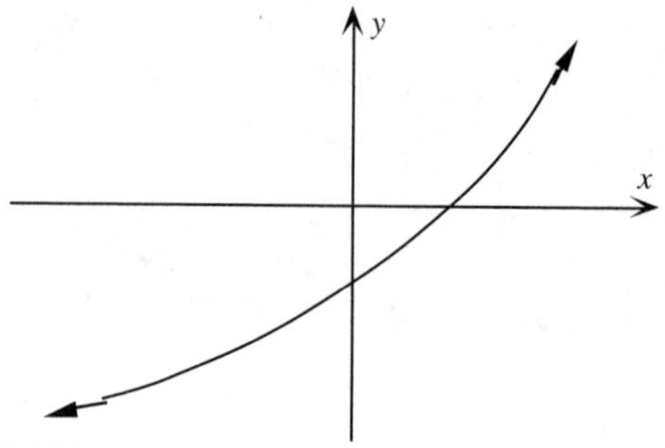

59.

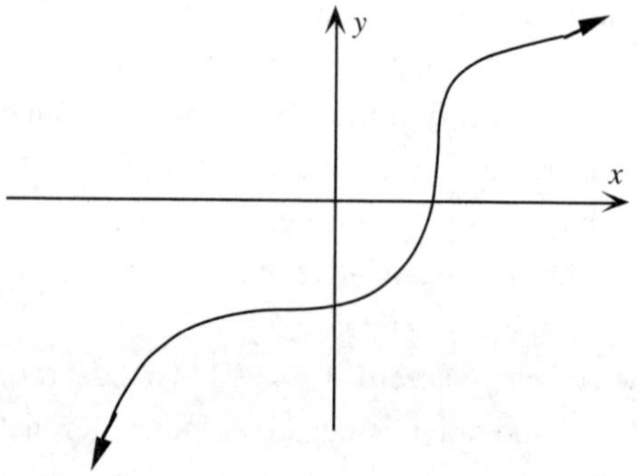

61. Not necessarily; it could be neither a relative maximum nor a relative minimum, as in the graph of $y = x^3$ at the origin.

63. Answers will vary.

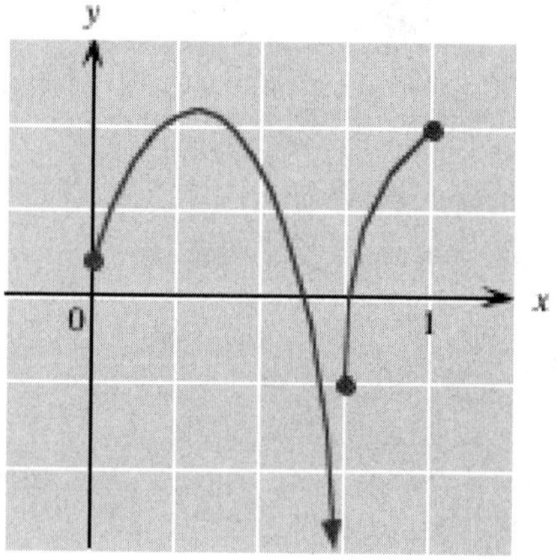

5.2

1. Solve for $y = 10 - x$ and substitute:

$P = x(10 - x) = 10x - x^2.$

Stationary points: Set $P'(x) = 0$ and solve for x:

$P'(x) = 10 - 2x$

$P'(x) = 0$ when $x = 5$. Since the graph of $P(x)$ is a parabola opening downward, $x = 5$ must be its vertex, hence gives the maximum. The corresponding value of y is $y = 10 - x = 5$. Hence, $x = y = 5$ and $P = 25$.

3. Solve for $y = 9/x$ and substitute:

$S = x + 9/x$

with domain $x > 0$.

Stationary points: Set $S'(x) = 0$ and solve for x:

$S'(x) = 1 - 9/x^2$

$S'(x) = 0$ when $x = 3$ ($x = -3$ is not in the domain). Testing points on either side of $x = 3$ we see that we have the minimum. Hence, $x = y = 3$ and $S = 6$.

5. Solve for $x = 10 - 2y$ and substitute:

$F(y) = (10 - 2y)^2 + y^2$

$= 100 - 40y + 5y^2.$

Stationary points: Set $F'(y) = 0$ and solve for y:

$F'(y) = -40 + 10y$

$F'(y) = 0$ when $y = 4$. Since the graph of F is a parabola opening upward, $y = 4$ must be its vertex, hence gives the minimum. Hence, $x = 2$, $y = 4$, and $F = 20$.

7. Since y appears in both constraints, we can solve for the other two variables in terms of y: $x = 30 - y$ and $z = 30 - y$. Substitute:

$P = (30 - y) \cdot y \cdot (30 - y)$

$= y(30 - y)^2.$

Since all the variables must be nonnegative, we have $0 \le y \le 30$ as the domain.

Stationary points: Set $P'(y) = 0$ and solve for y:

$P'(y) = (30 - y)^2 - 2y(30 - y)$

$= (30 - y)(30 - y - 2y)$

$= (30 - y)(30 - 3y)$

$P'(y) = 0$ when $y = 10$ or $y = 30$. Substituting these values and the other endpoint $y = 0$ into P, we find that the maximum occurs when $y = 10$. Thus, $x = 20$, $y = 10$, $z = 20$, and $P = 4000$.

9. Let x and y be the dimensions. Then we want to maximize $A = xy$ subject to $2x + 2y = 20$. Solve for $y = 10 - x$ and substitute:

$A = x(10 - x) = 10x - x^2.$

Stationary points: Set $A'(x) = 0$ and solve for x:

$A'(x) = 10 - 2x;$

$A'(x) = 0$ when $x = 5$. Since the graph of $A(x)$ is a parabola opening downward, $x = 5$ must be its vertex, hence gives the maximum. The corresponding value of y is $y = 10 - x = 5$. Hence, the rectangle should have dimensions 5×5.

11. Unknown: x, the number of MP3 players

Objective function: average cost

$$\bar{C}(x) = \frac{C(x)}{x} = \frac{25{,}000 + 20x + 0.001x^2}{x}$$

$$= \frac{25{,}000}{x} + 20 + 0.001x$$

with domain $x > 0$.

Stationary points: Set $\bar{C}'(x) = 0$ and solve for x:

$$\bar{C}'(x) = -\frac{25{,}000}{x^2} + 0.001 = 0$$

$x^2 = 25{,}000/0.001 = 25{,}000{,}000$

$x = 5000$

Testing points on either side shows that this gives a minimum and a consideration of the graph shows that this minimum is absolute. Thus, average cost is minimized when we manufacture 5000 MPs per day, giving an average cost of

$\bar{C}(5000) = \dfrac{25{,}000}{5000} + 20 + 0.001(5000)$

$= \$30$ per MP3 player.

13. Unknown: q, the number of pounds of pollutant removed per day

Objective function: average cost

$$\bar{C}(q) = \frac{C(q)}{q} = \frac{4000 + 100q^2}{q}$$

$$= \frac{4000}{q} + 100q$$

with domain $q > 0$.

Stationary points: Set $\bar{C}'(q) = 0$ and solve for q:

$$\bar{C}'(q) = -\frac{4000}{q^2} + 100 = 0$$

$$q^2 = 4000/100 = 40$$

$$q = \sqrt{40} \approx 6.32 \text{ pounds of pollutant per day.}$$

Testing points on either side shows that this gives a minimum and a consideration of the graph shows that this minimum is absolute. Thus, average cost is minimized when we remove about 6.32 pounds per day, giving an average cost of

$$\bar{C}(\sqrt{40}\,) = \frac{4000 + 100(40)}{\sqrt{40}}$$

$\approx \$1265$ per pound.

15. Net cost is

$$N = C(q) - 500q = 4000 + 100q^2 - 500q,$$

with $q \geq 0$.

Stationary points: Set $N'(q) = 0$ and solve for q:

$$N'(q) = 200q - 500$$

$N'(q) = 0$ when $q = 2.5$; $N'(q)$ is defined for all q. Testing the endpoint $q = 0$, the stationary point $q = 2.5$, and one more point to the right of 2.5, we see that the net cost is minimized when $q = 2.5$ pounds of pollutant per day.

17. Let x be the length of the east and west sides and let y be the length of the north and south sides. The area is $A = xy$ and the cost of the fence is $2 \cdot 4x + 2 \cdot 2y = 8x + 4y$. (We multiply by 2 because there are two sides of length x and two sides of length y.) So, our problem is to maximize $A = xy$ subject to $8x + 4y = 80$. Solve for y:

$$y = 20 - 2x$$

and substitute:

$$A(x) = x(20 - 2x) = 20x - 2x^2.$$

Since x and y must both be nonnegative, we have $0 \leq x \leq 10$.

Stationary points: Set $A'(x) = 0$ and solve for x:

$$A'(x) = 20 - 4x$$

$A'(x) = 0$ when $20 - 4x = 0$, $x = 5$; $A'(x)$ is always defined. Testing the endpoints 0 and 10 as well as the stationary point 5, we see that the maximum area occurs when $x = 5$. The corresponding value of y is $y = 10$, so the largest area possible is $5 \times 10 = 50$ square feet.

19. $R = pq = p(200{,}000 - 10{,}000p)$

$= 200{,}000p - 10{,}000p^2$.

For p and q to both be nonnegative, we must have $0 \leq p \leq 20$.

Stationary points: Set $R'(p) = 0$ and solve for p:

$$R'(p) = 200{,}000 - 20{,}000p$$

$R'(p) = 0$ when $p = 10$; $R'(p)$ is always defined. Testing the endpoints 0 and 20 as well as the stationary point 10, we see that the maximum revenue occurs when the price is $p = \$10$.

21. $R = pq = p\left(-\frac{4}{3}p + 80\right) = -\frac{4}{3}p^2 + 80p.$

For p and q to both be nonnegative, we must have $0 \leq p \leq 60$.

Stationary points: Set $R'(p) = 0$ and solve for p:

$$R'(p) = -\frac{8}{3}p + 80$$

$R'(p) = 0$ when $p = 30$; $R'(p)$ is always defined. Testing the endpoints 0 and 60 as well as the stationary point 30, we see that the maximum revenue occurs when the price is $p = \$30$.

23. (a) $R = pq = \dfrac{500{,}000}{q^{1.5}} \cdot q$

$= \dfrac{500{,}000}{q^{0.5}} = 500{,}000q^{-0.5}$.

We are given that $q \geq 5000$.

Stationary points: Set $R'(q) = 0$ and solve for q:

$R'(q) = -250{,}000q^{-1.5}$

$R'(q)$ is never 0; $R'(q)$ is defined for all $q > 0$. We test the endpoint $q = 5000$ and a point to the right: $R(5000) = 7071.07$ and $R(10{,}000) = 5000$. Thus, the revenue is decreasing and its maximum value occurs at the endpoint $q = 5000$. The corresponding price is $p = \$1.41$ per pound.

(b) As found in part (a), the maximum occurs at $q = 5000$ pounds.

(c) The maximum revenue is $R(5000) = \$7071.07$ per month.

25. We are given two points: $(p, q) = (25, 22)$ and $(14, 27.5)$. The equation of the line through these two points is

$q = -p/2 + 34.5$.

The revenue is

$R = pq = p(-p/2 + 34.5)$

$= -p^2/2 + 34.5p$.

For p and q to both be nonnegative we must have $0 \leq p \leq 69$.

Stationary points: Set $R'(p) = 0$ and solve for p:

$R'(p) = -p + 34.5$

$R'(p) = 0$ when $p = 34.5$; $R'(p)$ is defined for all p. Testing the endpoints 0 and 69 as well as the stationary point 34.5, we find that the revenue is maximized when the price is $p = 34.5¢$ per pound, for an annual (per capita) revenue of \$5.95.

27. $R = pq$ and $C = 25q$, so

$P = R - C$

$= pq - 25q$

$= p\left(-\frac{4}{3}p + 80\right) - 25\left(-\frac{4}{3}p + 80\right)$

$= -\frac{4}{3}p^2 + \frac{340}{3}p - 2000$.

For p and q to both be nonnegative, we must have $0 \leq p \leq 60$.

Stationary points: Set $P'(p) = 0$ and solve for p:

$P'(p) = -\frac{8}{3}p + \frac{340}{3}$

$P'(p) = 0$ when $p = 42.5$; $P'(p)$ is defined for all p. Testing the endpoints 0 and 60 as well as the stationary point 42.5, we find that the profit is maximized when the price is $p = \$42.50$ per ruby, for a weekly profit of $P(42.5) = \$408.33$.

29. (a) $R = pq$ and $C = 100q$, so

$P = R - C$

$= pq - 100q$

$= \dfrac{1000}{q^{0.3}} \cdot q - 100q$

$= 1000q^{0.7} - 100q$.

For p and q to be defined and nonnegative we need $q > 0$.

Stationary points: Set $P'(q) = 0$ and solve for q:

$P'(q) = 700q^{-0.3} - 100$

$P'(q) = 0$ when $q = 7^{1/0.3} \approx 656$; $P'(q)$ is defined for all $q > 0$. Testing the stationary point at approximately 656 and points on either side, we see that the profit is maximized when you sell 656 headsets, for a profit of $P(656) \approx \$28{,}120$.

(b) The corresponding price is $p \approx \$143$ per headset.

31. Let x be the length of one side of the square cut out of each corner, as in the figure:

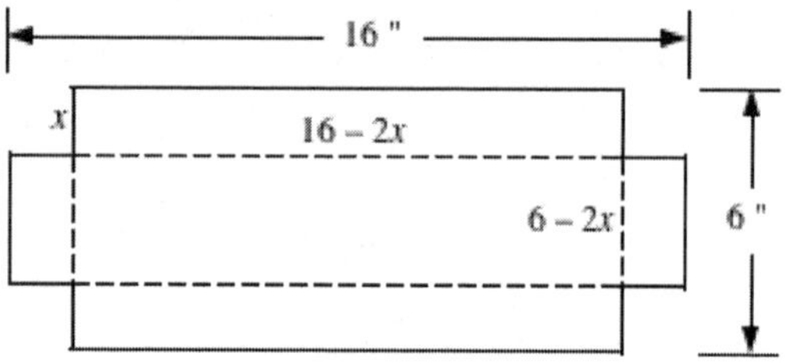

When the sides are folded up, the resulting box will have volume

$V = x(16 - 2x)(6 - 2x)$

$= 4x^3 - 44x^2 + 96x.$

For the sides to have nonnegative lengths, we must have $0 \le x \le 3$.

Stationary points: Set $V'(x) = 0$ and solve for x:

$V'(x) = 12x^2 - 88x + 96$

$= 2(x - 6)(6x - 8)$

$V'(x) = 0$ when $x = 4/3$ or $x = 6$, but $x = 6$ is outside of the domain. Testing the endpoints 0 and 3 and the stationary point 4/3, we find that the largest volume occurs when $x = 4/3"$. Thus, thee box with the largest volume has dimensions 13 1/3" × 3 1/3" × 1 1/3" and it has volume

$V(4/3) = 1600/27 \approx 59$ cubic inches.

33. Let x be the width and depth of the box and let y be the height. The amount of material used will be

$S = 2x^2 + 4xy,$

counting the top, bottom (each of which has area x^2) and four sides (each of which has area xy). We are told that the volume is 125 cm^3, so we must have

$x^2y = 125.$

So, our problem is to maximize

$S = 2x^2 + 4xy$ subject to

$x^2y = 125$

Also, $x > 0$ for x and y to be nonnegative. Solve for $y = 125/x^2$ and substitute:

$S = 2x^2 + 500/x.$

Stationary points: Set $S'(x) = 0$ and solve for x:

$S'(x) = 4x - 500/x^2$

$S'(x) = 0$ when $x^3 = 500/4 = 125$, so $x = 5$. (The corresponding value of y is $y = 125/x^2 = 5$ also.) Testing this stationary point and a point on either side, we see that the least material is used building a box of dimensions 5 × 5 × 5 cm.

35. Let l = length, w = width, and h = height. We want to maximize the volume

$V = lwh$

but we are restricted by

$l + w + h \le 62.$

Since we're looking for the largest volume, we can assume that

$l + w + h = 62.$

We are also told that $h = w$. Thus, our problem is to maximize

$V = lwh$ subject to

$l + w + h = 62$ and

$h = w.$

Substitute $h = w$ in the other constraint to get

$l + 2w = 62,$

then solve for l:

$l = 62 - 2w.$

Substitute to get

$V = (62 - 2w)w^2 = 62w^2 - 2w^3.$

For all dimensions to be nonnegative we need $0 \le w \le 31$.

Stationary points:

$V'(w) = 124w - 6w^2$

$V'(w) = 0$ when

$124w - 6w^2 = 0$

$2w(62 - 3w) = 0,$

so $w = 0$ or $w = 62/3$.

Singular points: None; $V'(w)$ is defined for all w.

Testing the endpoints 0 and 31 as well as the (other) stationary point 62/3, we see that the volume is maximized when $w = 62/3$. The corresponding values of the other dimensions are $l = h = 62/3$. Thus, the largest volume bag has dimensions $l = w = h \approx 20.67$ in, and volume $V \approx 8827$ in^3.

37. Let l = length, w = width, and h = height. We want to maximize the volume $V = lwh$ but we have the constraints $l + w = 45$ and $w + h = 45$. Solve for $l = 45 - w$ and $h = 45 - w$. Substitute to get $V = w(45 - w)^2$. For all dimensions to be nonnegative we need $0 \le w \le 45$. Stationary points: $V'(w) = (45 - w)^2 - 2w(45 - w) = (45 - w)(45 - w - 2w) = (45 - w)(45 - 3w)$; $V'(w) = 0$ when $w = 15$ or $w = 45$; $V'(w)$ is defined for all w. Testing the endpoints 0 and 45 as well as the stationary point 15, we see that the volume is maximized when $w = 15$. The corresponding values of the other dimensions are $l = h = 30$. Thus, the largest volume bag has dimensions $l = 30$ in, $w = 15$ in, and $h = 30$ in.

39. Let l = length, w = width, and h = height. We want to maximize the volume $V = lwh$ but we are restricted by $l + 2(w + h) \le 108$. Since we're looking for the largest volume, we can assume that $l + 2(w + h) = 108$. We are also told that $w = h$. Substitute to get $l + 2(2w) = 108$, or $l + 4w = 108$. Solve for $l = 108 - 4w$ and substitute to get $V = w^2(108 - 4w) = 108w^2 - 4w^3$. For all the dimensions to be nonnegative we need $0 \le w \le 27$.

Stationary points: $V'(w) = 216w - 12w^2 = 12w(18 - w)$; $V'(w) = 0$ when $w = 0$ or $w = 18$; $V'(w)$ is defined for all w. Testing the endpoints 0 and 27 as well as the stationary point 18, we see that the volume is maximized when $w = 18$. The corresponding values of the other dimensions are $h = 18$ and $l = 36$. Thus, the largest volume package has dimensions $l = 36$ in and $w = h = 18$ in, and has volume $V = 11{,}664$ in^3.

41. (a) Stationary points:

$R'(t) = 350(39)e^{-0.3t} - 0.3(350)(39t + 68)e^{-0.3t}$
$= 350(39 - 11.7t - 20.4)e^{-0.3t}$
$= 350(18.6 - 11.7t)e^{-0.3t}$.

$R'(t) = 0$ when $t \approx 1.6$; $R'(t)$ is defined for all t. Testing points on either side of the stationary point, we see that the maximum revenue occurs at $t \approx 1.6$ years, or year 2001.6.

(b) The maximum revenue is $R(1.6) \approx \$28{,}241$ million.

43. Let $C(t) = \dfrac{D(t)}{S(t)} = \dfrac{10 + t}{2.5e^{0.08t}} = 0.4(10 + t)e^{-0.08t}$.

Stationary points:

$C'(t) = 0.4e^{-0.08t} - 0.032(10 + t)e^{-0.08t}$
$= (0.4 - 0.32 - 0.032t)e^{-0.08t}$
$= (0.08 - 0.032t)e^{-0.08t}$

$C'(t) = 0$ when $t = 2.5$; $C'(t)$ is defined for all t. Testing points on either side of 2.5 we see that the maximum revenue occurs at $t = 2.5$ or midway through 1972, when $D(2.5)/S(2.5) \approx 4.09$. Midway through 1972 the number of new (approved) drugs per \$1 billion dollars of spending on research and development reached a high of around 4 approved drugs per \$1 billion.

45. $p = ve^{-0.05t} = (300{,}000 + 1000t^2)e^{-0.05t}$. $(t \ge 5)$

Endpoint: $t = 5$

Stationary points:

$p' = 2000te^{-0.05t} - 0.05(300{,}000 + 1000t^2)e^{-0.05t}$

$= (-15{,}000 + 2000t - 50t^2)e^{-0.05t}$

$p' = 0$ when

$-15{,}000 + 2000t - 50t^2 = 0$

$-50(t - 30)(t - 10) = 0$

so $t = 10$ or $t = 30$; p' is always defined. Testing $t = 5$, 10, 30, and 40, we see that the maximum discounted value occurs $t = 30$ years from now.

47. $R = (100 + 2t)(400{,}000 - 2500t)$
$= 40{,}000{,}000 + 550{,}000t - 5000t^2$
Stationary points:
$R' = 550{,}000 - 10{,}000t$
$R' = 0$ when $t = 55$; R' is defined for all t. Testing $t = 55$ and points on either side of it (or recognizing that the graph of R is a parabola), we see that the release should be delayed for 55 days.

49. $R = 500\sqrt{x}$, $C = 10{,}000 + 2x$, so the total profit is
$P = R - C = 500\sqrt{x} - (10{,}000 + 2x)$
and the average profit per copy is
$\bar{P} = 500x^{-1/2} - 10{,}000x^{-1} + 2.$
Stationary points:
$\bar{P}' = -250x^{-3/2} + 10{,}000x^{-2}$
$\bar{P}' = 0$ when
$x = (10{,}000/250)^2 = 1600$
$\bar{P}'$ is defined for all $x > 0$. Testing the stationary point 1600 and a point on either side, we see that the average profit is maximized at $x = 1600$ copies. For this many copies, the average profit is
$\bar{P}(1600) = \$8.25$/copy.
Since $P' = 250x^{-1/2} + 2$, the marginal profit is
$P'(1600) = \$8.25$/copy also.
At this value of x, average profit equals marginal profit; beyond this the marginal profit is smaller than the average and so the average declines.

51. We are being asked to find the extreme values of the derivative, $N'(t)$. Call this function
$M(t) = N'(t) = -435t^2 + 10{,}600t + 1300.$
Then, for stationary points,
$M'(t) = -870t + 10{,}600$
$M'(t) = 0$ when $t \approx 12.2$. Testing the endpoints 0 and 23 as well as the stationary point 12.2, we see that M has an absolute maximum of
$M(12.2) \approx 66{,}000$
and an absolute minimum of
$M(0) = 1300.$
Hence, N was increasing most rapidly in 1992 and increasing least rapidly in 1980.

53. Let $r(t) = c'(t) = -0.195t^3 + 6.8t - 22$. $r'(t) = -0.39t + 6.8$; $r'(t) = 0$ when $t \approx 17$. Testing the endpoints 8 and 30 as well as the stationary point 17, we see that $c'(t)$ has its maximum when $t = 17$ days. This means that the embryo's oxygen consumption is increasing most rapidly 17 days after the egg is laid.

55. We want to minimize $S = \pi r^2 + 2\pi rh$ subject to $\pi r^2 h = 5000$. Solve for $h = 5000/(\pi r^2)$ and substitute to get $S = \pi r^2 + 2\pi r \dfrac{5000}{\pi r^2} = \pi r^2 + \dfrac{10{,}000}{r}$. For the dimensions to be defined and nonnegative we need $r > 0$. $S'(r) = 2\pi r - \dfrac{10{,}000}{r^2}$; $S'(r) = 0$ when $r = \sqrt[3]{\dfrac{5000}{\pi}} = 10\sqrt[3]{\dfrac{5}{\pi}} \approx 11.7$. Testing the stationary point and a point on either side, we see that $S(r)$ has an absolute minimum at $r \approx 11.7$. The corresponding value of h is $h = 10\sqrt[3]{\dfrac{5}{\pi}}$ also. Hence, the bucket using the least plastic has dimensions $h = r \approx 11.7$ cm.

57. Let $y(x)$ be the annual yield per tree when there are x trees; $y(x) = 100 - (x - 50) = 150 - x$. If $Y(x)$ is the total annual yield from x trees, then $Y(x) = xy(x) = x(150 - x) = 150x - x^2$. $Y'(x) = 150 - 2x$; $Y'(x) = 0$ when $x = 75$. Since the graph of Y is a parabola opening downward, we know that this gives the maximum value of Y. Hence, the total annual yield is largest when there are 75 trees, or 25 additional trees beyond the 50 we already have.

59. We want to minimize $C = 20{,}000x + 365y$ subject to $x^{0.4}y^{0.6} = 1000$. Solve for y:

$$y = (1000x^{-0.4})^{1/0.6} = 1000^{5/3}x^{-2/3}$$

and substitute:

$$C = 20{,}000x + 365(1000^{5/3})x^{-2/3}.$$

The domain is $x > 0$.

Stationary points:

$$C'(x) = 20{,}000 - \frac{730}{3}1000^{5/3}x^{-5/3}$$

$C'(x) = 0$ when

$$x = \left(\frac{60{,}000}{730 \cdot 1000^{5/3}}\right)^{-3/5} \approx 71.$$

Testing the stationary point at 71 and a point on either side, we see that C has its minimum at $x \approx 71$. So, you should hire 71 employees.

61. We want to find the maximum of $N'(t)$. Here are the graphs of $N(t)$, $N'(t)$ and $N''(t)$:

Graph of $N(t)$:

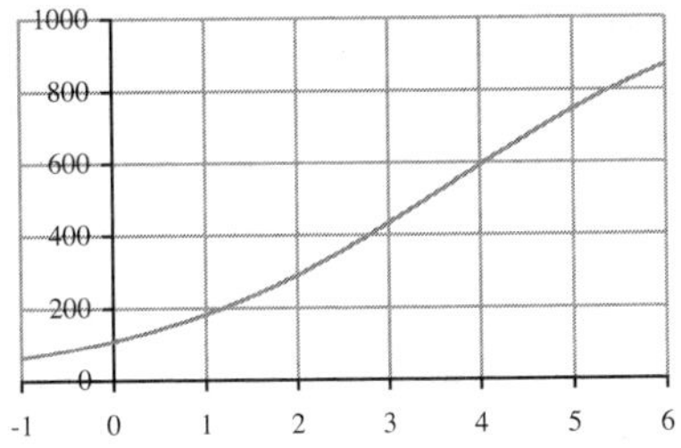

Graph of $N'(t)$:

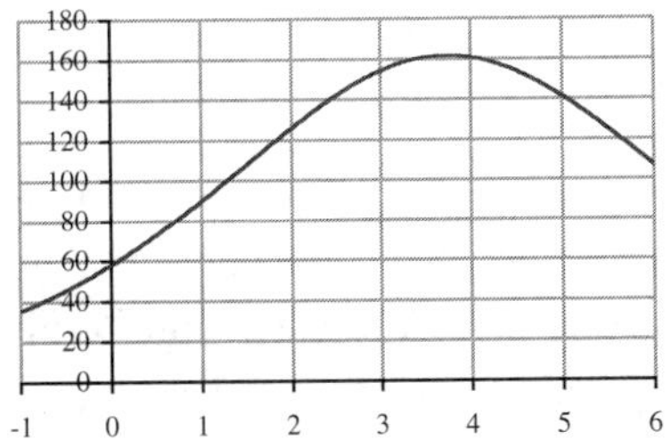

Graph of $N''(t)$:

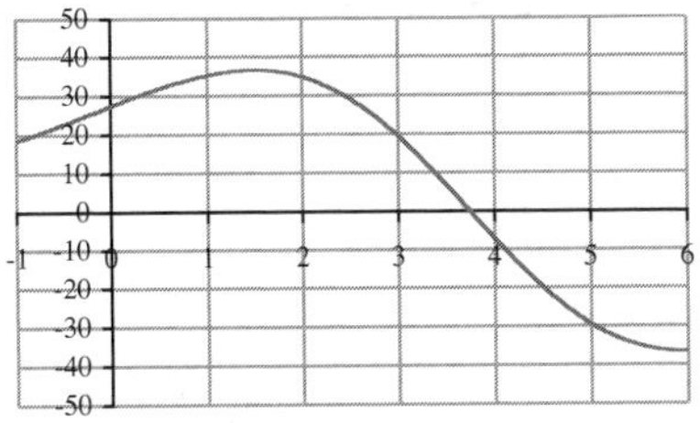

$N'(t)$ appears to have a maximum around $t = 3.5$. To find this value more accurately, trace to the point where the graph of $N''(t)$ crosses the t-axis: at approximately $t = 3.7$. The associated rate of sales is given by the y-coordinate:

$$N'(3.7) \approx 160$$

Thus, iPod sales were increasing most rapidly in the fourth quarter of 2003 ($t \approx 3.7$). The rate of increase was $N'(3.7) \approx 160$ thousand iPods per quarter.

Solution without technology:

Let $R(t) = N'(t) = \dfrac{9900(\ln 1.8)(1.8)^{-t}}{[1 + 9(1.8)^{-t}]^2} \approx \dfrac{5800(1.8)^{-t}}{[1 + 9(1.8)^{-t}]^2}$.

$$R'(t) = \frac{-3400(1.8)^{-t}[1 + 9(1.8)^{-t}]^2 + 61{,}000(1.8)^{-t}[1 + 9(1.8)^{t}](1.8)^{-t}}{[1 + 9(1.8)^{-t}]^4}$$

$$= \frac{(1.8)^{-t}[-3400 - 30{,}600(1.8)^{-t} + 61{,}000(1.8)^{-t}]}{[1 + 9(1.8)^{-t}]^3}$$

$$= \frac{(1.8)^{-t}[-3400 + 30{,}400(1.8)^{-t}]}{[1 + 9(1.8)^{-t}]^3};$$

$R'(t) = 0$ when $t = -\ln(3400/30{,}400)/\ln(1.8) \approx 3.7$. Testing the endpoints -1 and 6 and the stationary point at 3.7, we see that $R(t)$ has a maximum when $t \approx 3.7$. Hence, iPod sales were increasing most rapidly in the fourth quarter of 2003 ($t \approx 3.7$). The rate of increase was $N'(3.7) \approx 160$ thousand iPods per quarter.

63. Graph of derivative of $S(t)/R(t)$:

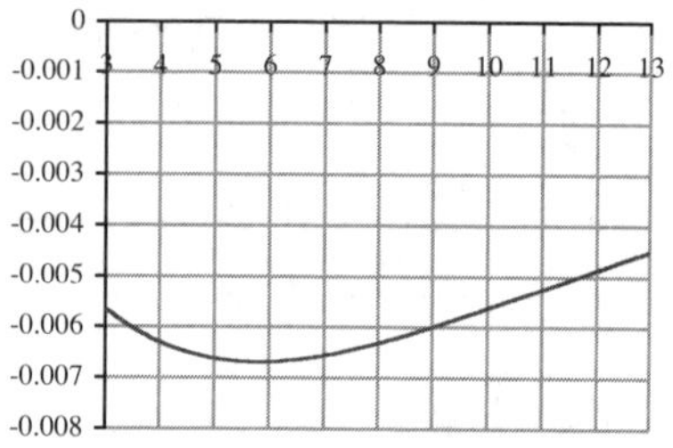

On zooming in we can determine that the minimum occurs at approximately $t = 6$, with value approximately -0.0067. The fraction of bottled water sales due to sparkling water was decreasing most rapidly in 1996. At that time it was decreasing at a rate of 0.67 percentage points per year.

65. $p = v(1.05)^{-t} = \dfrac{10{,}000(1.05)^{-t}}{1 + 500e^{-0.5t}}$.

Technology formula:

Excel:

```
10000*(1.05)^(-x)/(1+500*exp(-0.5*x))
```

TI-83/84:

```
10000*(1.05)^(-x)/(1+500*e^(-0.5*x))
```

Here are the graphs of $p(t)$ and $p'(t)$:

Graph of $p(t)$;

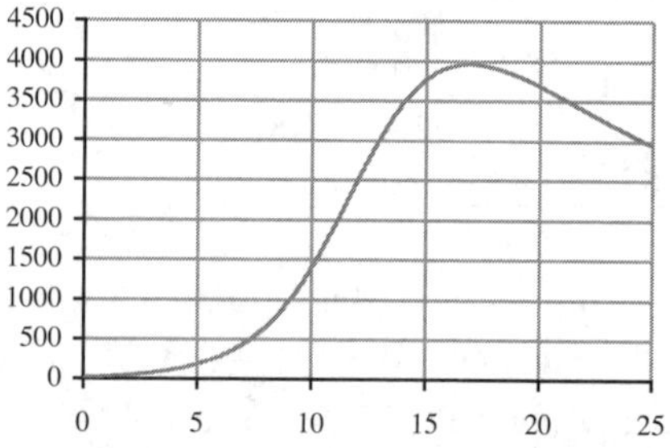

Graph of $p'(t)$:

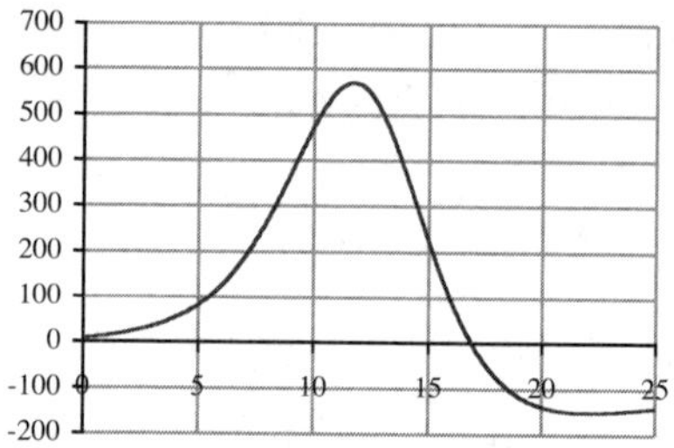

From the graph, we see that the maximum occurs betrween $t = 15$ and $t = 20$. The maximum is more accurately seem in the graph of $p'(t)$ where it crosses the t-axis. Zooming in on the graph of $p'(t)$, we see the following:

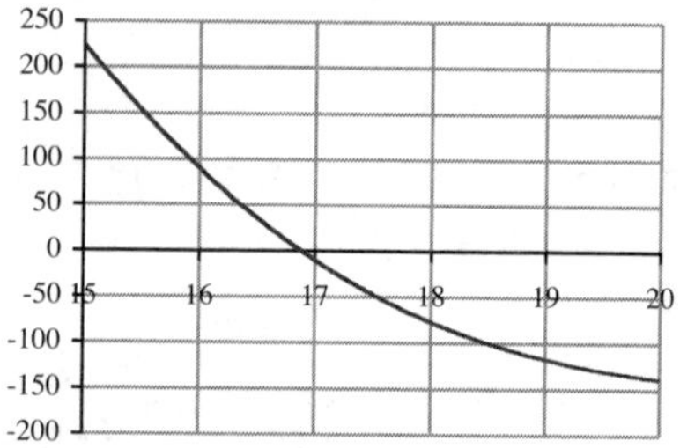

and so the maximum is very close to $t = 17$ years. To obtain the value, substitute $t = 17$ in the formula for $p(t)$ or else use the trace feature to see the y-coordinate of the highest point in the graph of $p(t)$: approximately \$3960.

67. $\dfrac{d(TR)}{dQ} = b - 2cQ$; $\dfrac{d(TR)}{dQ} = 0$ when $Q = \dfrac{b}{2c}$, which is (D).

69. The problem is uninteresting because the company can accomplish the objective by cutting away the entire sheet of cardboard, resulting in a box with surface area zero. Put another way, if it doesn't cut away everything, it can make the surface area be as close to zero as you like.

71. Not all absolute extrema occur at stationary points; some may occur at an end-point or singular point of the domain, as in Exercises 17, 18, 47 and 48.

73. The minimum of *dq*/*dp* is the fastest that the demand is dropping in response to increasing price.

5.3

1. $\frac{dy}{dx} = 6x$; $\frac{d^2y}{dx^2} = 6$

3. $\frac{dy}{dx} = -\frac{2}{x^2}$; $\frac{d^2y}{dx^2} = \frac{4}{x^3}$

5. $\frac{dy}{dx} = 1.6x^{-0.6} - 1$; $\frac{d^2y}{dx^2} = -0.96x^{-1.6}$

7. $\frac{dy}{dx} = -e^{-(x-1)} - 1$; $\frac{d^2y}{dx^2} = e^{-(x-1)}$

9. $\frac{dy}{dx} = -\frac{1}{x^2} - \frac{1}{x}$; $\frac{d^2y}{dx^2} = \frac{2}{x^3} + \frac{1}{x^2}$

11. (a) $s = 12 + 3t - 16t^2$

$v = \frac{ds}{dt} = 3 - 32t$

$a = \frac{dv}{dt} = -32$ ft/sec^2

(b) $a(2) = -32$ ft/sec^2

13. (a) $s = \frac{1}{t} + \frac{1}{t^2}$

$v = \frac{ds}{dt} = -\frac{1}{t^2} - \frac{2}{t^3}$

$a = \frac{dv}{dt} = \frac{2}{t^3} + \frac{6}{t^4}$ ft/sec^2

(b) $a(1) = 8$ ft/sec^2

15. (a) $s = \sqrt{t} + t^2$

$v = \frac{ds}{dt} = \frac{1}{2}t^{-1/2} + 2t$

$a = \frac{dv}{dt} = -\frac{1}{4}t^{-3/2} + 2$ ft/sec^2

(b) $a(4) = \frac{63}{32}$ ft/sec^2

17. (1, 0): changes from concave down to concave up

19. (1, 0): changes from concave down to concave up

21. None: always concave down

23. (−1, 0): changes from concave up to concave down; (1, 1): changes from concave down to concave up

For exercises 25–28, remember that a point of inflection of f corresponds to a relative extreme point of f' that is internal, not an endpoint.

25. Points of inflection at $x = -1$ (relative max of f') and $x = 1$ (relative min of f')

27. One point of inflection, at $x = -2$ (relative min of f'). Note that f' has a stationary point at $x = 1$ but not a relative extreme point there.

For exercises 29–34, remember that a point of inflection of f corresponds to a point at which f'' changes sign, from positive to negative or vice versa. This could be a point where its graph crosses the x-axis, or a point where its graph is broken: positive on one side of the break and negative on the other.

29. Points of inflection where the graph of f'' crosses the x-axis: at $x = -2$, $x = 0$, and $x = 2$.

31. Points of inflection where the graph of f'' crosses the x-axis at $x = -2$ and $x = 2$.

33. $f(x) = x^2 + 2x + 1$

$f'(x) = 2x + 2$; $f'(x) = 0$ when $x = -1$; $f'(x)$ is always defined. $f''(x) = 2$; $f''(x)$ is never 0 and is always defined.

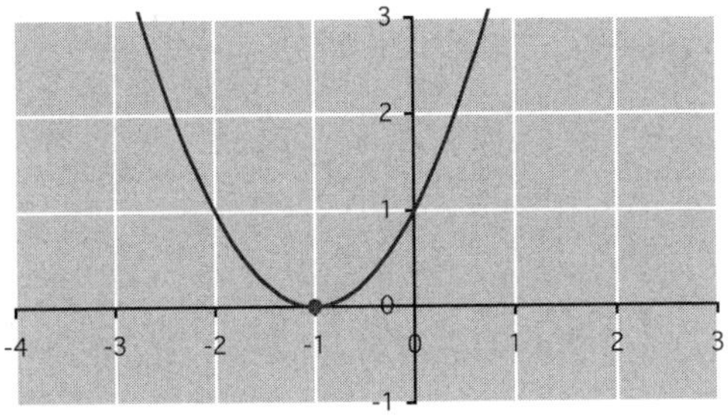

f has an absolute min at $(-1, 0)$ and no points of inflection.

35. $f(x) = 2x^3 + 3x^2 - 12x + 1$

$f'(x) = 6x^2 + 6x - 12 = 6(x^2 + x - 2)$;
$f'(x) = 0$ when $x^2 + x - 2 = 0$, $(x + 2)(x - 1) = 0$, $x = -2$ or $x = 1$; $f'(x)$ is always defined.
$f''(x) = 12x + 6$; $f''(x) = 0$ when $x = -1/2$; $f''(x)$ is always defined.

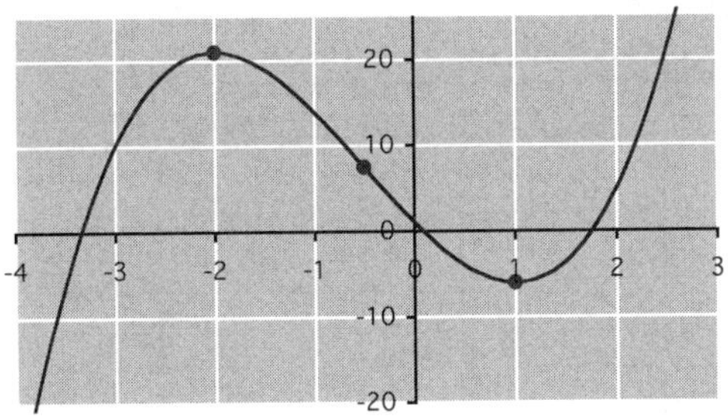

f has a relative max at $(-2, 21)$, a relative min at $(1,-6)$, and a point of inflection at $(-1/2, 15/2)$.

37. $g(x) = x^3 - 12x$, domain $[-4, 4]$

$g'(x) = 3x^2 - 12$; $g'(x) = 0$ when $x = \pm 2$; $g'(x)$ is always defined. $g''(x) = 6x$; $g''(x) = 0$ when $x = 0$; $g''(x)$ is always defined.

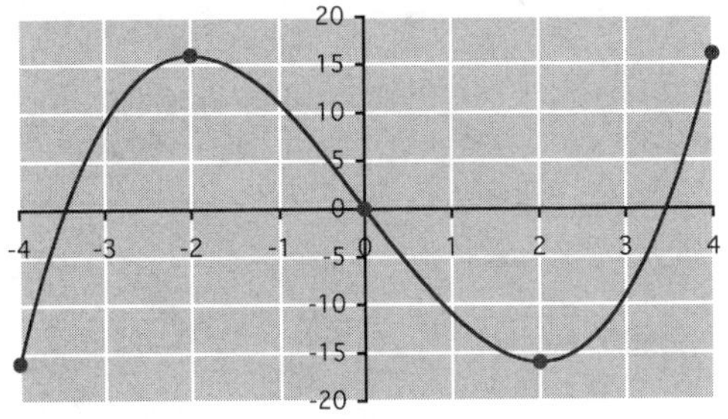

g has absolute mins at $(-4, -16)$ and $(2, -16)$, absolute maxes at $(-2, 16)$ and $(4, 16)$, and a point of inflection at $(0, 0)$.

39. $g(t) = \frac{1}{4}t^4 - \frac{2}{3}t^3 + \frac{1}{2}t^2$

$g'(t) = t^3 - 2t^2 + t = t(t^2 - 2t + 1)$; $g'(t) = 0$ when $t = 0$ or $t = 1$; $g'(t)$ is always defined.
$g''(t) = 3t^2 - 4t + 1$; $g''(t) = 0$ when $3t^2 - 4t + 1 = 0$, $(3t - 1)(t - 1) = 0$, $t = 1/3$ or $t = 1$.

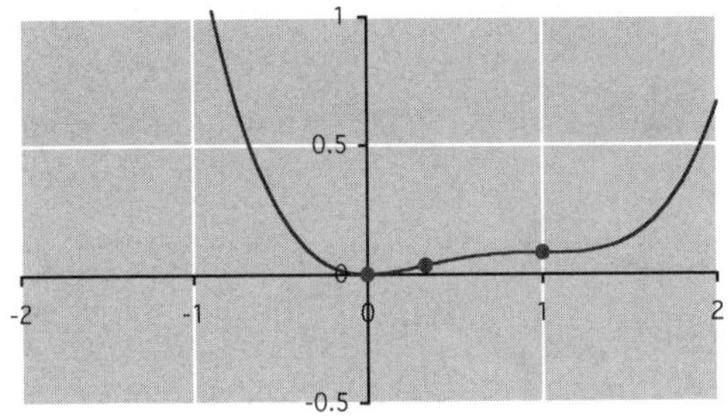

Note that g increases from the stationary point at $t = 0$ to the one at $t = 1$, then continues to increase to the right (as is clear from the graph or from a test point to the right of $t = 1$). So, g has an absolute min at $(0, 0)$ and points of inflection at $(1/3, 11/324)$ and $(1, 1/12)$.

41. $f(t) = \dfrac{t^2+1}{t^2-1}$, domain $[-2, 2]$, $t \neq \pm 1$

(Notice that $t = \pm 1$ are not in the domain of f.)

$$f'(t) = \frac{2t(t^2 - 1) - (t^2 + 1)(2t)}{(t^2 - 1)^2} = \frac{-4t}{(t^2 - 1)^2};$$

$f'(t) = 0$ when $t = 0$; $f'(t)$ is defined for all t in the domain of f.

$f''(t) = \dfrac{-4(t^2-1)^2 + 4t(2)(t^2-1)(2t)}{(t^2-1)^4} = \dfrac{(t^2-1)[-4(t^2-1)+16t^2]}{(t^2-1)^4} = \dfrac{12t^2+4}{(t^2-1)^3}$; $f''(t)$ is never 0; $f''(t)$ is defined for all t in the domain of f.

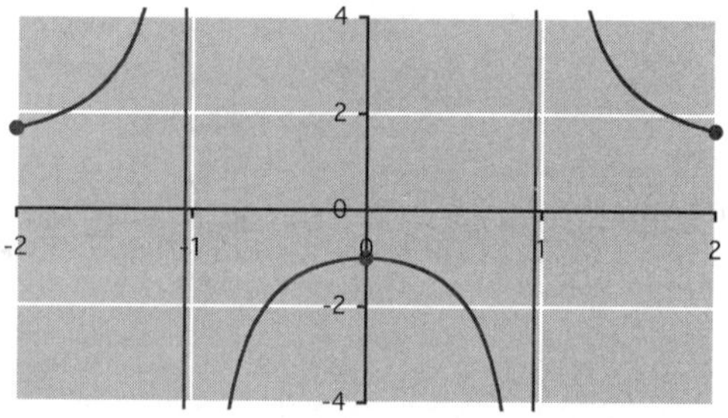

f has relative mins at $(-2, 5/3)$ and $(2, 5/3)$, a relative max at $(0, -1)$, and vertical asymptotes at $x = \pm 1$.

43. $f(x) = x + \dfrac{1}{x}$

Notice that the domain of f includes all numbers except 0.

$f'(x) = 1 - \dfrac{1}{x^2}$; $f'(x) = 0$ when $x^2 = 1$, $x = \pm 1$; $f'(x)$ is defined for all x in the domain of f. $f''(x) = \dfrac{2}{x^3}$; $f''(x)$ is never 0; $f''(x)$ is defined for all x in the domain of f.

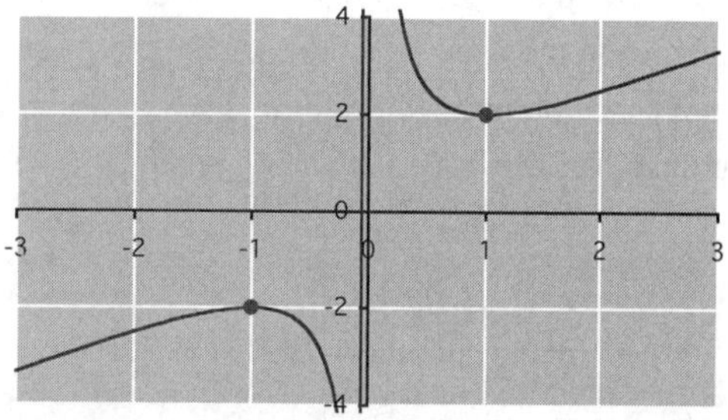

f has a relative min at $(1, 2)$, a relative max at $(-1, -2)$, no points of inflection, and a vertical asymptote at $y = 0$.

45. $k(x) = \dfrac{2x}{3} + (x+1)^{2/3}$

Technology format:

```
2*x/3+((x+1)^2)^(1/3)
```

Relative Extrema:

$k'(x) = \dfrac{2}{3} + \dfrac{2}{3}(x+1)^{-1/3} = \dfrac{2}{3} + \dfrac{2}{3(x+1)^{1/3}}$

Stationary points: $k'(x) = 0$ when

$\dfrac{2}{3} + \dfrac{2}{3(x+1)^{1/3}} = 0$

$\dfrac{2}{3} = -\dfrac{2}{3(x+1)^{1/3}}$

cross-multiply:

$6(x+1)^{1/3} = -6$

$(x+1)^{1/3} = -1$

$(x+1) = (-1)^3 = -1$

$x = -2$,

so we have a stationary point at $x = -2$.

Singular points: $k'(x)$ is undefined when $x = -1$, so we have a singular point at $x = -1$.

In addition we use a test point to the left of $x = -2$ and to the right of $x = -1$.

x	−3	−2	−1	0
$f(x)$	−0.41	−1/3	−2/3	1

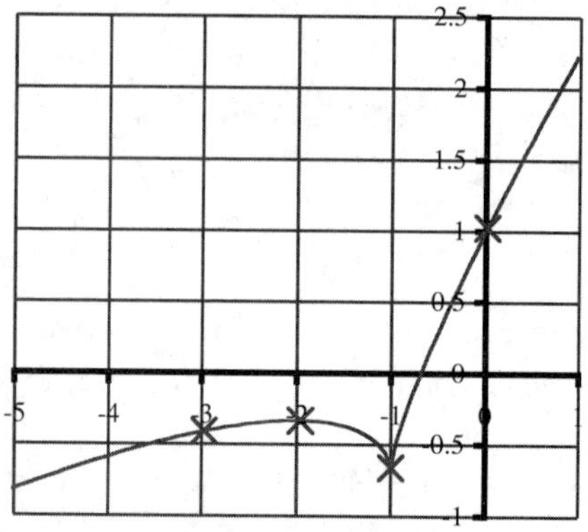

From the graph we see that we have a relative maximum at $(-2, -1/3)$ and a relative minimum at $(-1, -2/3)$.

Points of Inflection:

$$k'(x) = \frac{2}{3} + \frac{2}{3(x+1)^{1/3}}$$

$$k''(x) = -\frac{2}{9(x+1)^{4/3}}$$

$k''(x)$ is never zero and not defined when $x = -1$. So, the only candidate for a point of inflection is $x = -1$ which we see from the graph is not one. No points of inflection.

Behavior near points where the function is not defined:

The domain of this function consists of all real numbers, so there are no such points.

Behavior at infinity:

$$\lim_{x\to-\infty} k(x) = \lim_{x\to-\infty} [2x/3 + (x+1)^{2/3}] = -\infty$$

(computing $k(x)$ for large negative values of x gives large negative numbers).

$$\lim_{x\to+\infty} k(x) = \lim_{x\to+\infty} [2x/3 + (x+1)^{2/3}] = +\infty$$

Final sketch:

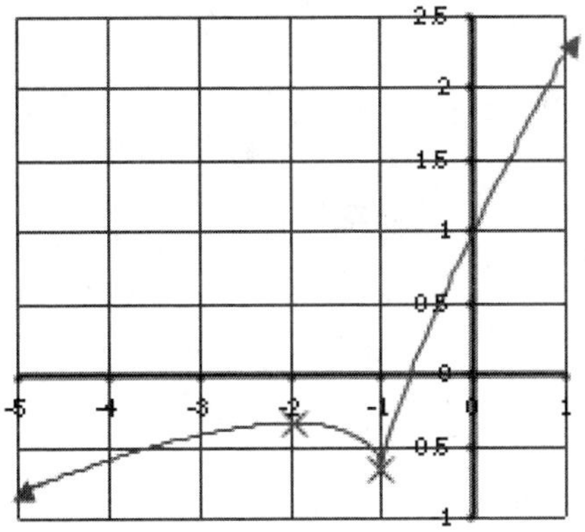

47. $g(x) = x^3/(x^2+3)$

$g'(x) = \dfrac{3x^2(x^2+3) - x^3(2x)}{(x^2+3)^2} = \dfrac{x^4 + 9x^2}{(x^2+3)^2}$; $g'(x) = 0$ when $x^4 + 9x^2 = 0$, $x^2(x^2+9) = 0$, $x = 0$; $g'(x)$ is defined for all x. $g''(x) =$

$$\frac{(4x^3 + 18x)(x^2+3)^2 - (x^4 + 9x^2)(2)(x^2+3)(2x)}{(x^2+3)^4}$$

$$= \frac{(x^2+3)(-6x^3 + 54x)}{(x^2+3)^4} = \frac{-6x(x^2-9)}{(x^2+3)^3};$$

$g''(x) = 0$ when $x = 0$ or $x = \pm 3$; $g''(x)$ is always defined.

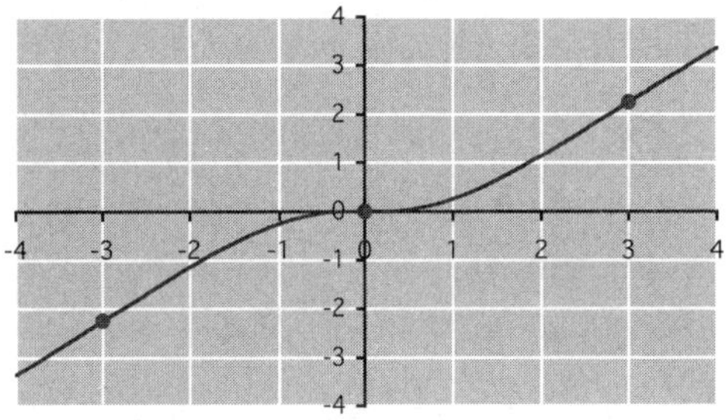

It is difficult to tell from the graph, but the points at $x = \pm 3$ are points of inflection. We can tell by computing, for example, $g''(2) = 60/7^3 > 0$ and $g''(4) = -168/19^3 < 0$, which shows that the concavity changes at $x = 3$. So, g has no extrema and points of inflection at (0, 0), (−3, −9/4), and (3, 9/4).

49. $f(x) = x - \ln x$, domain $(0, +\infty)$

$f'(x) = 1 - \dfrac{1}{x}$; $f'(x) = 0$ when $x = 1$; $f'(x)$ is defined for all x in the domain of f. $f''(x) = \dfrac{1}{x^2}$;

$f''(x)$ is never 0; $f''(x)$ is defined for all x in the domain of f.

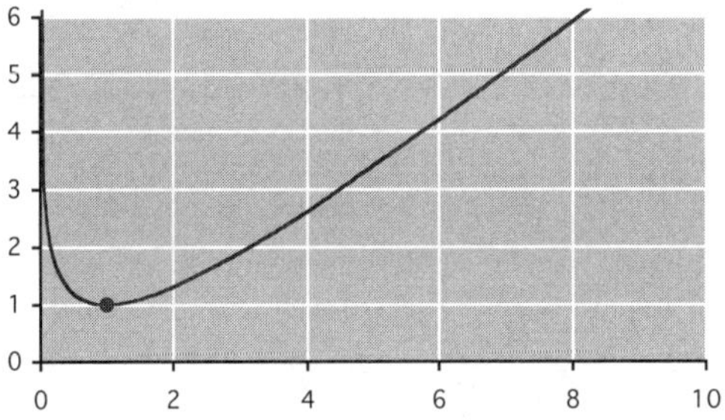

f has an absolute min at (1, 1) and a vertical asymptote at $x = 0$.

51. $f(x) = x^2 + \ln x^2$

The domain of f is all numbers except 0.

$f'(x) = 2x + \dfrac{2x}{x^2} = 2x + \dfrac{2}{x}$; $f'(x)$ is never 0; $f'(x)$ is defined for all x in the domain of f. $f''(x) = 2 - \dfrac{2}{x^2}$; $f'(x) = 0$ when $x^2 = 1$, $x = \pm 1$; $f''(x)$ is defined for all x in the domain of f.

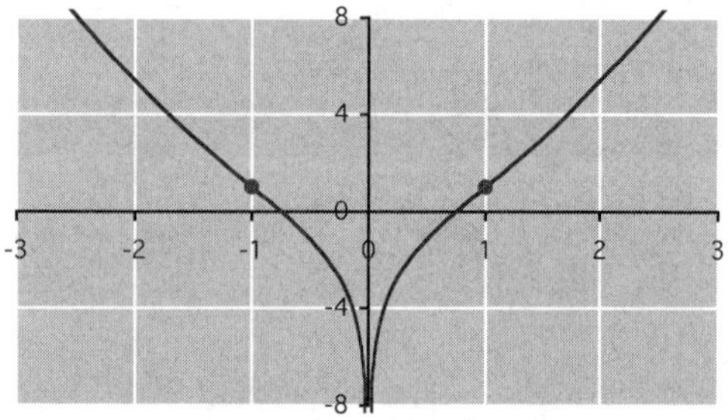

f has no relative extrema, points of inflection at $(1, 1)$ and $(-1, 1)$, and a vertical asymptote at $x = 0$.

53. $g(t) = e^t - t$, domain $[-1, 1]$

$g'(t) = e^t - 1$; $g'(t) = 0$ when $e^t = 1$, $t = \ln 1 = 0$; $g'(t)$ is always defined. $g''(t) = e^t$; $g''(t)$ is never 0; $g''(t)$ is always defined.

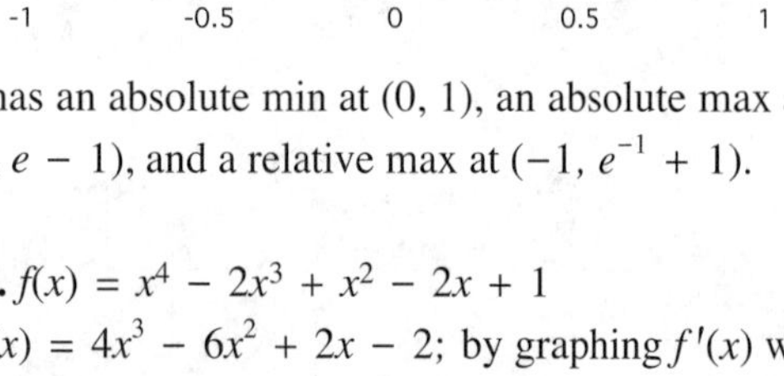

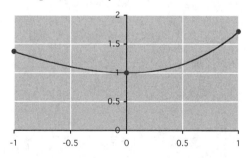

g has an absolute min at $(0, 1)$, an absolute max at $(1, e - 1)$, and a relative max at $(-1, e^{-1} + 1)$.

55. $f(x) = x^4 - 2x^3 + x^2 - 2x + 1$

$f'(x) = 4x^3 - 6x^2 + 2x - 2$; by graphing $f'(x)$ we see that $f'(x) = 0$ for $x \approx 1.40$; $f'(x)$ is always defined. $f''(x) = 12x^2 - 12x + 2$; $f''(x) = 0$ for $x = \dfrac{1}{2} \pm \dfrac{\sqrt{3}}{6}$ (by the quadratic formula), $x \approx 0.21$ or $x \approx 0.79$; $f''(x)$ is always defined.

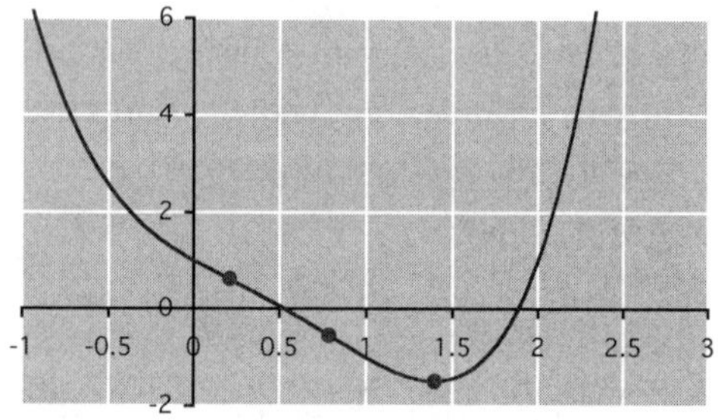

f has an absolute min at $(1.40, -1.49)$ and points of inflection at $(0.21, 0.61)$ and $(0.79, -0.55)$.

57. $f(x) = e^x - x^3$

$f'(x) = e^x - 3x^2$; by graphing $f'(x)$ we see that $f'(x) = 0$ for $x \approx -0.46$, $x \approx 0.91$, and $x \approx 3.73$; $f'(x)$ is always defined. $f''(x) = e^x - 6x$; by graphing $f''(x)$ we see that $f''(x) = 0$ for $x \approx 0.20$ and $x \approx 2.83$; $f''(x)$ is always defined.

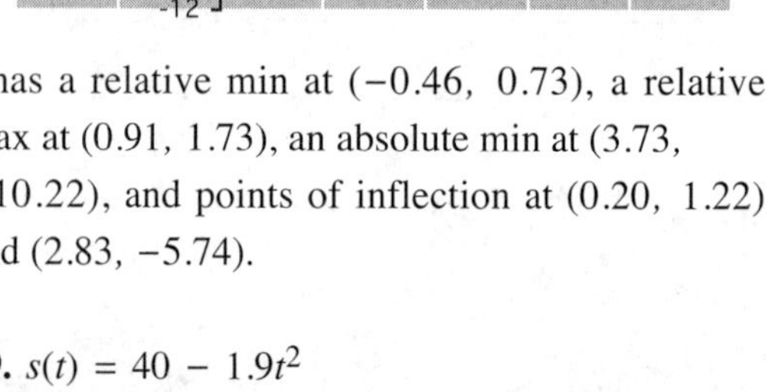

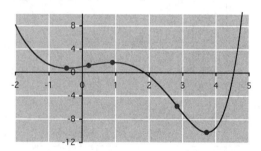

f has a relative min at $(-0.46, 0.73)$, a relative max at $(0.91, 1.73)$, an absolute min at $(3.73, -10.22)$, and points of inflection at $(0.20, 1.22)$ and $(2.83, -5.74)$.

59. $s(t) = 40 - 1.9t^2$

$v(t) = s'(t) = -3.8t$

$a(t) = s''(t) = -3.8 \text{ m/s}^2$

61. $s(t) = t^3 - t^2$
$v(t) = s'(t) = 3t^2 - 2t$
$a(t) = s''(t) = 6t - 2$ ft/s^2
$a(1) = 6(1) - 2 = 4$ ft/s^2
Since this is positive, the velocity is increasing.

63. $R(t) = 17t^2 + 100t + 2300$
$R'(t) = 34t + 100$
$R''(t) = 34$ million gals/yr^2
Accelerating by 34 million gals/yr^2

65. (a) $c(t) = -0.065t^3 + 3.4t^2 - 22t + 3.6$
$c(20) = -0.065(20)^3 + 3.4(20)^2 - 22(20) + 3.6$
≈ 400 ml
(b) $c'(t) = -0.195t^2 + 6.8t - 22$
$c'(20) = -0.195(20)^2 + 6.8(20) - 22$
$= 36$ ml/day
(c) $c''(t) = -0.39t + 6.8$
$c''(20) = -0.39(20) + 6.8 = -1$ ml/day^2

67. (a) Where the graph is steepest: 2 years into the epidemic.
(b) At the point at inflection 2 years into the epidemic: There the steepness stops increasing and starts to decline, so the rate of new infections starts to drop.

69. (a) 2000: the point where the graph is increasing and steepest.
(b) 2002: the point where the graph is decreasing and steepest.
(c) 1998: where the graph changed from concave down to concave up.

71. The graph of P is concave up when P'' is positive, and concave down when P'' is negative. From the graph of P'', we see that it is negative until about $t = 8$, at which time it turns positive. Therefore: The graph of P is concave up for $8 < t < 20$, concave down for $0 < t < 8$, and there is a point of inflection around $t = 8$ (when $P'' = 0$). Interpretation: From the graph of P' we see that P' has a minimum at around $t = 8$, meaning that the percentage of articles written by researchers in the U.S. was decreasing most rapidly at around $t = 8$ (1991).

73. (a) The graph of c' is concave down throughout the range [8, 30], and therefore has no points of inflection: Choice (B).
(b) At $t = 18$ (the point of inflection) the graph of c' has a maximum. Since c' measures the rate of change of daily oxygen consumption, it means that oxygen consumption is increasing at a maximum rate at around $t = 18$: Choice (B).
(c) For $t > 18$, the graph of c is increasing but concave down, so that oxygen consumption is increasing at a decreasing rate: Choice (A).

75. (a) $S'(n) \approx -1757(n - 180)^{-2.325}$; $S''(n) \approx 4085(n - 180)^{-3.325}$; $S''(n)$ is never zero and is always defined for $n > 180$. So, there are no points of inflection in the graph of S.
(b) Since the graph is concave up ($S''(n) > 0$ for $n > 180$), the derivative of S is increasing, and so the rate of *decrease* of SAT scores with increasing numbers of prisoners is diminishing. In other words, the apparent effect on SAT scores of increasing numbers of prisoners is diminishing.

77. (a) $\frac{dn}{ds} = 144.42 - 47.72s + 4.371s^2$; $\frac{d^2n}{ds^2} = -47.72 + 8.742s$; $\left.\frac{d^2n}{ds^2}\right|_{s=3} = -21.494$. Thus, for a firm with annual sales of \$3 million, the rate at which new patents are produced decreases with increasing firm size. This means that the returns (as measured in the number of new patents per increase of \$1 million in sales) are diminishing as the firm size increases.

(b) $\left.\frac{d^2n}{ds^2}\right|_{s=7} = 13.474$. Thus, for a firm with annual sales of \$7 million, the rate at which new patents are produced increases with increasing firm size by 13.474 new patents per \$1 million increase in annual sales.

(c) There is a point of inflection when $s \approx 5.4587$, so that in a firm with sales of \$5,458,700 per year, the number of new patents produced per additional \$1 million in sales is a minimum.

79. $M(t)/B(t) = (2.48t+6.87)/(15{,}700t+444{,}000)$

Graph:

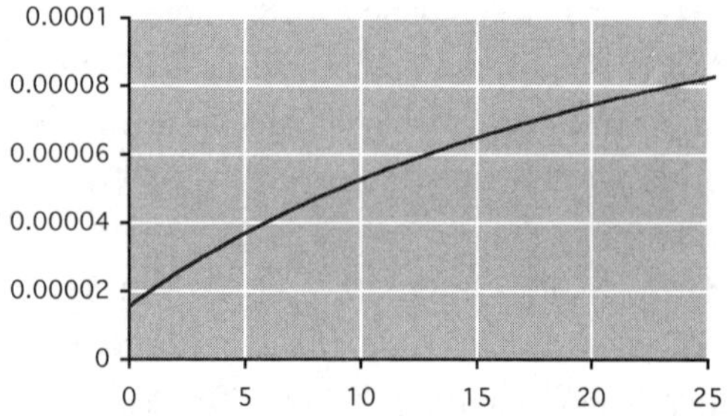

The graph is concave down, which means that the slope is decreasing. The correct choice is (C) because the number of manatees killed per boat is increasing (the slope is always positive), but the rate of increase (the slope) is decreasing.

81. $p = v(1.05)^{-t} = \frac{10{,}000(1.05)^{-t}}{1 + 500e^{-0.5t}}$.

Technology formula:

Excel:

```
10000*(1.05)^(-x)/(1+500*exp(-0.5*x))
```

TI-83/84:

```
10000*(1.05)^(-x)/(1+500*e^(-0.5*x))
```

Here are the graphs of $p(t)$ and $p'(t)$:

Graph of $p(t)$;

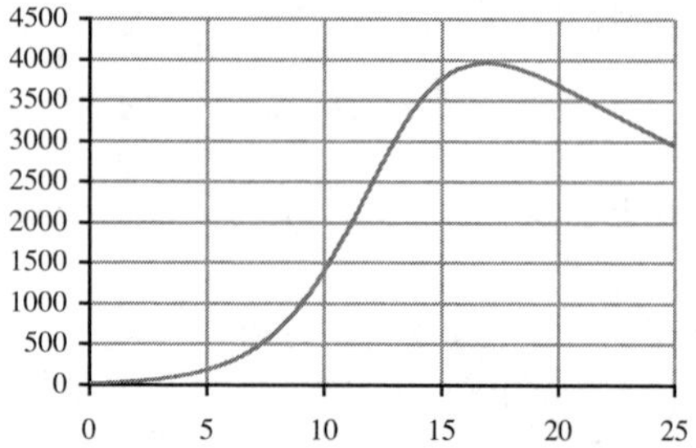

Graph of $p'(t)$:

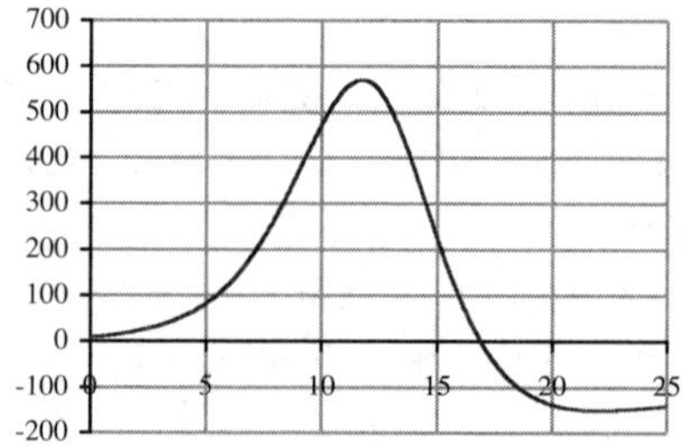

Graph of $p''(t)$:

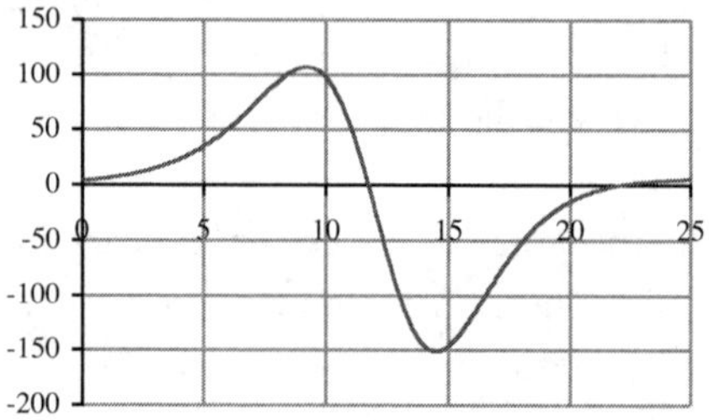

The greatest rate of increase of p occurs when the the derivative is greatest. This high point on the graph of $p'(t)$ is located accurately by determining where the graph of $p''(t)$ crosses the t-axis: at approximately $t = 12$. The value of the greatest rate of increase at this point is the y-coordinate of $p'(t)$ at $t = 12$, which we can determine from the graph of $p'(t)$ as approximately \$570 per year.

83. $p(t) = ve^{-0.05t} = (300{,}000 + 1000t^2)\, e^{-0.05t}$

Technology formula:

Excel:

```
(300000+1000*x^2)*exp(-0.05*x)
```

TI-84/85:

```
(300000+1000*x^2)*e^(-0.05*x)
```

Graph of $p(t)$:

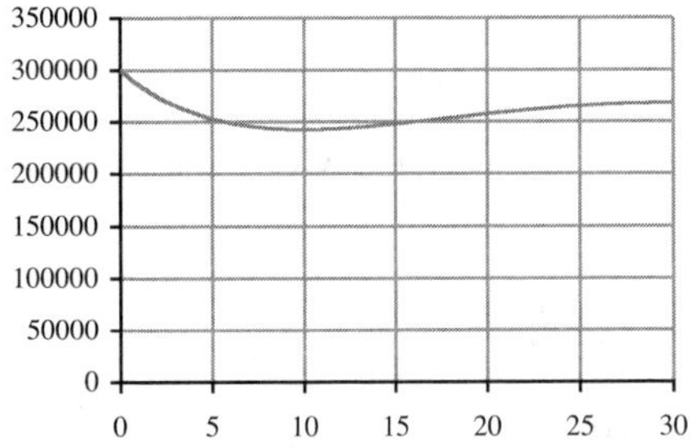

Graph of $p'(t)$:

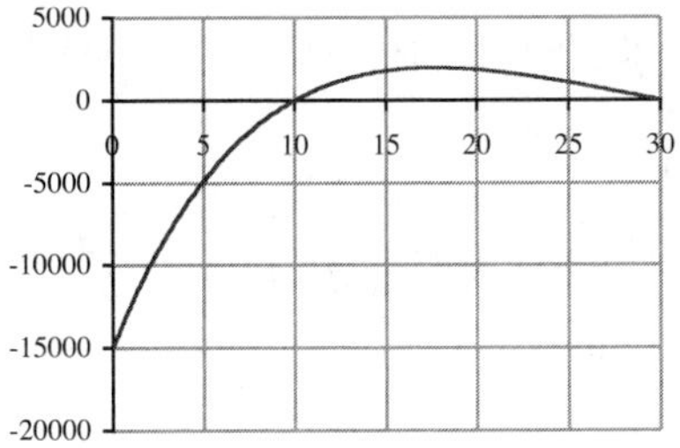

Graph of $p''(t)$:

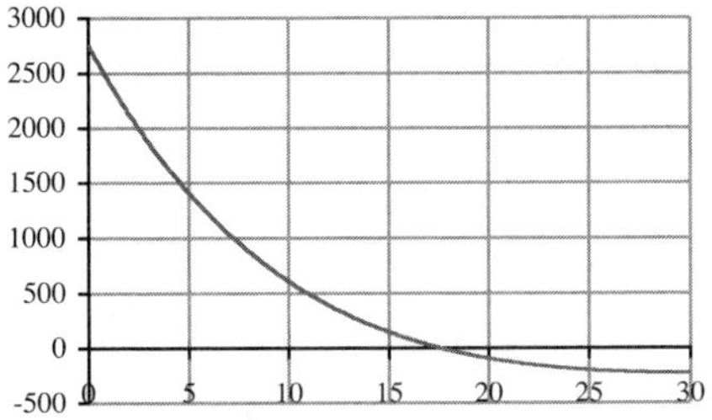

The greatest rate of increase of p occurs when the the derivative is greatest. This high point on the graph of $p'(t)$ is located accurately by determining where the graph of $p''(t)$ crosses the t-axis: at approximately $t = 17.7$.

$p(t)$ is decreasing most rapidly at the point where the derivative $p'(t)$ is a minimum, which occurs at $t = 0$ (see the graph of $p'(t)$).

85. Nonnegative

87. Daily sales were decreasing most rapidly in June 2002.

89.

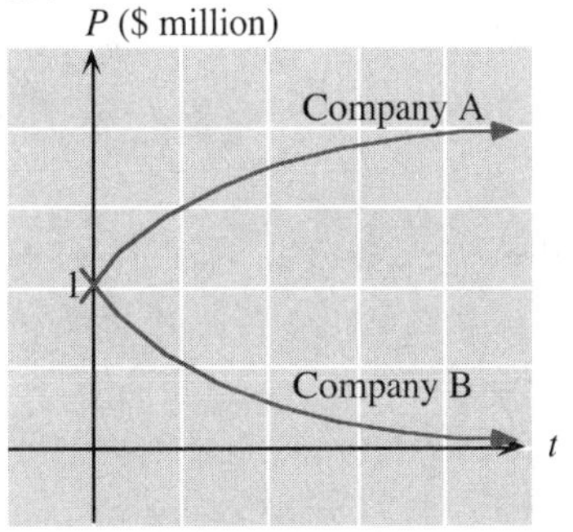

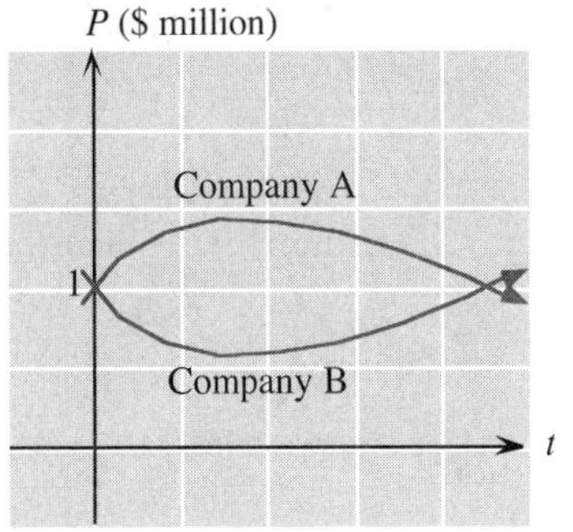

91. At a point of inflection, the graph of a function changes either from concave up to concave down, or vice-versa. If it changes from concave up to concave down, then the derivative changes from increasing to decreasing, and hence has a relative maximum. Similarly, if it changes from concave down to concave up, the derivative has a relative minimum.

5.4

1. The population P is currently 10,000 and its rate of change is 1000 per year:

$$P = 10{,}000 \text{ and } \frac{dP}{dt} = 1000.$$

3. Let R be the annual revenue of my company and let q be annual sales. R is currently \$7000 but and its rate of change is −\$700 each year. Find how fast q is changing:

$$R = 7000 \text{ and } \frac{dR}{dt} = -700.$$

Find $\frac{dq}{dt}$.

5. Let p be the price of a pair of shoes and let q be the demand for shoes.

$$\frac{dp}{dt} = 5. \text{ Find } \frac{dq}{dt}.$$

7. Let T be the average global temperature and let q be the number of Bermuda shorts sold per year.

$$T = 60 \text{ and } \frac{dT}{dt} = 0.1. \text{ Find } \frac{dq}{dt}.$$

9. (a) Changing quantities: the radius r and the area A.

The problem:

$$\frac{dA}{dt} = 1200. \text{ Find } \frac{dr}{dt} \text{ when } r = 10{,}000.$$

The relationship:

$$A = \pi r^2$$

$\frac{d}{dt}$ of both sides:

$$\frac{dA}{dt} = 2\pi r \frac{dr}{dt}.$$

Substitute:

$$1200 = 2\pi(10{,}000) \frac{dr}{dt}$$

so $\frac{dr}{dt} = 6/(100\pi) \approx 0.019$ km/sec.

(b) This time the problem is to find $\frac{dr}{dt}$ when A = 640,000. From part (a) we have the derived equation

$$\frac{dA}{dt} = 2\pi r \frac{dr}{dt}.$$

Since r appears in the derived equation but not A, we need to find r from $A = \pi r^2$:

$$640{,}000 = \pi r^2$$

$$r = \sqrt{640{,}000/\pi} = 800\sqrt{\pi}.$$

Substituting these values in the derived equation:

$$\frac{dA}{dt} = 2\pi r \frac{dr}{dt}.$$

$$1200 = 2\pi(800\sqrt{\pi}) \frac{dr}{dt}$$

$$\frac{dr}{dt} = 6/(8\sqrt{\pi}) \approx 0.4231 \text{ km/sec.}$$

11. Changing quantities: the volume V and the radius r.

The problem:

$$\frac{dV}{dt} = 3. \text{ Find } \frac{dr}{dt} \text{ when } r = 1.$$

The relationship:

$$V = \frac{4}{3}\pi r^3$$

$\frac{d}{dt}$ of both sides:

$$\frac{dV}{dt} = 4\pi r^2 \frac{dr}{dt}$$

Substitute:

$$3 = 4\pi(1)^2 \frac{dr}{dt}$$

$$\frac{dr}{dt} = \frac{3}{4\pi} \approx 0.24 \text{ ft/min}$$

13. Changing quantities: b = the distance of the base of the ladder from the wall and h = the height of the top of the ladder.

The problem:

$$\frac{db}{dt} = 10. \text{ Find } \frac{dh}{dt} \text{ when } b = 30$$

The relationship:

$b^2 + h^2 = 50^2$

$\frac{d}{dt}$ of both sides:

$2b\frac{db}{dt} + 2h\frac{dh}{dt} = 0.$

We need the value of h:

$30^2 + h^2 = 50^2$

$h = \sqrt{2500 - 900} = 40.$

Substitute:

$2(30)(10) + 2(40)\frac{dh}{dt} = 0$

$\frac{dh}{dt} = -600/80 = -7.5,$

so the top of the ladder is sliding down at 7.5 ft/sec.

15. Changing quantities: the average cost $\bar{C}$ and the number of CD players x.

The problem:

$x = 3000$ and $\frac{dx}{dt} = 100$. Find $\frac{d\bar{C}}{dt}$.

The relationship:

$\bar{C}(x) = 150{,}000x^{-1} + 20 + 0.0001x$ $\frac{d}{dt}$ of both sides:

$\frac{d\bar{C}}{dt} = -150{,}000x^{-2}\frac{dx}{dt} + 0.0001\frac{dx}{dt}$

Substitute:

$\frac{d\bar{C}}{dt} = -150{,}000(3000)^{-2}(100) + 0.0001(100)$

$\approx -1.66.$

The average cost is decreasing at a rate of \$1.66 per player per week.

17. Changing quantities: the number q of T shirts sold per month and the price p per T-shirt

The problem:

$p = 15$ and $\frac{dp}{dt} = 2$. Find $\frac{dq}{dt}$

The relationship: This is the given demand equation:

$q = 500 - 100p^{0.5}$

$\frac{d}{dt}$ of both sides:

$\frac{dq}{dt} = -50p^{-0.5}\frac{dp}{dt}.$

Substitute:

$\frac{dq}{dt} = -50(15)^{-0.5}(2) \approx -26.$

Monthly sales will drop at a rate of 26 T-shirts per month.

19. Changing quantities: The price p, the weekly demand q, and the weekly revenue R.

The problem:

$q = 50$, $p = 30¢$, and $\frac{dq}{dt} = -5.$

Find $\frac{dp}{dt}$ if $\frac{dR}{dt} = 0.$

The relationship:

$R = pq$ Revenue = price × quantity

$\frac{d}{dt}$ of both sides:

$\frac{dR}{dt} = \frac{dp}{dt}q + p\frac{dq}{dt}.$

Substitute:

$0 = \frac{dp}{dt}(50) + (30)(-5)$

$\frac{dp}{dt} = 150/50 = 3.$

You must raise the price by 3¢ per week.

21. Changing quantities: the number P of automobiles produced per year, the number x of employees, and the daily operating budget y.

The problem:

P is constant at 1000. $x = 150$ and $\frac{dx}{dt} = 10.$

Find $\frac{dy}{dt}.$

The relationship:

$P = 10x^{0.3}y^{0.7}$

$\frac{d}{dt}$ of both sides:

$$0 = 3x^{-0.7}y^{0.7}\frac{dx}{dt} + 7x^{0.3}y^{-0.3}\frac{dy}{dt}$$

(We are told that P is constant so its derivative is zero.)

The solution to the problem is a bit simpler if we first solve for $\frac{dy}{dt}$ before substituting values:

$$\frac{dy}{dt} = -\frac{3x^{-0.7}y^{0.7}}{7x^{0.3}y^{-0.3}} = -\frac{3y}{7x}\frac{dx}{dt}.$$

We need the value of y:

$$1000 = 10(150)^{0.3}y^{0.7}$$

$$y = (100/150^{0.3})^{1/0.7} = 100^{10/7}/150^{3/7}$$

Substitute:

$$\frac{dy}{dt} = -\frac{3(100^{1/0.7})/150^{3/7}}{7(150)}(10)$$

$$= -\frac{30(100^{10/7})}{7(150)^{10/7}} \approx -2.40.$$

The daily operating budget is dropping at a rate of \$2.40 per year.

23. Changing quantities: the number q of pounds of tuna that can be sold in one month, the price p in dollars per pound.

The problem:

$$q = 900 \text{ and } \frac{dq}{dt} = 100. \text{ Find } \frac{dp}{dt}.$$

The relationship: This is the demand equation:

$$pq^{1.5} = 50{,}000$$

$\frac{d}{dt}$ of both sides:

$$\frac{dp}{dt}q^{1.5} + 1.5pq^{0.5}\frac{dq}{dt} = 0$$

We will also need the value of p:

$$p(900)^{1.5} = 50{,}000$$

$$p = 50{,}000/(900)^{1.5}$$

Substitute:

$$\frac{dp}{dt}(900)^{1.5}+1.5[50{,}000/(900)^{1.5}](900)^{0.5}(100) = 0$$

$$\frac{dp}{dt} = -(75{,}000/9)/(900)^{1.5} \approx -0.31$$

The price is decreasing at a rate of approximately 31¢ per pound per month.

25. Changing quantities: Let x be the distance of the Mona Lisa from Montauk and let y be the distance of the Dreadnaught from Montauk. Let z be the distance between the two ships.

The problem:

$$x = 50, \frac{dx}{dt} = 30, y = 40, \text{ and } \frac{dy}{dt} = 20.$$

Find $\frac{dz}{dt}$.

The relationship:

$$z^2 = x^2 + y^2$$

$\frac{d}{dt}$ of both sides:

$$2z\frac{dz}{dt} = 2x\frac{dx}{dt} + 2y\frac{dy}{dt}$$

We need the value of z:

$$z^2 = 50^2 + 40^2 = 4100$$

$$z = \sqrt{4100}$$

Substitute:

$$2\sqrt{4100}\frac{dz}{dt} = 2(50)(30) + 2(40)(20)$$

$$= 4600$$

$$\frac{dz}{dt} = \frac{2300}{\sqrt{4100}} \approx 36 \text{ mph.}$$

27. Changing quantities: Let x be the distance of the batter from home base, and let h be the distance from third base as shown here:

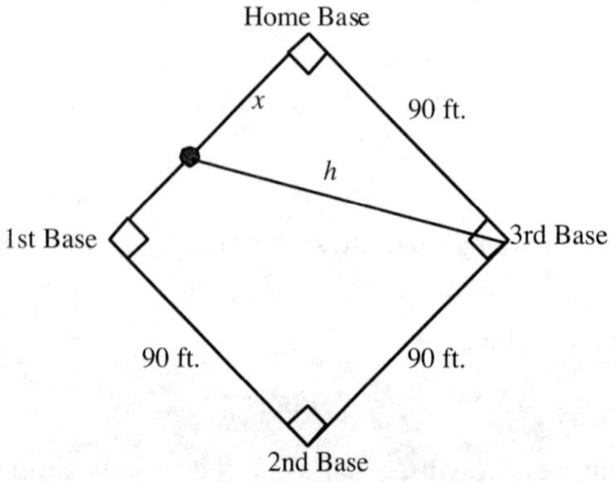

The problem:

Given that $\frac{dx}{dt} = 24$, find $\frac{dh}{dt}$ when $x = 45$.

Equation relating the changing quantities:

$$x^2 + 90^2 = h^2$$
$$x^2 + 8100 = h^2$$

Derived equation:

$$2x\frac{dx}{dt} = 2h\frac{dh}{dt}$$
$$\frac{dh}{dt} = \frac{x}{h}\frac{dx}{dt}$$
$$h^2 = 8100 + 45^2$$
$$= 8100 + 2025 = 10,125$$
$$h = \sqrt{10,125} \approx 100.62$$
$$\frac{dh}{dt} = \frac{x}{h}\frac{dx}{dt} = \frac{45}{\sqrt{10,125}}(24) \approx 10.7 \text{ ft/sec}$$

29. Changing quantities: the x-coordinate x and the y-coordinate y of a point on the graph.

The problem:

$$\frac{dx}{dt} = 4 \text{ and } y = 2. \text{ Find } \frac{dy}{dt}.$$

The relationship:

$$y = x^{-1}$$

$\frac{d}{dt}$ of both sides:

$$\frac{dy}{dt} = -x^{-2}\frac{dx}{dt}$$

We need the value of x:

$$2 = 1/x$$
$$x = 1/2.$$

Substitute:

$$\frac{dy}{dt} = -(1/2)^{-2}(4) = -16$$

The y coordinate is decreasing at a rate of 16 units per second.

31. Changing quantities: I and n.

The problem:

$$n = 13 \text{ and } \frac{dn}{dt} = \frac{1}{3}. \text{ Find } \frac{dI}{dn}.$$

The relationship:

$$I = 2928.8n^3 - 115,860n^2 + 1,532,900n - 6,760,800$$

$\frac{d}{dt}$ of both sides:

$$\frac{dI}{dt} = 8786.4n^2\frac{dn}{dt} - 231,720n\frac{dn}{dt} + 1,532,900\frac{dn}{dt}$$

Substitute:

$$\frac{dI}{dt} = 8786.4(13)^2(1/3)$$
$$- 231,720(13)(1/3) + 1,532,900(1/3)$$
$$\approx \$1814 \text{ per year.}$$

33. Changing quantitie: V, e, g.

The problem:

V is constant at 200. $g = 3.0$ and $\frac{dg}{dt} = -0.2$.

Find $\frac{de}{dt}$.

The relationship:

$$V = 3e^2 + 5g^3$$

$\frac{d}{dt}$ of both sides:

$$0 = 6e\frac{de}{dt} + 15g^2\frac{dg}{dt}$$

We need the value of e:

$$200 = 3e^2 + 5(3.0)^3$$
$$e = \sqrt{65/3} \approx 4.655$$

Substitute:

$$0 = 6(4.655)\frac{de}{dt} + 15(3.0)^2(-0.2)$$
$$\frac{de}{dt} = \frac{27}{27.93} \approx 0.97.$$

Their prior experience must increase at a rate of approximately 0.97 years every year.

35. Changing quantities: V, h

The problem:

$$\frac{dV}{dt} = 100 \text{ and } V = 200\pi. \text{ Find } \frac{dh}{dt}.$$

The relationship:

The given formula expresses V in terms of both h and r. To get the relationship between V and h we need to know how r is related to h. Looking at the vessel from the side, we can see that, for any given value of h, the corresponding radius r must satisfy

$r/h = 30/50$

the ratio at the brim of the vessel. So,

$r = \frac{3}{5} h.$

Substituting into $V = \frac{1}{3} \pi r^2 h$ we get the relationship we want:

$V = \frac{3}{25} \pi h^3$

$\frac{d}{dt}$ of both sides:

$$\frac{dV}{dt} = \frac{9}{25} \pi h^2 \frac{dh}{dt}$$

We need the value of h:

$200\pi = \frac{3}{25} \pi h^3$

$h = (5000/3)^{1/3}$

Substitute:

$$100 = \frac{9\pi}{25} \left(\frac{5000}{3}\right)^{2/3} \frac{dh}{dt}$$

$$\frac{dh}{dt} = \frac{2500}{9\pi} \left(\frac{3}{5000}\right)^{2/3} \approx 0.63 \text{ m/sec.}$$

37. Changing quantities: V and h.

The problem:

r is constant at 2. $V = 4t^2 - t$ and $h = 2$.

Find $\frac{dh}{dt}$.

The relationship:

From $V = \pi r^2 h = 4\pi h$ we get

$h = \frac{V}{4\pi}$

$\frac{d}{dt}$ of both sides:

$$\frac{dh}{dt} = \frac{1}{4\pi} \frac{dV}{dt} = \frac{1}{4\pi}(8t - 1).$$

We need the value of t when $h = 2$. We first find

$V = 4\pi h = 8\pi$

so $8\pi = 4t^2 - t$

$4t^2 - t - 8\pi = 0$

$$t = \frac{1 \pm \sqrt{1 + 128\pi}}{8}$$

by the quadratic formula. We take the positive solution

$$t = \frac{1 + \sqrt{1 + 128\pi}}{8},$$

where the volume is rising. Substituting:

$$\frac{dh}{dt} = \frac{\sqrt{1 + 128\pi}}{4\pi} \approx 1.6 \text{ cm/sec.}$$

39. Changing quantities: q and x.

The problem:

$x = 30{,}000$ and $\frac{dx}{dt} = 2000$. Find q and $\frac{dq}{dt}$.

The relationship:

$q = 0.3454 \ln x - 3.047$

$\frac{d}{dt}$ of both sides:

$$\frac{dq}{dt} = \frac{0.3454}{x} \frac{dx}{dt}$$

Substitute:

$$\frac{dq}{dt} = \frac{0.3454}{30{,}000}(2000) \approx 0.0230.$$

We are also asked for the value of q (the number of computers per household) so we substitute $x = 30{,}000$ in the original equation:

$q = 0.3454 \ln(30{,}000) - 3.047 \approx 0.5137$

So, there are approximately 0.5137 computers per household, increasing at a rate of 0.0230 computers per household per year.

41. Changing quantities: S and n.

The problem:

$n = 475$ and $\frac{dn}{dt} = 35$. Find S and $\frac{dS}{dt}$.

The relationship:

$$S = 904 + \frac{1326}{(n-180)^{1.325}}$$

$S(475) \approx 904.71$ from the formula.

Derived relationship:

$$\frac{dS}{dt} = 1326(-1.325)(n - 180)^{-2.325} \frac{dn}{dt}.$$

Substitute:

$$\frac{dS}{dt} = -1756.95(475 - 180)^{-2.325}(35)$$
$$\approx -0.11$$

The average SAT score was 904.71, decreasing at a rate of 0.11 per year.

43. $r = 1.1$ and $\frac{dr}{dt} = 0.05$. Find $\frac{d}{dt}d(r)$. Note that $r = 1.1$ is in the range where $d(r) = -40r + 74$ and that we stay in that range because we are interested in the slope at that point. Thus, $\frac{d}{dt}d(r) = -40\frac{dr}{dt} = -40(0.05) = -2$. The divorce rate is decreasing by 2 percentage points per year.

45. The section is called "related rates" because the goal is to compute the rate of change of a quantity based on a knowledge of the rate of change of a related quantity. The relationship between the quantities gives a relationship between their rates of change.

47. Answers may vary. For example: A rectangular solid has dimensions 2 cm $\times$5 cm $\times$10 cm and each side is expanding at a rate of 3 cm/second. How fast is the volume increasing?

49. $R = pq$, so $R' = p'q + pq'$, where the derivatives are with respect to time t. Divide by $R = pq$ to get $\frac{R'}{R} = \frac{p'q}{pq} + \frac{pq'}{pq} = \frac{p'}{p} + \frac{q'}{q}$ as claimed.

51. The derived equation is linear in the unknown rate X. This follows from the chain rule, since if Q is a quantity and $f(Q)$ is any expression in Q, we have $\frac{d}{dt}f(Q) = f'(Q)\frac{dQ}{dt}$, which is linear in the derivative $\frac{dQ}{dt}$. The presence of other variables may add terms not containing $\frac{dQ}{dt}$ but those maintain the linearity.

53. Let x = my grades and y = your grades. If $dx/dt = 2\ dy/dt$, then $dy/dt = (1/2)\ dx/dt$

5.5

1. $E = -\frac{dq}{dp} \cdot \frac{p}{q}$

$= -(-20)\frac{p}{1000 - 20p} = \frac{20p}{1000 - 20p}$

When $p = 30$, $E = 1.5$: The demand is going down 1.5% per 1% increase in price at that price level. $E = 1$ when

$\frac{20p}{1000 - 20p} = 1$

$20p = 1000 - 20p$

$p = 25$

Revenue is maximized when $p = \$25$; weekly revenue at that price is $R = pq = 25(1000 - 20 \cdot 25) = \$12{,}500$.

3. (a) $E = -\frac{dq}{dp} \cdot \frac{p}{q}$

$= -(-2)(100 - p)\frac{p}{(100 - p)^2}$

$= \frac{2p}{100 - p}.$

When $p = 30$, $E = 6/7$. The demand is going down 6% per 7% increase in price at that price level. Thus, a price increase is in order.

(b) $E = 1$ when

$\frac{2p}{100 - p} = 1$

$2p = 100 - p$

$p = 100/3$

Revenue is maximized when $p = 100/3 \approx \$33.33$.

(c) Demand would be $(100- 100/3)^2 =$ $(200/3)^2 \approx 4444$ cases per week.

5. (a) $E = -\frac{dq}{dp} \cdot \frac{p}{q}$

$= -(-4p + 33)\cdot\frac{p}{-2p^2 + 33p}$

$= \frac{4p - 33}{-2p + 33}$

(b) $E(10) = \frac{4(10) - 33}{-2(10) + 33} = \frac{7}{13} \approx 0.54$

Interpretation: The demand for $E=mc^2$ T-shirts is going down by about 0.54% per 1% increase in the price.

(c) Revenue is maximized when $E = 1$:

$\frac{4p - 33}{-2p + 33} = 1$

$4p - 33 = -2p + 33$

$6p = 66$

$p = \$11$

Therefore, the club should charge \$11 per T shirt to maximize revenue.

At this price, the total revenue is given by

$R = pq = p(-2p^2 + 33p)$

$= (11)(-2(11)^2 + 33(11))$

$= \$1331$

7. (a) $E = -\frac{dq}{dp} \cdot \frac{p}{q}$

$= -(-2.2)\cdot\frac{p}{9900 - 2.2p}$

$= \frac{2.2p}{9900 - 2.2p}$

$E(2900) = \frac{2.2(2900)}{9900 - 2.2(2900)}$

≈ 1.81

Thus, the demand is elastic at the given tuition level, showing that a decrease in tuition will result in an increase in revenue.

(c) Revenue is maximized when $E = 1$:

$\frac{2.2p}{9900 - 2.2p} = 1$

$2.2p = 9900 - 2.2p$

$4.4p = 9900$

$p = 9900/4.4 = \$2250$ per student, and this will result in an enrollment of about

$q = 9900 - 2.2(2250) = 4950$ students, giving a revenue of about

$pq = 2250 \times 4950 = \$11{,}137{,}500.$

9. (a) $E = -\frac{dq}{dp} \cdot \frac{p}{q}$

$= -(-6p + 1)100e^{-3p^2+p} \frac{p}{100e^{-3p^2+p}}$

$= p(6p - 1).$

When $p = 3$, $E = 51$: The demand is going down 51% per 1% increase in price at that price level. Thus, a large price decrease is advised.

(b) $E = 1$ when

$p(6p - 1) = 1$

$6p^2 - p - 1 = 0$

$(3p + 1)(2p - 1) = 0$

$p = 1/2.$

(We reject the solution $p = -1/3$ because we must have $p > 0$.) Revenue is maximized when $p =$ ¥0.50.

(c) Demand would be $q = 100e^{-3/4+1/2} \approx 78$ paint-by-number sets per month.

11. (a) $E = -\frac{dq}{dp} \cdot \frac{p}{q}$

$= -m \frac{p}{mp + b} = -\frac{mp}{mp + b}$

(b) $E = 1$ when

$-\frac{mp}{mp + b} = 1$

$-mp = mp + b$

$p = -\frac{b}{2m}$

13. (a) $E = -\frac{dq}{dp} \cdot \frac{p}{q}$

$= -(-r)\frac{k}{p^{r+1}} \frac{p}{k/p^r} = r$

(b) E is independent of p.

(c) If $r = 1$ the revenue is not affected by the price. If $r > 1$ the revenue is always elastic, whereas if $r < 1$ the revenue is always inelastic. This is an unrealistic model because there should be a price at which the revenue is a maximum.

15. (a) We have two data points: $(p, q) =$ (2.00, 3000) and (4.00, 0). The line through these two points is

$q = -1500p + 6000.$

(b) $E = -\frac{dq}{dp} \cdot \frac{p}{q}$

$= -(-1500) \frac{p}{-1500p + 6000}$

$= \frac{1500p}{-1500p + 6000}$

$E = 1$ when

$\frac{1500p}{-1500p + 6000} = 1$

$1500p = -1500p + 6000$

$p = \$2$ per hamburger.

This gives a total weekly revenue of

$R = pq$

$= 2(-1500 \cdot 2 + 6000) = \$6000.$

17. $E = \frac{dq}{dx} \cdot \frac{x}{q}$

$= 0.01(-0.0156x + 1.5)$

$\cdot \frac{x}{0.01(-0.0078x^2 + 1.5x + 4.1)}$

$= \frac{-0.0156x^2 + 1.5x}{-0.0078x^2 + 1.5x + 4.1}.$

When $x = 20$, $E \approx 0.77$: At a family income level of \$20,000, the fraction of children attending a live theatrical performance is increasing by 0.77% per 1% increase in household income.

19. (a) $E = \frac{dq}{dx} \cdot \frac{x}{q}$

$= \frac{0.3454}{x} \cdot \frac{x}{0.3454 \ln x - 3.047}$

$= \frac{0.3454}{0.3454 \ln x - 3.047}.$

When $x = 60{,}000$, $E \approx 0.46$. The demand for computers is increasing by 0.46% per 1% increase in household income.

(b) E decreases as income increases because the denominator of E gets larger. **(c)** Unreliable; it predicts a likelihood greater than 1 at incomes of

\$123,000 and above. ($0.3454 \ln x - 3.047 = 1$ when $x = e^{4.047/0.3454} \approx 123{,}000$) In a model appropriate for large incomes, one would expect the curve to level off at or below 1.

(d) E approaches 0 as x goes to infinity, so for very large x we have $E \approx 0$.

21. The income elasticity of demand is

$$\frac{dQ}{dY} \cdot \frac{Y}{Q} = a\beta P^{\alpha} Y^{\beta-1} \frac{Y}{aP^{\alpha}Y^{\beta}} = \beta.$$

An increase in income of $x\%$ will result in an increase in demand of $\beta x\%$.

23. (a) The data points $(p, q) = (3.00, 407)$ and $(5.00, 223)$ give us the exponential function $q = 1000e^{-0.3p}$.

(b) $E = -\frac{dq}{dp} \cdot \frac{p}{q}$

$$= 300e^{-0.3p} \frac{p}{1000e^{-0.3p}} = 0.3p$$

At $p = \$3$, $E = 0.9$; at $p = \$4$, $E = 1.2$; and at $p = \$5$, $E = 1.5$.

(c) $E = 1$ when $0.3p = 1$, so $p = \$3.33$.

(d) We first find the price that produces a demand of 200 pounds:

$$1000e^{-0.3p} = 200$$
$$e^{-0.3p} = 0.2$$
$$p = -(\ln 0.2)/0.3 = \$5.36.$$

Selling at a lower price would increase demand, but you cannot sell more than 200 pounds anyway so your revenue would go down. On the other hand, if you set the price higher than \$5.36 the decrease in sales will outweigh the increase in price, which we know because the elasticity at $p = 5.36$ is $E \approx 1.6 > 1$. You should therefore set the price at \$5.36 per pound.

25. the price is lowered

27. Start with $R = pq$ and differentiate with respect to p to obtain

$$\frac{dR}{dp} = q + p\frac{dq}{dp}.$$

For a stationary point, $dR/dp = 0$, so

$$q + p\frac{dq}{dp} = 0.$$

Rearranging gives

$$p\frac{dq}{dp} = -q, \text{ and hence}$$
$$-\frac{dq}{dp} \cdot \frac{p}{q} = 1, \text{ or}$$
$$E = 1,$$

showing that stationary points of R correspond to points of unit elasticity.

29. The distinction is best illustrated by an example. Suppose that q is measured in weekly sales and p is the unit price in dollars. Then the quantity $-dq/dp$ measures the drop in weekly sales per \$1 increase in price. The elasticity of demand E, on the other hand, measures the *percentage* drop in sales per *one percent* increase in price. Thus, $-dq/dp$ measures absolute change, while E measures fractional, or percentage, change.

Chapter 5 Review Exercises

1. $f(x) = 2x^3-6x+1$ on $[-2,+\infty)$

End points: -2

Stationary points:

$f'(x) = 6x^2 - 6$

$f'(x) = 0$ when

$6x^2 - 6 = 0$

$6(x^2 - 1) = 0$

$x = \pm 1$

Singular points:

$f'(x)$ is defined for all x, so no singular points.

Thus, we have stationary points at $x = \pm 1$ and the endpoint $x = -2$. We test a point to the right of 1 as well:

x	-2	-1	1	2
$f(x)$	-3	5	-3	5

The graph must increase from $x = -2$ to $x = -1$, decrease to $x = 1$, and increase from then on.

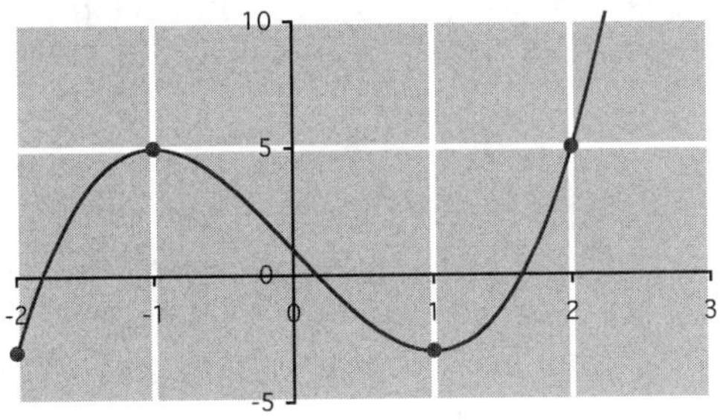

This gives absolute mins at $(-2, -3)$ and $(1, -3)$ and a relative max at $(-1, 5)$.

2. $f(x) = x^3-x^2-x-1$ on $(-\infty, \infty)$

End points: None

Stationary points:

$f'(x) = 3x^2 - 2x - 1 = (3x + 1)(x - 1)$

$f'(x) = 0$ when $(3x + 1)(x - 1) = 0$

$x = -1/3$ or 1.

Singular points: None

x	$y = x^3-x^2-x-1$
-2 (Test point)	-11
$-1/3$	$-22/27$
0 (Test point)	-1
1	-2
2 (Test point)	1

This shows a relative maximum at $(-1/3, -22/27)$, and a relative minimum at $(1, -2)$.

3. $g(x) = x^4-4x$ on $[-1, 1]$

End points: -1, 1

Stationary Points:

$g'(x) = 4x^3-4$

$= 4(x^3- 1)$

$= 4(x-1)(x^2+x+1)$

$g'(x) = 0$ when

$4(x-1)(x^2+x+1) = 0$

$x = 1$

Singular points: None

x	$y = x^4- 4x$
-1	5
1	-3

From the chart we see that g has an absolute maximum at $(-1, 5)$ and an absolute minimum at $(1, -3)$.

4. $f(x) = \dfrac{x+1}{(x-1)^2}$ on $[-2, 1) \cup (1, 2]$

End points: -2, 2

Stationary points:

$$f'(x) = \frac{(x-1)^2 - 2(x+1)(x-1)}{(x-1)^4}$$

$$= \frac{x - 1 - 2(x+1)}{(x-1)^3} = \frac{-x-3}{(x-1)^3}$$

$f'(x) = 0$ when $x = -3$, which is outside of the domain.1 So, no stationary points.

Singular points:

$f'(x)$ is undefined at $x = 1$, which is not in the domain of f.

Thus, there are no critical points in the domain of f. We test the endpoints and another point on each side of the point $x = 1$:

x	-2	-1	1.5	2
$f(x)$	-0.11	0	10	3

The graph must increase from $x = -2$ until the vertical asymptote at $x = 1$. It then decreases on the other side of $x = 1$ until $x = 2$.

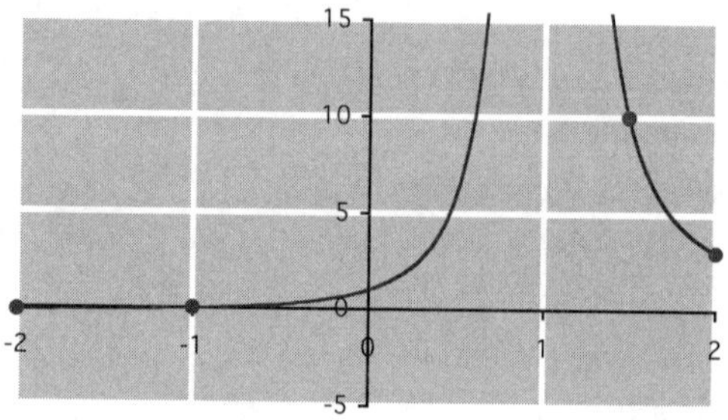

This gives an absolute min at $(-2, -1/9)$ and a relative min at $(2, 3)$.

5. $g(x) = (x-1)^{2/3}$

End points: None.

Stationary points:

$$g'(x) = \frac{2}{3}(x-1)^{-1/3}$$

$g'(x)$ is never 0, so no stationary points.

Singular points:

$g'(x)$ is not defined at $x = 1$.

Thus, we have a single critical point at $x = 1$. We test a point on either side:

x	0	1	2
$g(x)$	1	0	1

The graph must decrease until $x = 1$ and then increase.

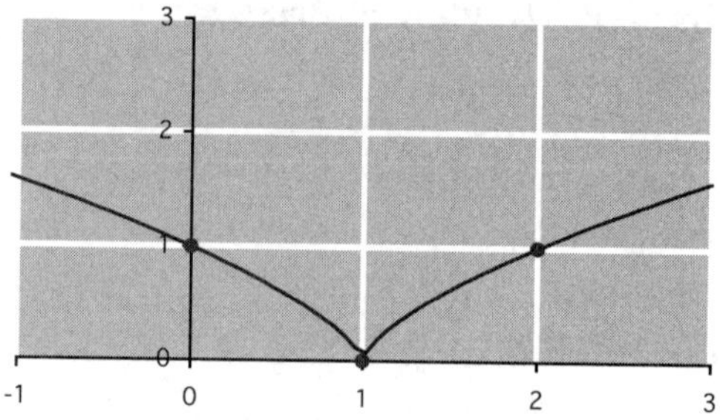

This gives an absolute min at $(1, 0)$.

6. $g(x) = x^2 + \ln x$ on $(0, +\infty)$

End points: None

Stationary points:

$$g'(x) = 2x + 1/x$$

$g'(x) = 0$ when

$$2x + 1/x = 0,$$

which cannot happen for $x > 0$. Thus, no stationary points.

Singular points:

$g'(x)$ is defined for all x in the domain of g.

Thus, g has no critical points in its domain.

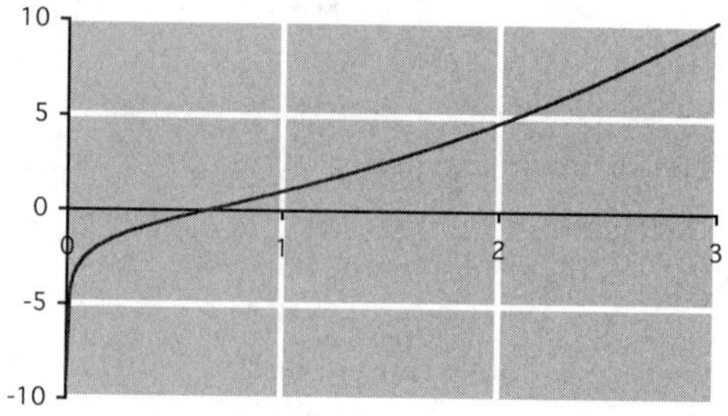

g has no extrema.

7. $h(x) = \dfrac{1}{x} + \dfrac{1}{x^2}$

End points: The domain of h is all $x \neq 0$. Thus, there are no end-points of the domain.

Stationary points:

$$h'(x) = -\frac{1}{x^2} - \frac{2}{x^3}$$

$h'(x) = 0$ when

$2x^2 = -x^3$

$x^3 + 2x^2 = 0$

$x^2(x + 2) = 0$

$x = -2$ ($x = 0$ is not in the domain)

Singular points:

$h'(x)$ is defined for all x in the domain of h, so there are no singular points.

Thus, h has a stationary point at $x = -2$. We test points on either side of -2 and points to the right of $x = 0$:

x	-3	-2	-1	1	2
$h(x)$	-0.22	-0.25	0	2	0.75

The graph decreases to $x = -2$ then increases approaching the vertical asymptote at $x = 0$. On the other side of the asymptote it decreases.

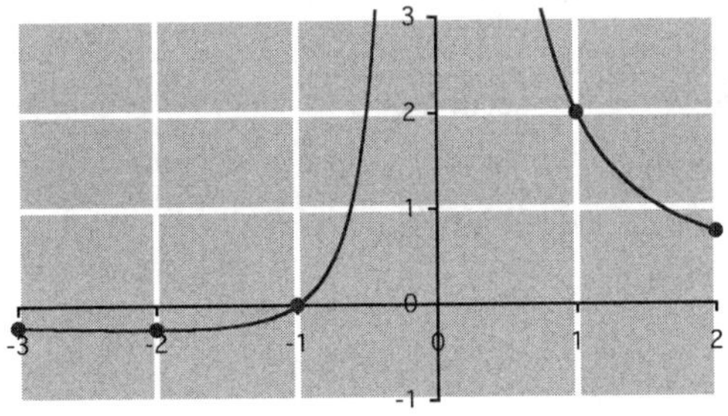

h has an absolute min at $(-2, -1/4)$.

8. $h(x) = e^{x^2} + 1$

End points: None

Stationary points:

$h'(x) = 2xe^{x^2}$

$h'(x) = 0$ when $x = 0$

Singular points:

$h'(x)$ is always defined, so there are no singular points.

We test points on either side of the stationary point at $x = 0$:

x	-1	0	1
$h(x)$	3.72	2	3.72

The graph decreases until $x = 0$ then increases.

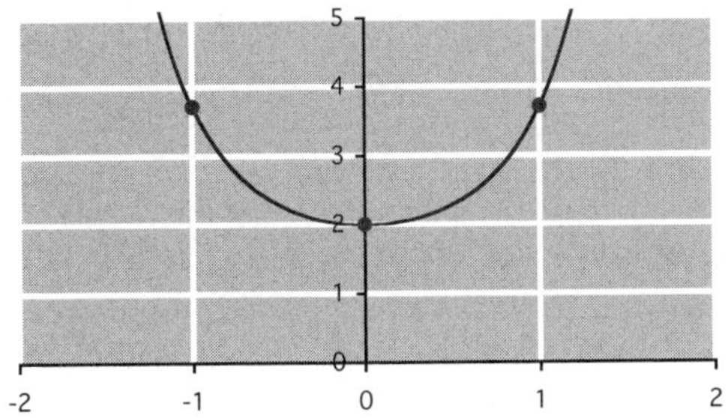

h has an absolute min at $(0, 2)$.

9. Relative max at $x = 1$, point of inflection at $x = -1$.

10. Relative min at $x = 2$, point of inflection at $x = -1$.

11. Relative max at $x = -2$: the derivative goes from positive to negative, so the f must go from increasing to decreasing; relative min at $x = 1$: f goes from decreasing to increasing; point of inflection at $x = -1$: a min of f' gives a point of inflection of f.

12. Relative min at $x = -2.5$ and $x = 1.5$: in both cases f goes from decreasing to increasing; relative max at $x = 3$: f goes from increasing to decreasing; point of inflection at $x = 2$: a max of f' gives a point of inflection of f.

13. One point of inflection, at $x = 0$. Note that f'' is not defined at $x = 0$, but does change from negative to positive there.

14. Points of inflection at $x = 0$,where the graph of f'' crosses the x-axis, and $x = 1$, where f'' is not defined but changes sign.

15. $s = \frac{2}{3t^2} - \frac{1}{t}$

$v = \frac{ds}{dt} = -\frac{4}{3t^3} + \frac{1}{t^2}$

$a = \frac{dv}{dt} = \frac{4}{t^4} - \frac{2}{t^3}$ m/sec^2

(b) At time $t = 1$, acceleration is

$a = \frac{4}{(1)^4} - \frac{2}{(1)^3} = 2$ m/sec^2

16. (a) $s = \frac{4}{t^2} - \frac{3t}{4}$

$v = \frac{ds}{dt} = -\frac{8}{t^3} + \frac{3}{4}$

$a = \frac{dv}{dt} = \frac{24}{t^4}$ m/sec^2

(b) At time $t = 2$, acceleration is

$a = \frac{24}{2^4} = \frac{24}{16} = 1.5$ m/sec^2

17. $f'(x) = 3x^2 - 12$

Stationary Points:

$f'(x) = 0$ when $3(x^2 - 4) = 0$, $x = \pm 2$; Singular Points: None; $f'(x)$ is always defined.

Possible points of inflection:

$f''(x) = 6x$

$f''(x) = 0$ when $x = 0$; $f''(x)$ is always defined.

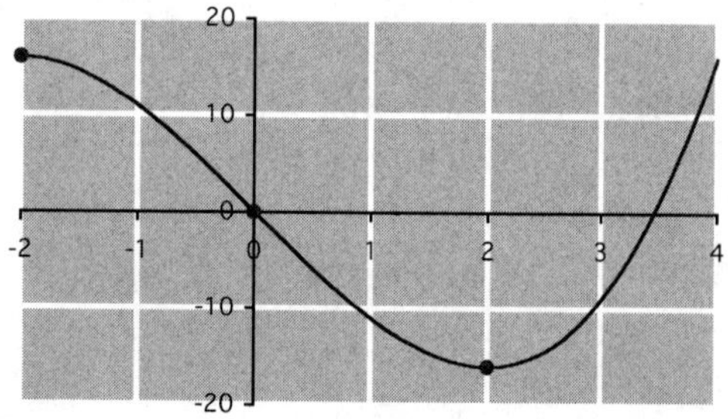

f has a relative max at $(-2, 16)$, an absolute min at $(2, -16)$, and a point of inflection at $(0, 0)$. It has no horizontal or vertical asymptotes.

18. $g(x) = x^4 - 4x$ on $[-1, 1]$

End points: $-1, 1$

Stationary Points:

$g'(x) = 4x^3 - 4$

$= 4(x^3 - 1)$

$= 4(x-1)(x^2+x+1)$

$g'(x) = 0$ when

$4(x-1)(x^2+x+1) = 0$

$x = 1$

Singular points: None

x	$y = x^4 - 4x$
-1	5
1	-3

From the chart we see that g has an absolute maximum at $(-1, 5)$ and an absolute minimum at $(1, -3)$

Possible points of inflection:

$g''(x) = 12x^2 = 0$ when $x = 0$

However, the second derivative is never negative, meaning that the curve is never concave down. Therefore, there are no points of inflection.

Graph:

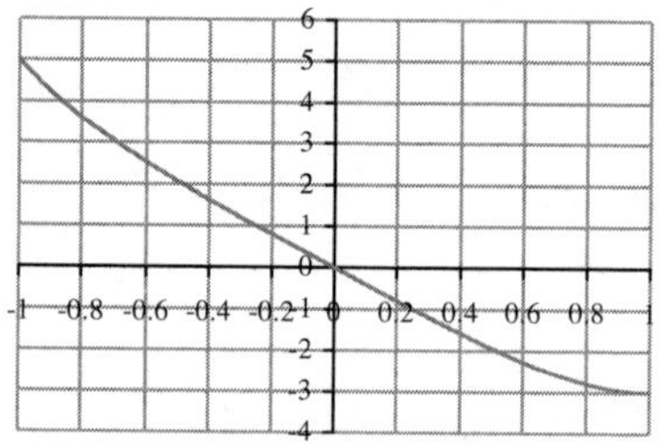

19. The domain of f includes all numbers except 0.

Stationary points:

$f'(x) = \frac{2x \cdot x^3 - (x^2 - 3)(3x^2)}{x^6}$

$= \frac{-x^4 + 9x^2}{x^6} = \frac{-x^2 + 9}{x^4}$

$f'(x) = 0$ when $x = \pm 3$

Singular points: None; $f'(x)$ is defined for all x in the domain of f.

Inflection points:

$$f''(x) = \frac{-2x \cdot x^4 - (-x^2 + 9)(4x^3)}{x^8}$$

$$= \frac{2x^5 - 36x^3}{x^8} = \frac{2(x^2 - 18)}{x^5}$$

$f''(x) = 0$ when $x = \pm\sqrt{18} = \pm 3\sqrt{2}$; $f''(x)$ is defined for all x in the domain of f.

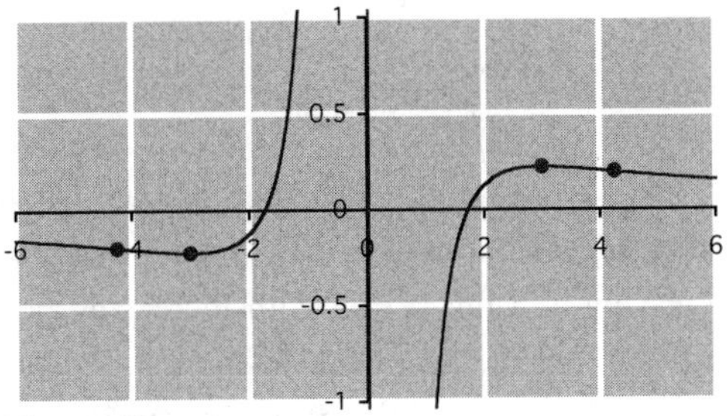

f has a relative min at $(-3, -2/9)$, a relative max at $(3, 2/9)$, and points of inflection at $(-3\sqrt{2}, -5\sqrt{2}/36)$ and $(3\sqrt{2}, 5\sqrt{2}/36)$. f has a vertical asymptote at $x = 0$ and a horizontal asymptote at $y = 0$.

20. $f(x)$ is defined for all x. $f'(x) = \frac{2}{3}(x - 1)^{-1/3} + \frac{2}{3}$; $f'(x) = 0$ when $(x - 1)^{-1/3} = -1$, $x - 1 = (-1)^{-3} = -1$, $x = 0$; $f'(x)$ is not defined when $x = 1$. $f''(x) = -\frac{2}{9}(x - 1)^{-4/3}$; $f''(x)$ is never 0; $f''(x)$ is not defined when $x = 1$.

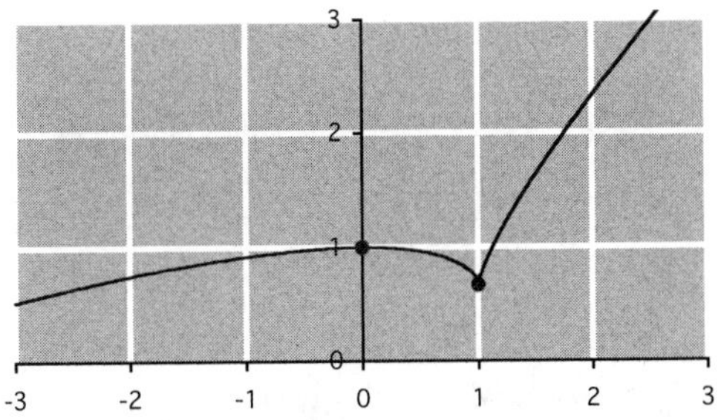

f has a relative max at $(0, 1)$ and a relative min at $(1, 2/3)$. It has no points of inflection and no horizontal or vertical asymptotes.

21. The domain of g is $x \geq 0$.

Stationary points:

$$g'(x) = \sqrt{x} + (x - 3)\frac{1}{2\sqrt{x}} = \frac{3\sqrt{x}}{2} - \frac{3}{2\sqrt{x}}$$

$g'(x) = 0$ when

$\sqrt{x} = \frac{1}{\sqrt{x}}$, $x = 1$;

Singular points: $g'(x)$ is not defined when $x = 0$, so $x = 0$ is a singular point.

Inflection points: $g''(x) = \frac{3}{4\sqrt{x}} + \frac{3}{4x^{3/2}}$;

$g''(x)$ is never 0; $g''(x)$ is not defined when $x = 0$.

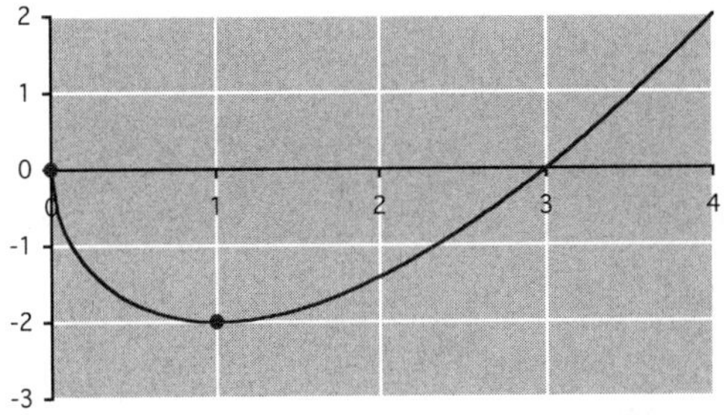

g has a relative max at $(0, 0)$ and an absolute min at $(1, -2)$.

22. The domain of g is $x \geq 0$.

$$g'(x) = \sqrt{x} + (x + 3)\frac{1}{2\sqrt{x}} = \frac{3\sqrt{x}}{2} + \frac{3}{2\sqrt{x}}$$

$g'(x)$ is never 0; $g'(x)$ is not defined when $x = 0$.

$g''(x) = \frac{3}{4\sqrt{x}} - \frac{3}{4x^{3/2}}$; $g''(x) = 0$ when $\sqrt{x} = x^{3/2}$, $\sqrt{x} - x\sqrt{x} = 0$, $x = 1$; $g''(x)$ is not defined when $x = 0$.

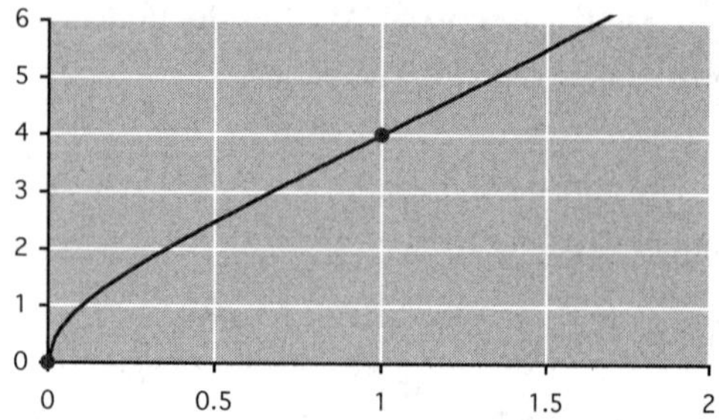

It is difficult to see from the graph, but g has a point of inflection where $x = 1$: we can calculate that $g''(0.5) < 0$ and $g''(2) > 0$. So, g has a relative min at (0, 0) and a point of inflection at (1, 4).

23. Objective: Maximize $R = pq = p(-p^2 + 33p + 9)$

$= -p^3 + 33p^2 + 9p$

End-points: 18, 28

Stationary points:

$$\frac{dR}{dp} = -3p^2 + 66p + 9 = 0$$

$$\text{when } p = \frac{-66 \pm \sqrt{66^2 - 4(-3)(9)}}{2(-3)}$$

$p \approx -0.14$ or 22.14

We reject the negative value and obtain the following table

p	18	22.14	28
$R = -p^3 + 33p^2 + 9p$	5022	5522.61	4172

So the maximum revenue of \$5522.61 occurs when p = \$22.14 per book

24. Profit $P = R - C = pq - (9q + 100) = p(-p^2 + 33p + 9) - 9(-p^2 + 33p + 9) - 100 = -p^3 + 42p^2 - 288p - 181$

25. $P' = -3p^2 + 84p - 288$

$= -3(p^2 - 28p + 96)$

$= -3(p - 24)(p - 4)$

$P' = 0$ when $p = 4$ or 24. To see which, if either, gives us the maximum profit, we test points on either side:

p	0	4	24	30
$P(p)$	−181	−725	3275	1979

P decreases to a low at $p = 4$ then increases to a maximum at $p = 24$, after which it decreases again. So, the company should charge \$24 per copy.

26. $P(24) = \$3275$

27. For maximum revenue, the company should charge \$22.14 per copy. At this price, the cost is decreasing at a linear rate with increasing price, while the revenue is not decreasing (its derivative is zero). Thus, the profit is increasing with increasing price, suggesting that the maximum profit will occur at a higher price. This is, in fact, what we just found.

28. Start by labeling the edges of the box:

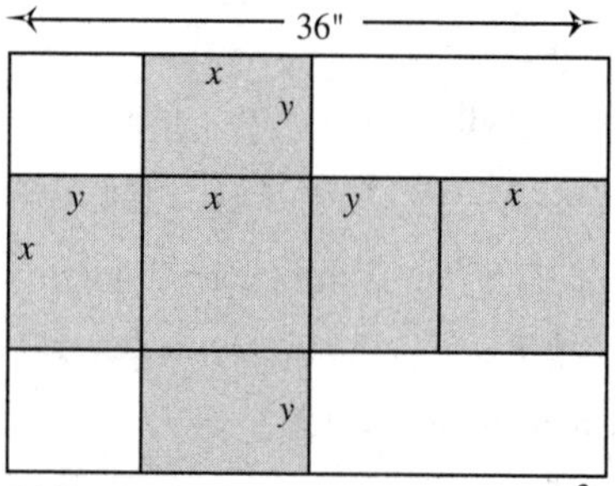

Objective: Maximize $V = xxy = x^2y$

subject to:

$2x + 2y = 36$

or $y = 18 - x$

Substitute into the objective

$V = x^2(18-x) = 18x^2 - x^3$

$$\frac{dV}{dx} = 36x - 3x^2 = 3x(12 - x)$$

Stationary point occurs at $x = 0, 12$

$x = 12$ in gives the maximum volume of

$V = 18(12)^2 - (12)^3 = 864$ cubic inches

The height of the box is

$y = 18 - x = 18 - 12 = 6$ in

29. $E = -(-2p + 33)\dfrac{p}{-p^2 + 33p + 9}$

$= \dfrac{2p^2 - 33p}{-p^2 + 33p + 9}$

30. $E(20) \approx 0.52$ and $E(25) \approx 2.03$. When the price is \$20, demand is dropping at a rate of 0.52% per 1% increase in the price; when the price is \$25, demand is dropping at a rate of 2.03% per 1% increase in the price.

31. $E = 1$ when $2p^2 - 33p = -p^2 + 33p + 9$

$3p^2 - 66p - 9 = 0$

$p^2 - 22p - 3 = 0$

so $p \approx \$22.14$ per book (using the quadratic formula; the other solution is negative so we reject it).

32. $E = -\dfrac{dq}{dp} \cdot \dfrac{p}{q}$

$= -(-2p+1)1000e^{-p^2+p} \cdot \dfrac{p}{1000e^{-p^2+p}}$

$= -(-2p+1)p = 2p^2 - p$

33. $E(2) = 2(2)^2 - 2 = 6$

Interpretation: The demand is dropping at a rate of 6% per 1% increase in the price.

34. For maximum revenue, $E = 1$:

$2p^2 - p = 1$

$2p^2 - p - 1 = 0$

$(2p+1)(p-1) = 0$

$p = \$1.00$ (reject the negative solution)

For the revenue,

$R = pq = p1000e^{-p^2+p}$

$R(1) = 1000e^{-1+1} = \$1000$ per month

35. Weekly sales are growing fastest when the rate of change s' is a maximum. From the graph of s', we see that this occurred at about week 5 (see also the graph of s'' which becomes zero at that point).

36. This point (a maximum in the graph of s') corresponds to a point of inflection on the graph of s. On the graph of s', that point is given by the maximum, and by the t-intercept in the graph of s''.

37. The graph appears to level of around s = 10,500; If weekly sales continue as predicted by the model, they will level off at around 10,500 books per week in the long-term.

38. The graph of s' appears to level off at $s' = 0$. If weekly sales continue as predicted by the model, the rate of change of sales approaches zero in the long-term.

39. Let h be the distance between Marjory Duffin and John O'Hagan. Let y be Marjory Duffin's distance from the corner, and let x be John O'Hagan's distance. Then

$x^2 + y^2 = h^2$

$2x\dfrac{dx}{dt} + 2y\dfrac{dy}{dt} = 2h\dfrac{dh}{dt}$

(a) $2(2)(5) + 2(2)(5) = 2(2\sqrt{2})\dfrac{dh}{dt}$

$40 = 4\sqrt{2}\dfrac{dh}{dt}$

$\dfrac{dh}{dt} = \dfrac{10}{\sqrt{2}}$ ft/sec

(b) $2(1)(5) + 2(1)(5) = 2\sqrt{2}\dfrac{dh}{dt}$

$20 = 2\sqrt{2}\dfrac{dh}{dt}$

$\dfrac{dh}{dt} = \dfrac{10}{\sqrt{2}}$ ft/sec

(c) $2(h)(5) + 2(h)(5) = 2h\sqrt{2}\frac{dh}{dt}$

$20h = 2h\sqrt{2}\frac{dh}{dt}$

$\frac{dh}{dt} = \frac{10}{\sqrt{2}}$ ft/sec

(d) Since the answer to part (c) is independent of h, it also must hold as $h \to 0$, giving the same answer as parts (a) through (c)

40. The combined area is $A = hw + \frac{1}{4}\pi h^2$. Given that $w = 1$, $\frac{dw}{dt} = 0.5$, and $h = 3$, we want to find $\frac{dh}{dt}$. Using the fact that A is constant, we differentiate: $0 = \frac{dh}{dt} w + h \frac{dw}{dt} + \frac{1}{2}\pi h \frac{dh}{dt}$. Substitute: $0 = \frac{dh}{dt} + 3(0.5) + \frac{1}{2}\pi(3)\frac{dh}{dt} = 1.5 + (1 + 3\pi/2)\frac{dh}{dt}$, $\frac{dh}{dt} = -1.5/(1 + 3\pi/2) \approx -0.263$ inches per second.

Chapter 6
6.1

1. $\int x^n\,dx = \dfrac{x^{n+1}}{n+1} + C;\ n = 5$

$\int x^5\,dx = \dfrac{x^6}{6} + C$

3. $\int 6\,dx = 6\int 1\,dx = 6x + C$

5. $\int x^n\,dx = \dfrac{x^{n+1}}{n+1} + C;\ n = 1$

$\int x\,dx = \dfrac{x^2}{2} + C$

7. $\int (x^2 - x)\,dx = \int x^2\,dx - \int x\,dx$

$= \dfrac{x^3}{3} - \dfrac{x^2}{2} + C$

9. $\int (1 + x)\,dx = \int 1\,dx + \int x\,dx$

$= x + \dfrac{x^2}{2} + C$

11. $\int x^n\,dx = \dfrac{x^{n+1}}{n+1} + C;\ n = -5$

$\int x^{-5}\,dx = \dfrac{x^{-4}}{(-4)} + C = -\dfrac{x^{-4}}{4} + C$

13. $\int (x^{2.3} + x^{-1.3})\,dx$

$= \int x^{2.3}\,dx + \int x^{-1.3}\,dx$

$= \dfrac{x^{3.3}}{3.3} - \dfrac{x^{-0.3}}{0.3} + C$

15. $\int (u^2 - 1/u)\,du$

$= \int u^2\,du - \int u^{-1}\,du$

$= \dfrac{u^3}{3} - \ln|u| + C$

17. $\int \sqrt{x}\,dx = \int x^{1/2}\,dx$

$= \dfrac{x^{3/2}}{3/2} + C = \dfrac{2x^{3/2}}{3} + C$

19. $\int (3x^4 - 2x^{-2} + x^{-5} + 4)\,dx$

$= 3\int x^4\,dx - 2\int x^{-2}\,dx + \int x^{-5}\,dx + \int 4\,dx$

$= \dfrac{3x^5}{5} + 2x^{-1} - \dfrac{x^{-4}}{4} + 4x + C$

21. $\int \left(\dfrac{2}{u} + \dfrac{u}{4}\right) du$

$= \int (2u^{-1} + (1/4)u)\,du$

$= 2\ln|u| + \dfrac{1}{4}\dfrac{u^2}{2} + C$

$= 2\ln|u| + u^2/8 + C$

23. $\int \left(\dfrac{1}{x} + \dfrac{2}{x^2} - \dfrac{1}{x^3}\right) dx$

$= \int (x^{-1} + 2x^{-2} - x^{-3})\,dx$

$= \ln|x| + 2\frac{x^{-1}}{-1} - \frac{x^{-2}}{-2} + C$

$= \ln|x| - \frac{2}{x} + \frac{1}{2x^2} + C$

25. $\int (3x^{0.1} - x^{4.3} - 4.1)\, dx$

$= \frac{3x^{1.1}}{1.1} - \frac{x^{5.3}}{5.3} - 4.1x + C$

27. $\int \left(\frac{3}{x^{0.1}} - \frac{4}{x^{1.1}}\right) dx$

$= \int (3x^{-0.1} - 4x^{-1.1})\, dx$

$= 3\frac{x^{0.9}}{0.9} - 4\frac{x^{-0.1}}{-0.1} + C$

$= \frac{x^{0.9}}{0.3} + \frac{40}{x^{0.1}} + C$

29. $\int \left(5.1t - \frac{1.2}{t} + \frac{3}{t^{1.2}}\right) dt$

$= \int (5.1t - 1.2t^{-1} + 3t^{-1.2})\, dt$

$= \frac{5.1t^2}{2} - 1.2 \ln|t| + 3\frac{t^{-0.2}}{-0.2} + C$

$= 2.55t^2 - 1.2 \ln|t| - \frac{15}{t^{0.2}} + C$

31. $\int (2e^x + 5/x + 1/4)\, dx$

$= 2e^x + 5\ln|x| + \frac{x}{4} + C$

33. $\int \left(\frac{6.1}{x^{0.5}} + \frac{x^{0.5}}{6} - e^x\right) dx$

$= \int \left(6.1x^{-0.5} + \frac{x^{0.5}}{6} - e^x\right) dx$

$= 6.1\frac{x^{0.5}}{0.5} + \frac{x^{1.5}}{6 \cdot 1.5} - e^x + C$

$= 12.2x^{0.5} + \frac{x^{1.5}}{9} - e^x + C$

35. $\int (2^x - 3^x)\, dx$

$= \frac{2^x}{\ln 2} - \frac{3^x}{\ln 3} + C$

37. $\int 100(1.1^x)\, dx$

$= \frac{100(1.1^x)}{\ln(1.1)} + C$

39. $\int \frac{x+2}{x^3}\, dx = \int \left(\frac{x}{x^3} + \frac{2}{x^3}\right) dx$

$= \int \left(\frac{1}{x^2} + \frac{2}{x^3}\right) dx$

$= \int (x^{-2} + 2x^{-3})\, dx$

$= -x^{-1} - x^{-2} = -\frac{1}{x} - \frac{1}{x^2} + C$

41. $f'(x) = x$, so

$f(x) = \int x\, dx = \frac{x^2}{2} + C.$

$f(0) = 1$, so

$\frac{0^2}{2} + C = 1$

$C = 1$

So, $f(x) = \frac{x^2}{2} + 1.$

43. $f'(x) = e^x - 1$, so

$f(x) = \int (e^x - 1)\, dx = e^x - x + C.$

$f(0) = 0$, so

$e^0 - 0 + C = 0$

$C = -1$

So, $f(x) = e^x - x - 1$.

45. $C'(x) = 5 - \frac{x}{10{,}000}$

$C(x) = \int \left(5 - \frac{x}{10{,}000}\right) dx$

$= 5x - \frac{x^2}{20{,}000} + K$

$C(0) = 20{,}000$, so

$0 - 0 + K = 20{,}000.$

$K = 20{,}000$

$C(x) = 5x - \frac{x^2}{20{,}000} + 20{,}000.$

47. $C'(x) = 5 + 2x + \frac{1}{x}$

$C(x) = \int \left(5 + 2x + \frac{1}{x}\right) dx$

$= 5x + x^2 + \ln x + K$

(Note that $x > 0$,so there is no need to write $|x|$.)

$C(1) = 1000$, so $5 + 1 + \ln 1 + K = 1000$, $K = 994$. $C(x) = 5x + x^2 + \ln x + 994$.

49. $v(t) = t^2 + 1$

(a) $s(t) = \int v(t)\, dt = \int (t^2 + 1)\, dt = \frac{t^3}{3} + t + C$

(b) $0 + 0 + C = 1$ so

$s = \frac{t^3}{3} + t + 1$

51. $a(t) = -32$

$v(t) = \int a(t)\, dt = \int (-32)\, dt$

$= -32t + C$

$v(0) = 0$ is given, so

$0 = 0 + C$

$C = 0$

and so $v(t) = -32t$.

After 10 seconds,

$v(10) = -32(0) = -320$

so the stone is traveling 320 ft/s downward.

53. (a) $v(t) = \int a(t)\, dt = \int (-32)\, dt = -32t + C$

At time $t = 0$ $v = 16$ ft/sec, so

$16 = -32(0) + C$

$C = 16$

giving

$v(t) = -32t + 16$

(b) $s(t) = \int v(t)\, dt = \int (-32t + 16)\, dt$

$= -16t^2 + 16t + C$

At time $t = 0$ $s = 185$ ft, so

$185 = -16(0)^2 + 16(0) + C$

$C = 185$

giving

$s(t) = -16t^2 + 16t + 185$

It reaches its zenith when $v(t) = 0$

$-32t + 16 = 0$

$t = 16/32 = 0.5$ sec

Its height at that moment is

$s(0.5) = -16(0.5)^2 + 16(0.5) + 185 = 189$ feet,

4 feet above the top of the tower.

55. $a(t) = -32$

$v(t) = \int a(t)\, dt = \int (-32)\, dt = -32t + C.$

$v(0) = v_0$ is given, so

$v_0 = -32(0) + C$

$C = v_0$

Thus, $v(t) = -32t + v_0$.

The projectile has zero velocity when

$v(t) = 0$

$0 = -32t + v_0$

$t = v_0/32.$

This is when it reaches its highest point.

57. By Exercise 56, the ball reaches a maximum height of $v_0^2/64$ feet. Thus,

$v_0^2/64 = 20$

$v_0 = (1280)^{1/2} \approx 35.78$ ft/s

59. (a) By Exercise 56, the chalk reaches a maximum height of $v_0^2/64$ feet. Thus,

$v_0^2/64 = 100$

$v_0 = (6400)^{1/2} = 80$ ft/s

(b) As in Exercise 56,

$v(t) = -32t + v_0$

and so

$$s(t) = \int v(t)\,dt = \int (-32t + v_0)\,dt$$

$= -16t^2 + v_0 t + C.$

If we take the starting height as 0,

$0 + 0 + C = 0,$

so $C = 0$ and

$s(t) = -16t^2 + v_0 t.$

$= -16t^2 + 100t.$

The chalk strikes the ceiling when

$s(t) = 100, -16t^2 + 100t = 100$

$16t^2 - 100t + 100 = 0$

$4(4t - 5)(t - 5) = 0$

$t = 1.25$ or 5.

We take the first solution, which is the first time it strikes the ceiling. Now,

$v(t) = -32t + v_0 = -32t + 100,$

so the velocity when it strikes the ceiling is

$v(1,25) = 60$ ft/s.

(c) Start with $s(0) = 100$ and $v(0) = -60$. As before,

$v(t) = -32t + v_0 = -32t - 60.$

Now,

$$s(t) = \int v(t)\,dt = \int (-32t - 60)\,dt$$

$= -16t^2 - 60t + C. 0 - 0 + C = 100,$

so $s(t) = -16t^2 - 60t + 100.$

Now we find when $s(t) = 0$:

$-16t^2 - 60t + 100 = 0$

$-4(4t - 5)(t + 5) = 0$

$t = 1.25$ or -5

We take the positive solution and say that it takes 1.25 seconds to hit the ground.

61. Let v_0 be the speed at which Prof. Strong throws and let w_0 be the speed at which Prof. Weak throws. We have

$v_0^2/64 = 2w_0^2/64$

so

$v_0 = w_0\sqrt{2}.$

Thus, Prof. Strong throws $\sqrt{2} \approx 1.414$ times as fast as Prof. Weak.

63. $I(t) = \int 1000\,dt = 1000t + C$

$30{,}000 = I(0) = C$

So, $I(t) = 30{,}000 + 1000t$

$I(13) = \$43{,}000$

65. (a) $m = (100 - 65)/10 = 3.5$, so

$H'(t) = 3.5t + 65$ billion dollars per year.

(b) $H(t) = \int (3.5t + 65)\,dt$

$= \frac{3.5}{2}t^2 + 65t + C$

$H(0) = 700 = C$

Thus, $H(t) = 1.75t^2 + 65t + 700$ billion dollars

67. $S(t) = \int (17t^2 + 100t + 2300)\,dt$

$= \frac{17}{3}t^3 + 50t^2 + 2300t + C$

$S(3) = 0 = 7503 + C$

$C = -7503$

So, $S(t) = \frac{17}{3}t^3 + 50t^2 + 2300t - 7503.$

$S(13) = \frac{17}{3}(13)^3 + 50t^{(13)} + 2300(13) - 7503$

$\approx 43{,}000$ million gallons

69. They differ by a constant,

$G(x) - F(x) = \text{Constant}.$

71. Antiderivative; marginal

73. $\int f(x)\,dx$ represents the total cost of manufacturing x items. The units of $\int f(x)\,dx$ are the product of the units of $f(x)$ and the units of x.

75. $\int [f(x) + g(x)]\,dx$ is, by definition, an antiderivative of $f(x) + g(x)$. Let $F(x)$ be an antiderivative of $f(x)$ and let $G(x)$ be an antiderivative of $g(x)$. Then, because the derivative of $F(x) + G(x)$ is $f(x) + g(x)$ (by the rule for sums of derivatives), $F(x) + G(x)$ is an antiderivative of $f(x) + g(x)$. In symbols,

$$\int [f(x) + g(x)]\,dx = F(x) + G(x) + C$$

$$= \int f(x)\,dx + \int g(x)\,dx$$

the sum of the indefinite integrals.

77. $\int x \cdot 1\,dx = \int x\,dx = \frac{x^2}{2} + C,$

whereas

$$\int x\,dx \cdot \int 1\,dx = \left(\frac{x^2}{2} + D\right) \cdot (x + E),$$

which is not the same as $\frac{x^2}{2} + C$, no matter what values we choose for the constants C, D and E.

79. derivative; indefinite integral; indefinite integral; derivative

6.2

1. $u = 3x-5$

$du/dx = 3$

$dx = \frac{1}{3}du$

$$\int (3x-5)^3\,dx = \int u^3\,\tfrac{1}{3}du$$

$$= \frac{1}{3}\frac{u^4}{4} + C = u^4/12 + C$$

$$= (3x-5)^4/12 + C$$

3. $\int (3x-5)^3\,dx = \dfrac{(3x-5)^4}{(3)(4)} + C$

$$= (3x-5)^4/12 + C$$

5. $u = -x,\ du/dx = -1,\ dx = -du;\ \int e^{-x}\,dx = -$

$$\int e^u\,du = -e^u + C = -e^{-x} + C$$

7. $\int e^{-x}\,dx = \dfrac{1}{-1}e^{-x} + C = -e^{-x} + C$

9. $u = (x+1)^2,\ du/dx = 2(x+1),\ dx = \dfrac{du}{2(x+1)}$

$$\int (x+1)e^{(x+1)^2}\,dx = \int (x+1)e^u\,\frac{du}{2(x+1)}$$

$$= \int \tfrac{1}{2}e^u\,du = \tfrac{1}{2}e^u + C = \tfrac{1}{2}e^{(x+1)^2} + C$$

11. $u = 3x + 1,\ du = 3dx,\ dx = \frac{1}{3}\,du$

$$\int (3x + 1)^5\,dx = \int u^5\,\frac{1}{3}\,du = \frac{u^6}{18} + C$$

$$= \frac{(3x + 1)^6}{18} + C$$

13. $u = -2x + 2,\ du = -2\,dx,\ dx = -\frac{1}{2}\,du$

$$\int (-2x + 2)^{-2}\,dx = -\int u^{-2}\,\frac{1}{2}\,du = \frac{u^{-1}}{2} + C$$

$$= \frac{(-2x + 2)^{-1}}{2} + C$$

15. $u = 3x - 4,\ du = 3\,dx,\ dx = \frac{1}{3}\,du$

$$\int 7.2\sqrt{3x-4}\,dx = \int 7.2\sqrt{u}\,\frac{1}{3}\,du$$

$$= 2.4\int u^{1/2}\,du = 2.4\,\frac{u^{3/2}}{3/2} + C$$

$$= 1.6(3x - 4)^{3/2} + C$$

17. $u = 0.6x + 2,\ du = 0.6\,dx,\ dx = \frac{1}{0.6}\,du$

$$\int 1.2e^{(0.6x+2)}\,dx = \int 1.2e^u\,\frac{1}{0.6}\,du = 2e^u + C$$

$$= 2e^{(0.6x+2)} + C$$

19. $u = 3x^2 + 3,\ du = 6x\,dx,\ dx = \dfrac{1}{6x}\,du$

$$\int x(3x^2 + 3)^3\,dx = \int xu^3\,\frac{1}{6x}\,du$$

$$= \frac{1}{6}\int u^3\,du = \frac{u^4}{12} + C = \frac{(3x^2 + 3)^4}{24} + C$$

21. $u = x^2 + 1,\ du = 2x\,dx,\ dx = \dfrac{1}{2x}\,du$

$$\int x(x^2 + 1)^{1.3}\,dx = \int xu^{1.3}\,\frac{1}{2x}\,du$$

$$= \frac{1}{2}\int u^{1.3}\,du = \frac{u^{2.3}}{4.6} + C$$

$$= \frac{(x^2 + 1)^{2.3}}{4.6} + C$$

23. $u = 3.1x - 2,\ du = 3.1\ dx,\ dx = \frac{1}{3.1}\ du$

$$\int (1 + 9.3e^{3.1x-2})\ dx = \int dx + \int 9.3e^{3.1x-2}\ dx$$

$$= x + \int 9.3e^{u} \frac{1}{3.1}\ du = x + 3e^{u} + C$$

$$= x + 3e^{3.1x-2} + C$$

25. $u = 3x^2 - 1,\ du = 6x\ dx,\ dx = \frac{1}{6x}\ du$

$$\int 2x\sqrt{3x^2 - 1}\ dx = \int 2x\sqrt{u}\frac{1}{6x}\ du$$

$$= \frac{1}{3}\int u^{1/2}\ du = \frac{u^{3/2}}{9/2} + C$$

$$= \frac{2}{9}(3x^2 - 1)^{3/2} + C$$

27. $u = -x^2 + 1,\ du = -2x\ dx,\ dx = -\frac{1}{2x}\ du$

$$\int xe^{-x^2+1}\ dx = -\int xe^{u}\frac{1}{2x}\ du = -\frac{1}{2}\int e^{u}\ du$$

$$= -\frac{1}{2}e^{u} + C = -\frac{1}{2}e^{-x^2+1} + C$$

29. $u = -(x^2 + 2x),\ du = -(2x + 2)\ dx,$

$dx = -\frac{1}{2(x + 1)}\ du$

$$\int (x + 1)e^{-(x^2+2x)}\ dx$$

$$= -\int (x + 1)e^{u}\frac{1}{2(x + 1)}\ du$$

$$= -\frac{1}{2}\int e^{u}\ du = -\frac{1}{2}e^{u} + C$$

$$= -\frac{1}{2}e^{-(x^2+2x)} + C$$

31. $u = x^2 + x + 1,\ du = (2x + 1)\ dx,$

$dx = \frac{1}{2x + 1}\ du$

$$\int \frac{-2x - 1}{(x^2 + x + 1)^3}\ dx = \int \frac{-2x - 1}{u^3}\cdot\frac{1}{2x + 1}\ du$$

$$= -\int u^{-3}\ du = -\frac{u^{-2}}{-2} + C$$

$$= \frac{(x^2 + x + 1)^{-2}}{2} + C$$

33. $u = 2x^3 + x^6 - 5,\ du = (6x^2 + 6x^5)\ dx$

$dx = \frac{1}{6(x^2 + x^5)}\ du$

$$\int \frac{x^2 + x^5}{\sqrt{2x^3 + x^6 - 5}}\ dx$$

$$= \int \frac{x^2 + x^5}{\sqrt{u}}\cdot\frac{1}{6(x^2 + x^5)}\ du = \frac{1}{6}\int u^{-1/2}\ du$$

$$= \frac{1}{6}\cdot\frac{u^{1/2}}{1/2} + C = \frac{1}{3}(2x^3 + x^6 - 5)^{1/2} + C$$

$$= \frac{1}{3}\sqrt{2x^3 + x^6 - 5} + C$$

35., $du = dx$

$$\int x(x - 2)^5\ dx = \int xu^5\ dx$$

To remove the remaining x, solve for x in terms of u in the expression for u:

$u = x - 2$, so $x = u + 2$

The above integral is then

$$\int xu^5\ dx = \int (u + 2)u^5\ du$$

$$= \int (u^6 + 2u^5)\ du = \frac{1}{7}u^7 +$$

$$\frac{1}{3}u^6 + C = \frac{1}{7}(x - 2)^7 + \frac{1}{3}(x - 2)^6 + C$$

37. $u = x + 1,\ du = dx$

$$\int 2x\sqrt{x + 1}\ dx = \int 2x\sqrt{u}\ dx$$

To remove the remaining x, solve for x in terms of u in the expression for u:

$u = x + 1$, so $x = u - 1$

The above integral is then

$$\int 2x\sqrt{u}\,dx = \int 2(u - 1)\sqrt{u}\,du$$

$$= 2\int (u^{3/2} - u^{1/2})\,du$$

$$= 2\frac{u^{5/2}}{5/2} - 2\frac{u^{3/2}}{3/2} + C$$

$$= \frac{4}{5}(x + 1)^{5/2} - \frac{4}{3}(x + 1)^{3/2} + C$$

39. $u = 1 - e^{-0.05x}$, $du = 0.05e^{-0.05x}\,dx$

$dx = \frac{1}{0.05e^{-0.05x}}\,du$

$$\int \frac{e^{-0.05x}}{1 - e^{-0.05x}}\,dx$$

$$= \int \frac{e^{-0.05x}}{u} \cdot \frac{1}{0.05e^{-0.05x}}\,du$$

$$= 20\int \frac{1}{u}\,du = 20\ln|u| + C$$

$$= 20\ln|1 - e^{-0.05x}| + C$$

41. $u = -\frac{1}{x}$, $du = \frac{1}{x^2}\,dx$, $dx = x^2\,du$

$$\int \frac{3e^{-1/x}}{x^2}\,dx = \int \frac{3e^u}{x^2}x^2\,du = \int 3e^u\,du$$

$$= 3e^u + C = 3e^{-1/x} + C$$

43. $\int \frac{e^x + e^{-x}}{2}\,dx = \int \frac{e^x}{2} + \frac{e^{-x}}{2}\,dx$

$$= \frac{e^x}{2} + \int \frac{e^{-x}}{2}\,dx$$

let $u = -x$, $du = -dx$, $dx = -du$

$$\frac{e^x}{2} + \int \frac{e^{-x}}{2}\,dx = \frac{e^x}{2} - \int \frac{e^u}{2}\,du$$

$$= \frac{e^x}{2} - \frac{e^u}{2} + C = \frac{e^x - e^{-x}}{2} + C$$

45. $u = e^x + e^{-x}$, $du = (e^x - e^{-x})\,dx$

$dx = \frac{1}{e^x - e^{-x}}\,du$

$$\int \frac{e^x - e^{-x}}{e^x + e^{-x}}\,dx$$

$$= \int \frac{e^x - e^{-x}}{u} \cdot \frac{1}{e^x - e^{-x}}\,du$$

$$= \int \frac{1}{u}\,du = \ln|u| + C = \ln|e^x + e^{-x}| + C$$

47. $\int [(2x - 1)e^{2x^2-2x} + xe^{x^2}]\,dx$

$$= \int (2x - 1)e^{2x^2-2x}\,dx + \int xe^{x^2}\,dx\,.$$

For the first integral, let

$u = 2x^2 - 2x$, $du = (4x - 2)\,dx$,

$dx = \frac{1}{2(2x - 1)}\,du$

$$\int (2x - 1)e^{2x^2-2x}\,dx$$

$$= \int (2x - 1)e^u \frac{1}{2(2x - 1)}\,du$$

$$= \frac{1}{2}\int e^u\,du = \frac{1}{2}e^u + C = \frac{1}{2}e^{2x^2-2x} + C.$$

For the second integral, let

$u = x^2$, $du = 2x\,dx$, $dx = \frac{1}{2x}\,du$

$$\int xe^{x^2}\,dx = \int xe^u \frac{1}{2x}\,du = \frac{1}{2}\int e^u\,du$$

$$= \frac{1}{2}e^u + C = \frac{1}{2}e^{x^2} + C.$$

So, $\int [(2x-1)e^{2x^2-2x} + xe^{x^2}]\,dx$

$= \frac{1}{2}e^{2x^2-2x} + \frac{1}{2}e^{x^2} + C.$

49. Let $u = ax + b,\ du = a\,dx,\ dx = \frac{1}{a}\,du$

$\int (ax+b)^n\,dx = \int u^n \frac{1}{a}\,du$

$= \frac{u^{n+1}}{a(n+1)} + C = \frac{(ax+b)^{n+1}}{a(n+1)} + C$

(if $n \neq -1$)

51. Let $u = ax + b,\ du = a\,dx,\ dx = \frac{1}{a}\,du$

$\int \sqrt{ax+b}\,dx = \int \sqrt{u}\,\frac{1}{a}\,du = \frac{u^{3/2}}{(3/2)a} + C$

$= \frac{2}{3a}(ax+b)^{3/2} + C$

53. $\int e^{ax+b}\,dx = \frac{1}{a}e^{ax+b} + C$

$\int e^{-x}\,dx = \frac{1}{-1}e^{-x} + C = -e^{-x} + C$

55. $\int e^{ax+b}\,dx = \frac{1}{a}e^{ax+b} + C$

$\int e^{2x-1}\,dx = \frac{1}{2}e^{2x-1} + C$

57. $\int (ax+b)^n\,dx = \frac{(ax+b)^{n+1}}{a(n+1)} + C$

$\int (2x+4)^2\,dx = \frac{(2x+4)^3}{2(3)} + C = \frac{(2x+4)^3}{6} + C$

59. $\int (ax+b)^{-1}\,dx = \frac{1}{a}\ln\_ax+b\_ + C$

$\int \frac{1}{5x-1}\,dx = \int (5x-1)^{-1}\,dx$

$= \frac{1}{5}\ln\_5x-1\_ + C$

61. $\int (ax+b)^n\,dx = \frac{(ax+b)^{n+1}}{a(n+1)} + C$

$\int (1.5x)^3\,dx = \frac{(1.5x)^4}{1.5(4)} = \frac{1}{6}(1.5x)^4 + C$

63. $\int c^{ax+b}\,dx = \frac{1}{a\ln c}c^{ax+b} + C$

. $\int 1.5^{3x}\,dx = \frac{1.5^{3x}}{3\ln(1.5)} + C$

65. $\int c^{ax+b}\,dx = \frac{1}{a\ln c}c^{ax+b} + C$

$\int (2^{3x+4} + 2^{-3x+4})\,dx = \frac{2^{3x+4}}{3\ln 2} + \frac{2^{-3x+4}}{-3\ln 2}$

$= \frac{2^{3x+4} - 2^{-3x+4}}{3\ln 2} + C$

67. $f'(x) = x(x^2+1)^3$

So, $f(x) = \int x(x^2+1)^3\,dx$

$u = x^2 + 1,\ du = 2x\,dx,\ dx = \frac{1}{2x}\,du$

$\int x(x^2+1)^3\,dx = \int xu^3 \frac{1}{2x}\,du = \frac{1}{2}\int u^3\,du$

$= \frac{1}{8}u^4 + C = \frac{1}{8}(x^2+1)^4 + C.$

$f(0) = 0$, so

$\frac{1}{8}(0+1)^4 + C = 0$

$C = -\frac{1}{8}$.

and so $f(x) = \frac{1}{8}(x^2 + 1)^4 - \frac{1}{8}$

69. $f'(x) = xe^{x^2-1}$

So, $f(x) = \int xe^{x^2-1}\,dx$

$$u = x^2 - 1,\ du = 2x\,dx,\ dx = \frac{1}{2x}\,du$$

$$\int xe^{x^2-1}\,dx = \int xe^u \frac{1}{2x}\,du = \frac{1}{2}\int e^u\,du$$

$$= \frac{1}{2}e^u + C = \frac{1}{2}e^{x^2-1} + C.$$

$f(1) = 1/2$, so

$\frac{1}{2}e^{1-1} + C = 1/2$

$C = 0$

and so $f(x) = \frac{1}{2}e^{x^2-1}$

71. $C'(x) = 5 + 1/(x+1)^2$

So, $C(x) = \int [5 + (x + 1)^{-2}]\,dx$

$$= \int 5\ dx + = \int (x + 1)^{-2}\,dx$$

$$= 5x - (x + 1)^{-1} + K.$$

(We used the shortcut formula

$$\int (ax+b)^n\,dx = \frac{(ax+b)^{n+1}}{a(n+1)} + K$$

to do the second integal)

$C(1) = 1000$, so

$5 - 2^{-1} + K = 1000$

$K = 995.5$.

Thus, $C(x) = 5x - \frac{1}{(x + 1)} + 995.5$

73. Total number of articles is

$$N(t) = \int \frac{7e^{0.2t}}{5 + e^{0.2t}}\,dt$$

$$u = 5 + e^{0.2t},\ \frac{du}{dt} = 0.2e^{0.2t},\ dt = \frac{du}{0.2e^{0.2t}}$$

$$N(t) = \int \frac{7e^{0.2t}}{5 + e^{0.2t}}\,dt = \int \frac{7e^{0.2t}}{u}\cdot\frac{du}{0.2e^{0.2t}}$$

$$= \int 35.\frac{du}{u} = 35\ \ln|u| + C$$

$$= 35\ \ln(5 + e^{0.2t}) + C$$

At time $t = 0$, $N = 0$, so

$0 = 35\ \ln(5 + e^0) + C = 35\ln 6 + C$

$C = -35\ln 6 \approx -63$

so $N(t) = 35\ \ln(5 + e^{0.2t}) - 63$

(b) Since 2003 is represented by $t = 20$, we get

$N(10) = 35\ \ln(5 + e^{0.2(20)}) - 63 \approx 80$ thousand articles

75. (a) $s = \int v(t)\,dt = \int [t(t^2 + 1)^4 + t]\,dt$

$$= \int t(t^2 + 1)^4\,dt + \frac{1}{2}t^2$$

For the integral on the left, take

$$u = t^2 + 1,\ du = 2t\,dt,\ dt = \frac{1}{2t}\,du$$

$$\int t(t^2 + 1)^4\,dt = \int tu^4 \frac{1}{2t}\,du$$

$$= \frac{1}{2}\int u^4\,du = \frac{1}{10}u^5 + C$$

$$= \frac{1}{10}(t^2 + 1)^5 + C.$$

So, $s = \frac{1}{10}(t^2 + 1)^5 + \frac{1}{2}t^2 + C$

(b) $s(0) = 1$ gives

$\frac{1}{10}(0 + 1)^5 + 0 + C = 1$

$C = \frac{9}{10}$

So, $s = \frac{1}{10}(t^2 + 1)^5 + \frac{1}{2}t^2 + \frac{9}{10}$

77. $S(t)$

$= \int (17(t - 1990)^2 + 100(t - 1990) + 2300)\, dt$

$= \frac{17}{3}(t - 1990)^3 + 50(t - 1990)^2 + 2300(t - 1990) + C.$

(We used the shortcut formula to evaluate each term.)

$S(1993) = 0$ gives

$$0 = \frac{17}{3}3^3 + 50(3^2) + 2300(3) + C$$

$$0 = 7503 + C$$

$$C = -7503$$

So, $S(t) \approx \frac{17}{3}(t - 1990)^3 + 50(t - 1990)^2 + 2300(t - 1990) - 7503.$

79. None; the substitution $u = x$ simply replaces the letter x throughout by the letter u, and thus does not change the integral at all. For instance, the integral $\int x(3x^2 + 1)\, dx$ becomes $\int u(3u^2 + 1)\, du$ if we substitute $u = x$.

81. The purpose of substitution is to introduce a new variable that is defined in terms of the variable of integration. One cannot say $u = u^2 + 1$, because u is not a new variable. Instead, define $w = u^2 + 1$ (or any other letter different from u).

83. The integral $\int x(x^2 + 1)\, dx$ can be solved by the substitution $u = x^2 + 1$: $\int x(x^2 + 1)\, dx = \frac{1}{2}\int u\, du = \frac{1}{4}u^2 + C = \frac{1}{4}(x^2 + 1)^2 + C.$

85. The substitution $u = -x$ leads to $\int e^{-u^2}\, du$, which is just the original integral. The substitution $u = x^2$ leads to $\int e^{-u}\frac{1}{2x}\, du = \int e^{-u}\frac{1}{2\sqrt{u}}\, du$ which is no easier to evaluate. The substitution $u = -x^2$ is similar.

6.3

1. $\Delta x = \frac{b-a}{n} = \frac{5-0}{5} = 1$

Left Sum = $[f(0) + f(1) + f(2) + f(3) + f(4)]\Delta x$
$\approx [14 + 6 + 2 + 2 + 6](1) = 30$

3. $\Delta x = \frac{b-a}{n} = \frac{9-1}{4} = 2$

Left Sum = $[f(1) + f(3) + f(5) + f(7)]\Delta x$
$\approx [0 + 4 + 4 + 3](2) = 22$

5. $\Delta x = \frac{b-a}{n} = \frac{3.5-1}{5} = 0.5$

Left Sum = $[f(1) + f(1.5) + f(2) + f(2.5) + f(3)]\Delta x$
$\approx [-1 + (-1) + (-2) + (-1) + 1](0.5) = -2$

7. $\Delta x = \frac{b-a}{n} = \frac{3-1}{3} = 1$

Left Sum = $[f(0) + f(1) + f(2)]\Delta x$
$\approx [0 + 0 + 0](1) = 0$

9. $f(x) = 4x - 1$, $\Delta x = \frac{b-a}{n} = \frac{2-0}{4} = 0.5$

Left Sum = $[f(0) + f(0.5) + f(1) + f(1.5)](0.5)$
$= 4$

11. $f(x) = x^2$, $\Delta x = \frac{b-a}{n} = \frac{2-(-2)}{4} = 1$

Left Sum = $[f(-2) + f(-1) + f(0) + f(1)](1) = 6$

13. $f(x) = 1/(1 + x)$, $\Delta x = \frac{b-a}{n} = \frac{1-0}{5} = 0.2$

Left Sum = $[f(0) + f(0.2) + f(0.4) + f(0.6) + f(0.8)](0.2) \approx 0.7456$

15. $f(x) = e^{-x}$, $\Delta x = \frac{b-a}{n} = \frac{10-0}{5} = 2$

Left Sum = $[f(0) + f(2) + f(4) + f(6) + f(8)](2) \approx 2.3129$

17. $f(x) = e^{-x^2}$, $\Delta x = \frac{b-a}{n} = \frac{10-0}{4} = 2.5$

Left Sum = $[f(0) + f(2.5) + f(5) + f(7.5)](2.5) \approx 2.5048$

19. The area is a square of height 1 and width 1, so has area $1 \times 1 = 1$.

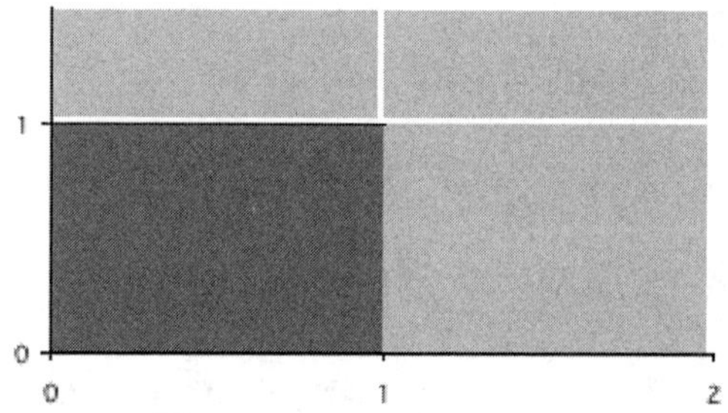

21. The area is a triangle of height 1 and base 1, so has area $\frac{1}{2}(1 \times 1) = \frac{1}{2}$.

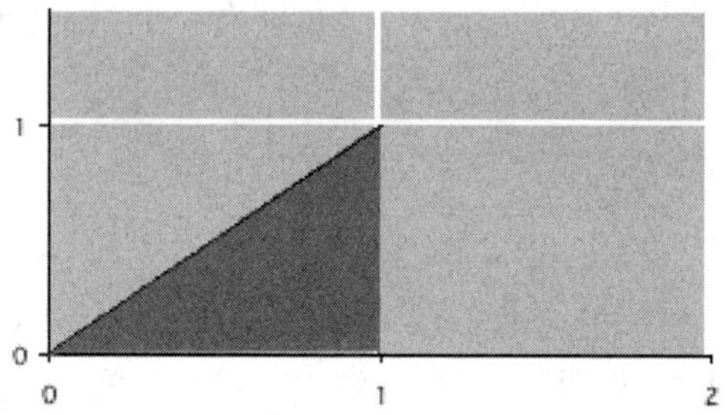

23. The area is a triangle of height $\frac{1}{2}$ and base 1, so has area $\frac{1}{2}\left(\frac{1}{2} \times 1\right) = \frac{1}{4}$.

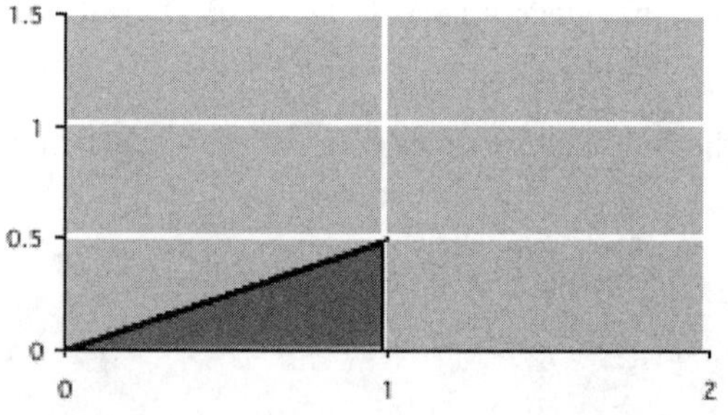

25. The area is a triangle of height 2 and base 2, so has area $\frac{1}{2}(2 \times 2) = 2$.

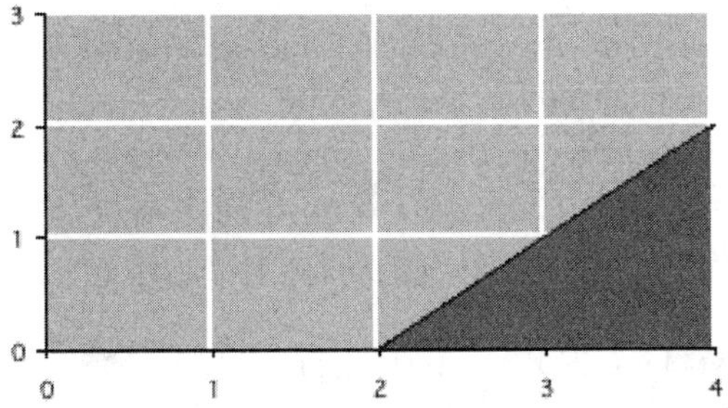

27. Since the are below the x axis is the same as the area above, the integral is 0.

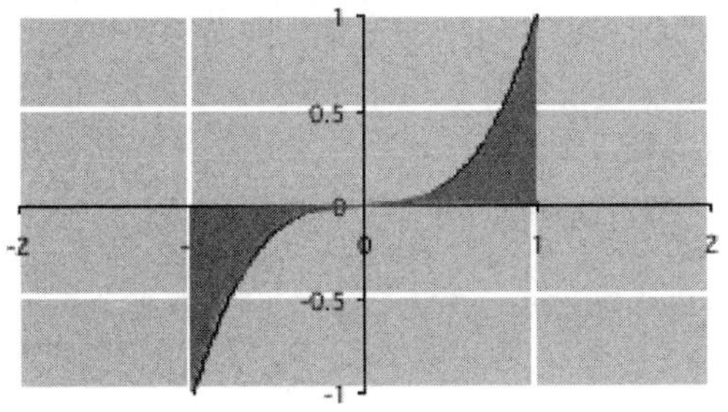

29. If counting grid squares, note that each grid square has an area of 1×0.5 = 0.5. Instead of counting grid squares, we average the left and right Riemann sums.

Left Sum = (1 + 1 + 1.5 + 2)(1) = 5.5

Right Sum = (1 + 1.5 + 2 + 2)(1) = 5.5

Total Change = Area under graph over [1, 5] = Average of left- and right- sums = (5.5+6.5)/2 = 6

31. Left Sum = (–1 – 0.5 + 0 + 0.5)(1) = –1

Right Sum = (– 0.5 + 0 + 0.5 + 1)(1) = 1

Total Change = Net area over [2, 6] = Average of left- and right- sums = (1+(–1))/2 = 0

33. Note that each grid square has an area of 0.5×0.5 = 0.25. Note that the areas corresponding to [–1, 0] and [0, 1] cancel out, so we are left with the area above [1, 2], which is 2 grid squares, or 0.5

35. Technology formula:

TI-83/84: `4*√(1-x^2)`

Excel: `4*SQRT(1-x^2)`

n = 10: 3.3045; n = 100: 3.1604; n = 1000: 3.1436

37. Technology formula:

`2*x^1.2/(1+3.5*x^4.7)`

n = 10: 0.0275; n = 100: 0.0258; n = 1000: 0.0256

39. $C'(x) = 20 - \frac{x}{200}$

$$\int_0^5 C'(x)\,dx \approx [C'(0) + C'(1) + C'(2) + C'(3) + C'(4)](1) = \$99.95$$

41. $R(t) = 17t^2 + 100t + 2300$

$$\int_5^{10} R(t)\,dt \approx [R(5) + R(6) + R(7) + R(8) + R(9)](1) = 19355 \text{ million gallons} \approx 19 \text{ billion gallons}$$

43. 2000–2003 is represented by the interval [2, 5] in the graph.

$$\text{Total value of sales} = \int_2^6 s(t)\,dt \text{ billion dollars}$$

$$\approx [f(2) + f(3) + f(4)](1)$$

$$\approx [4.5 + 7.5 + 10.5] = \$22.5 \text{ billion}$$

45. (a) Left sum ≈ (1 + 1.5 + 2 + 2.5 + 3 + 3.5 + 4.5 + 5)(2) = 46,

or about 46,000 articles

Right Sum ≈ (1.5 + 2 + 2.5 + 3 + 3.5 + 4.5 + 5 + 5.5)(2) = 55
or about 55,000 articles

(b) From the "Before we go on" discussion at the end of Example 4 we can estimate $\int_0^{16} r(t)\,dt$ as

$$\int_0^{16} r(t)\,dt \approx \text{Average of left- and right-sums}$$

$$= \frac{46+55.5}{2} = 50.5$$

Interpretation: Since $r(t)$ is given in thousands of articles per year, we conclude that a total of about 50,500 articles in *Physics Review* were written by researchers in Europe in the 16-year period beginning 1983.

47. The total change in number of students $c(t)$ from China who took the GRE exams is given by the definite integral of its rate of change:

$$\text{Total change} = \int_2^4 c'(t)\,dt$$

= Area above interval [2, 4]

Left sum = (35 + 32.5 + 30 + 20)(0.5) = 58.75

Right Sum = (32.5 + 30 + 20 + 17.5)(0.5) = 50

Average = $\frac{58.75 + 50}{2}$ = 54.375, or about 54,000 students

49. The total change in payroll $p(t)$ is given by the definite integral of its rate of change:

$$\text{Total change} = \int_0^4 p'(t)\,dt$$

= Area above t-axis – Area below t-axis

Each grid square on the graph has an area of 0.5×0.5 = 0.25.

Adding boxes gives us (–0.25) + (–0.25) + (–0.25) + (–0.125) + 0.125 + 0.125 + (–0.125) + (-0.25) = –1

(Alternatively, you can average the left-and right Riemann sums to obtain the same answer.)

Thus, the total change was about –\$1 billion.

51. $\int_8^{14} W(t)\,dt \approx [W(8) + W(9) + W(10) + W(11) + W(12) + W(13)](1) = 8160.$

This represents the total number of wiretaps authorized by U.S. courts from 1998 through 2003. Note that $t = 8$ represents the beginning of 1998 and $t = 14$ represents the beginning of 2004, or the end of 2003.

53. $v(t) = 40t^2$ ft/s

Height of rocket 2 seconds after launch

$$= \int_0^2 v(t)\,dt \approx [v(0) + v(0.2) + v(0.4) + \ldots + v(1.8)](0.2)$$

= 91.2 ft

55. Here is a method of computing the Riemann sum on the Web Site: Go to

Chapter 6 → Tools → Numerical Integration Utility

Enter `-1.5*x^2-0.9*x+1200` for $f(x)$, –10 and 10 for the left and right end-points, and 100 for the number of subdivisions. Then press "Left Sum" to obtain the left Riemann sum: 23,001.6 ≈ 23,000 to the nearest \$1000. Therefore, $\int_{-10}^{10} R(t)\,dt \approx \$23{,}000$. The median household income rose a total of approximately \$23,000 from 1980 to 2000.

57. Yes. The Riemann sum gives an estimated area of (0 + 15 + 18 + 8 + 7 + 16 + 20)(5) = 420 square feet.

59. (a) Graph:

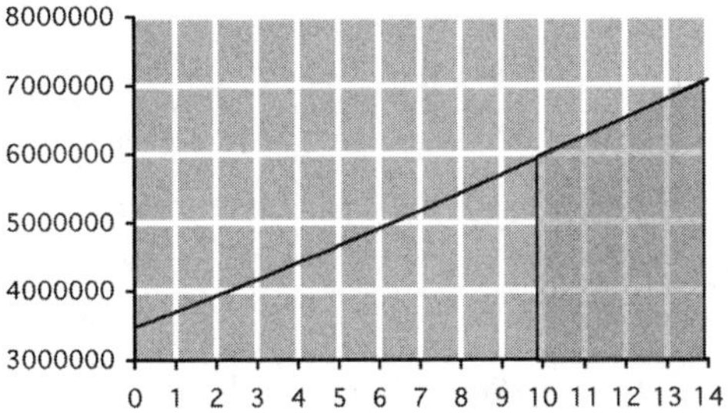

The area represents the total amount earned by households in the period 2000 through 2003, in millions of dollars.

(b) $\int_{10}^{14} A(t)\, dt \approx 26{,}000{,}000$. The total amount earned by households from 2000 through 2003 was approximately \$26 trillion.

61. (a) $p(x) = \frac{1}{5.2\sqrt{2\pi}} e^{-(x-72.6)^2/54.08}$, $\int_{60}^{100} p(x)\, dx \approx$ 0.994, so approximately 99.4% of students obtained between 60 and 100.

(b) $\int_{0}^{30} p(x)\, dx \approx 0$ (to at least 15 decimal places)

63. Stays the same: The graph is a horizontal line, and all Riemann sums give the exact area.

65. Increases: The left sum underestimates the function, by less as n increases.

67. The area under the curve and above the x axis equals the area above the curve and below the x axis.

69. Answers will vary. One example: Let r(t) be the rate of change of net income at time t. If r(t) is negative, then the net income is decreasing, so the change in net income, represented by the definite integral of r(t), is negative.

71. Answers may vary. For example:

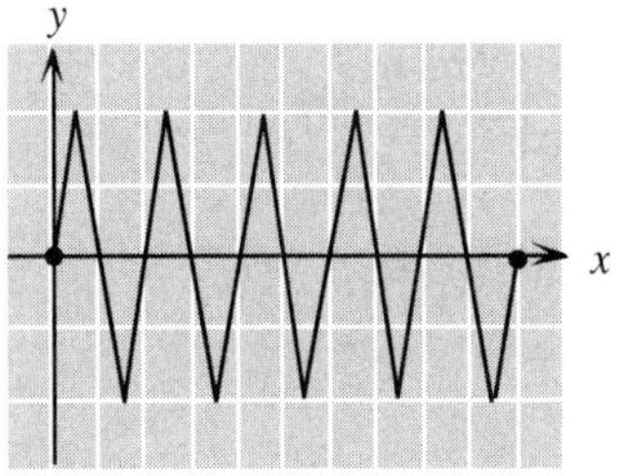

73. The total cost is $c(1) + c(2) + \ldots + c(60)$, which is represented by the Riemann sum approximation of $\int_{1}^{61} c(t)\, dt$ with $n = 60$.

75. $[f(x_1) + f(x_2) + \ldots + f(x_n)]\Delta x = \sum_{k=1}^{n} f(x_k)\Delta x$

77. If increasing n by a factor of 10 does not change the value of the answer when rounded to three decimal places, then the answer is (likely) accurate to three decimal places.

6.4

1. $\int_{-1}^{1}(x^2 + 2)\,dx = \left[\frac{x^3}{3} + 2x\right]_{-1}^{1}$

$= \frac{1}{3} + 2 - \left(\frac{-1}{3} - 2\right) = \frac{14}{3}$

3. $\int_{0}^{1}(12x^5 + 5x^4 - 6x^2 + 4)\,dx$

$= [2x^6 + x^5 - 2x^3 + 4x]_0^1$

$= 2 + 1 - 2 + 4 - (0) = 5$

5. $\int_{-2}^{2}(x^3 - 2x)\,dx = \left[\frac{x^4}{4} - x^2\right]_{-2}^{2}$

$= \frac{16}{4} - 4 - \left(\frac{16}{4} - 4\right) = 0$

7. $\int_{1}^{3}\left(\frac{2}{x^2} + 3x\right)dx = \int_{1}^{3}(2x^{-2} + 3x)\,dx$

$= \left[-2x^{-1} + \frac{3}{2}x^2\right]_1^3$

$= -\frac{2}{3} + \frac{27}{2} - \left(-2 + \frac{3}{2}\right) = \frac{40}{3}$

9. $\int_{0}^{1}(2.1x - 4.3x^{1.2})\,dx = \left[1.05x^2 - \frac{4.3}{2.2}x^{2.2}\right]_0^1$

$= 1.05 - \frac{4.3}{2.2} - (0) \approx -0.9045$

11. $\int_{0}^{1}2e^x\,dx = [2e^x]_0^1 = 2e - 2$

$= 2(e - 1)$

13. $\int_{0}^{1}\sqrt{x}\,dx = \int_{0}^{1}x^{1/2}\,dx = \left[\frac{2}{3}x^{3/2}\right]_0^1$

$= \frac{2}{3} - 0 = \frac{2}{3}$

15. $\int_{0}^{1}2^x\,dx = \left[\frac{2^x}{\ln 2}\right]_0^1 = \frac{2}{\ln 2} - \frac{1}{\ln 2} = \frac{1}{\ln 2}$

17. Let $u = 3x + 1$, $du = 3dx$, $dx = \frac{1}{3}du$; when $x = 0$, $u = 1$; when $x = 1$, $u = 4$;

$\int_{0}^{1}18(3x + 1)^5\,dx = \int_{1}^{4}18u^5\,\frac{1}{3}\,du = [u^6]_1^4$

$= 4^6 - 1 = 4095$

In Exercises 19–25, we use the shortcut integration formulas rather than substitution.

19. $\int_{-1}^{1}e^{2x-1}\,dx = \left[\frac{1}{2}e^{2x-1}\right]_{-1}^{1} = \frac{1}{2}e^1 - \frac{1}{2}e^{-3}$

$= \frac{1}{2}(e - e^{-3})$

21. $\int_{0}^{2}2^{-x+1}\,dx = \left[-\frac{2^{-x+1}}{\ln 2}\right]_0^2$

$= -\frac{2^{-1}}{\ln 2} - \left(-\frac{2}{\ln 2}\right) = \frac{3}{2\ln 2}$

23. $\int_{0}^{50}e^{-0.02x-1}\,dx = \left[-\frac{e^{-0.02x-1}}{0.02}\right]_0^{50}$

$= -50e^{-2} - (-50e^{-1}) = 50(e^{-1} - e^{-2})$

25. $\int_{-1.1}^{1.1}e^{x+1}\,dx = [e^{x+1}]_{-1.1}^{1.1} = e^{2.1} - e^{-0.1}$

27. Let $u = 2x^2 + 1$, $du = 4x\,dx$, $dx = \frac{1}{4x}\,du$; when $x = -\sqrt{2}$, $u = 5$; when $x = \sqrt{2}$, $u = 5$;

$$\int_{-\sqrt{2}}^{\sqrt{2}} 3x\sqrt{2x^2 + 1}\,dx = \int_0^0 3x\sqrt{u}\,\frac{1}{4x}\,du$$

$$= \frac{3}{4}\int_0^0 u^{1/2}\,du = \left[\frac{1}{2}u^{3/2}\right]_0^0 = 0 - 0 = 0$$

29. Let $u = x^2 + 2$, $du = 2x\,dx$, $dx = \frac{1}{2x}\,du$; when $x = 0$, $u = 2$; when $x = 1$, $u = 3$;

$$\int_0^1 5xe^{x^2+2}\,dx = \int_2^3 5xe^u\,\frac{1}{2x}\,du = \frac{5}{2}\int_2^3 e^u\,du$$

$$= \left[\frac{5}{2}e^u\right]_2^3 = \frac{5}{2}(e^3 - e^2)$$

31. Let $u = x^3 - 1$, $du = 3x^2\,dx$, $dx = \frac{1}{3x^2}\,du$; when $x = 2$, $u = 7$; when $x = 3$, $u = 26$;

$$\int_2^3 \frac{x^2}{x^3 - 1}\,dx = \int_7^{26} \frac{x^2}{u}\cdot\frac{1}{3x^2}\,du = \frac{1}{3}\int_7^{26}\frac{1}{u}\,du$$

$$= \left[\frac{1}{3}\ln|u|\right]_7^{26} = \frac{1}{3}(\ln 26 - \ln 7) \text{ or } \frac{1}{3}\ln\left(\frac{26}{7}\right)$$

33. Let $u = -x^2$, $du = -2x\,dx$, $dx = -\frac{1}{2x}\,du$; when $x = 0$, $u = 0$; when $x = 1$, $u = -1$;

$$\int_0^1 x(1.1)^{-x^2}\,dx = -\int_0^{-1} x(1.1)^u\,\frac{1}{2x}\,du$$

$$= -\frac{1}{2}\int_0^{-1}(1.1)^u\,du = \left[-\frac{(1.1)^u}{2\ln 1.1}\right]_0^{-1}$$

$$= -\frac{(1.1)^{-1}}{2\ln 1.1} - \left(-\frac{1}{2\ln 1.1}\right) = \frac{0.1}{2.2\ln 1.1}$$

35. Let $u = 1/x$, $du = -1/x^2\,dx$, $dx = -x^2\,du$; when $x = 1$, $u = 1$; when $x = 2$, $u = 1/2$;

$$\int_1^2 \frac{e^{1/x}}{x^2}\,dx = -\int_1^{1/2}\frac{e^u}{x^2}x^2\,du = -\int_1^{1/2} e^u\,du$$

$$= [-e^u]_1^{1/2} = -e^{1/2} - (-e) = e - e^{1/2}$$

37. Let $u = x + 1$, $du = dx$; when $x = 0$, $u = 1$; when $x = 2$, $u = 3$; $x = u - 1$;

$$\int_0^2 \frac{x}{x + 1}\,dx = \int_1^3 \frac{x}{u}\,du = \int_1^3 \frac{u - 1}{u}\,du$$

$$= \int_1^3\left(1 - \frac{1}{u}\right)du = [u - \ln|u|]_1^3$$

$$= 3 - \ln 3 - (1 - \ln 1) = 2 - \ln 3$$

39. Let $u = x - 2$, $du = dx$; when $x = 1$, $u = -1$; when $x = 2$, $u = 0$; $x = u + 2$;

$$\int_1^2 x(x - 2)^5\,dx = \int_{-1}^0 xu^5\,du = \int_{-1}^0 (u + 2)u^5\,du$$

$$= \int_{-1}^0 (u^6 + 2u^5)\,du = \left[\frac{u^7}{7} + \frac{u^6}{3}\right]_{-1}^0$$

$$= 0 + 0 - \left(\frac{-1}{7} + \frac{1}{3}\right) = -\frac{4}{21}$$

41. Let $u = 2x + 1$, $du = 2\,dx$, $dx = \frac{1}{2}\,du$; when $x = 0$, $u = 1$; when $x = 1$, $u = 3$; $x = \frac{1}{2}(u - 1)$;

$$\int_0^1 x\sqrt{2x + 1}\,dx = \int_1^3 x\sqrt{u}\,\frac{1}{2}\,du$$

$$= \frac{1}{2}\int_1^3 \frac{1}{2}(u - 1)u^{1/2}\,du = \frac{1}{4}\int_1^3 (u^{3/2} - u^{1/2})\,du$$

$$= \left[\frac{1}{10}u^{5/2} - \frac{1}{6}u^{3/2}\right]_1^3$$

$$= \frac{3^{5/2}}{10} - \frac{3^{3/2}}{6} - \left(\frac{1}{10} - \frac{1}{6}\right)$$

$$= \frac{3^{5/2}}{10} - \frac{3^{3/2}}{6} + \frac{1}{15}$$

43. Graph:

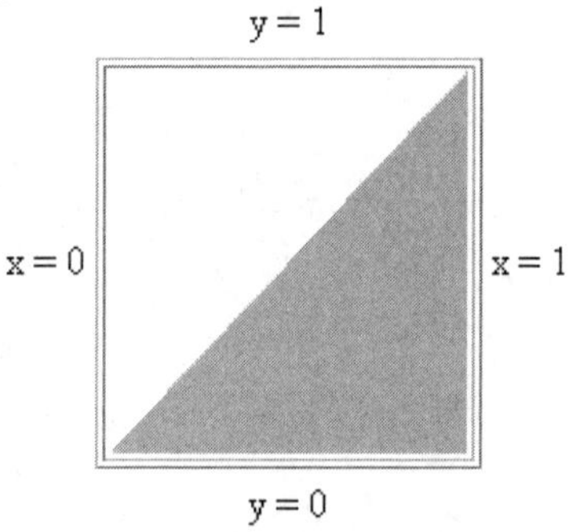

The line $y = x$ crosses the x axis at $x = 0$, so we can calculate the area as

$$\int_0^1 x\,dx = \left[\frac{x^2}{2}\right]_0^1 = \frac{1}{2}$$

45. Graph:

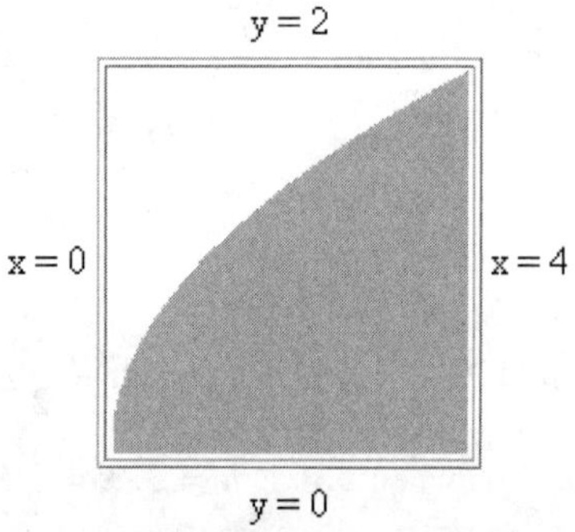

The curve $y = \sqrt{x}$ touches the x axis at $x = 0$ only, so we can calculate the area as

$$\int_0^4 \sqrt{x}\,dx = \int_0^4 x^{1/2}\,dx = \left[\frac{2}{3}x^{3/2}\right]_0^4$$

$$= \frac{16}{3} - 0 = \frac{16}{3}$$

47. Graph:

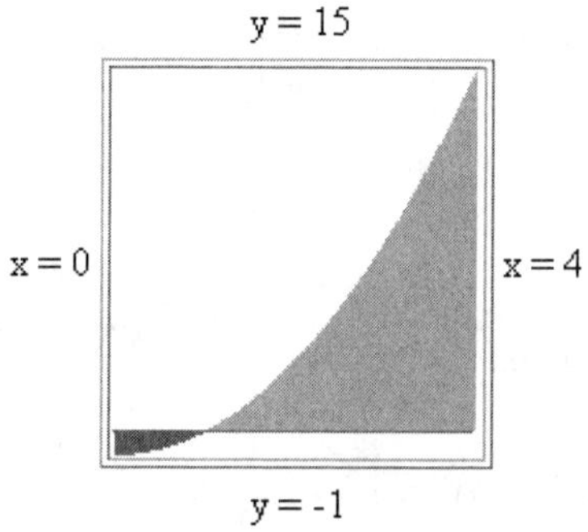

The curve $y = x^2 - 1$ crosses the x axis where $x^2 - 1 = 0$, so at $x = \pm 1$. We compute two integrals:

$$\int_0^1 (x^2 - 1)\,dx = \left[\frac{x^3}{3} - x\right]_0^1$$

$$= \frac{1}{3} - 1 - 0 = -\frac{2}{3}$$

and

$$\int_1^4 (x^2 - 1)\,dx = \left[\frac{x^3}{3} - x\right]_1^4$$

$$= \frac{64}{3} - 4 - \left(\frac{1}{3} - 1\right) = \frac{54}{3}.$$

The total area is the sum of the absolute values, so $\frac{2}{3} + \frac{54}{3} = \frac{56}{3}$.

49. Graph:

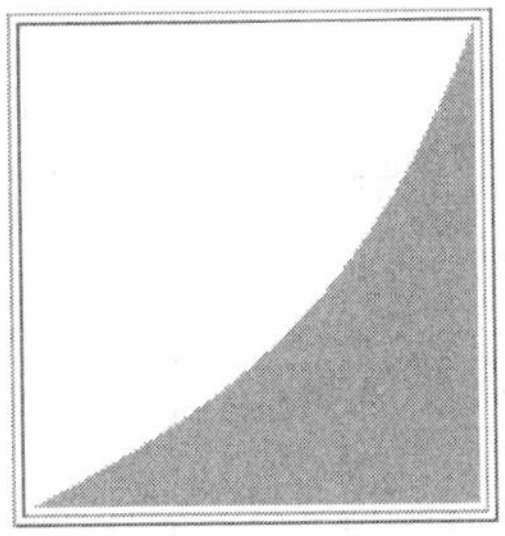

The curve $y = xe^{x^2}$ crosses the x axis at $x = 0$, so we can calculate the area as $\int_0^{(\ln 2)^{1/2}} xe^{x^2}\,dx$.

Let $u = x^2$, $du = 2x\,dx$, $dx = \frac{1}{2x}\,du$; when $x = 0$, $u = 0$; when $x = (\ln 2)^{1/2}$, $u = \ln 2$.

$$\int_0^{(\ln 2)^{1/2}} xe^{x^2}\,dx = \int_0^{\ln 2} xe^u \frac{1}{2x}\,du$$

$$= \frac{1}{2}\int_0^{\ln 2} e^u\,du = \left[\frac{1}{2}e^u\right]_0^{\ln 2} = \frac{1}{2}e^{\ln 2} - \frac{1}{2}$$

$$= \frac{1}{2}\cdot 2 - \frac{1}{2} = \frac{1}{2}$$

51. $C'(x) = 5 + \frac{x^2}{1000}$

$$\text{Change in cost} = \int_{10}^{100} C'(x)\,dx$$

$$= \int_{10}^{100}\left(5 + \frac{x^2}{1000}\right)dx = \left[5x + \frac{x^3}{3000}\right]_{10}^{100}$$

$$= 500 + \frac{1000}{3} - \left(50 + \frac{1}{3}\right) = \$783$$

53. $v(t) = 60 - e^{-t/10}$ mph

$$\text{Total distance traveled} = \int_1^6 v(t)\,dt$$

$$= \int_1^6 (60 - e^{-t/10})\,dt = [60t + 10e^{-t/10}]_1^6$$

$$= 360 + 10e^{-6/10} - (60 + 10e^{-1/10})$$

≈ 296 miles

55. $\text{Total sales} = \int_5^{10} (17t^2 + 100t + 2300)\,dt$

$$= \left[\frac{17}{3}t^3 + 50t^2 + 2300t\right]_5^{10}$$

$$= \left[\frac{17}{3}(10)^3 + 50(10)^2 + 2300(10)\right] - \left[\frac{17}{3}(5)^3 + 50(5)^2 + 2300(5)\right]$$

$\approx 20{,}000$ million gallons

57. $\text{Total change} = \int_{-10}^{10} R(t)\,dt$

$$= \int_{-10}^{10} (-1.5t^2 - 0.9t + 1200)\,dt$$

$$= [-0.5t^3 - 0.45t^2 + 1200t]_{-10}^{10}$$

$$= [-0.5(10)^3 - 0.45(10)^2 + 1200(10)] - [-0.5(-10)^3 - 0.45(-10)^2 + 1200(-10)]$$

$= \$23{,}000$

59. Total consumption

$$= \int_8^{10} (-0.065t^3 + 3.4t^2 - 22t + 3.6)\,dt$$

$$= \left[\frac{-0.065}{4}t^4 + \frac{3.4}{3}t^3 - 11t^2 + 3.6t\right]_8^{10}$$

≈ 68 milliliters

61. If $S(t)$ is the weekly sales rate, we have $S(t) = 50(1 - 0.05)^t$ T-shirts per week after t weeks. The total sales over the coming year will be

$$\int_0^{52} 50(0.95)^t\, dt = \left[\frac{50(0.95)^t}{\ln 0.95}\right]_0^{52}$$

$$= \frac{50}{\ln 0.95}[(0.95)^{52} - 1] \approx 907 \text{ T-shirts.}$$

63. $\int_0^{10} (1 - e^{-t})\, dt = [t + e^{-t}]_0^{10}$

$= 10 + e^{-10} - (0 + e^0) \approx 9$ gallons

65. Change in cost $= \int_0^x m(t)\, dt$

$= C(x) - C(0)$ by the FTC,

so

$$C(x) = C(0) + \int_0^x m(t)\, dt.$$

$C(0)$ is the *fixed cost.*

67. (a) $\frac{N}{1 + Ab^{-x}} = \frac{Nb^x}{(1 + Ab^{-x})b^x}$

$= \frac{Nb^x}{b^x + A} = \frac{Nb^x}{A + b^x}$

(b) Let $u = A + b^x$, $du = (\ln b)b^x\, dx$,

$$dx = \frac{1}{(\ln b)b^x}\, du.$$

$$\int \frac{N}{1 + Ab^{-x}}\, dx = \int \frac{Nb^x}{A + b^x}\, dx$$

$$= \int \frac{Nb^x}{u} \cdot \frac{1}{(\ln b)b^x}\, du = \frac{N}{\ln b}\int \frac{1}{u}\, du$$

$$= \frac{N \ln u}{\ln b} + C = \frac{N \ln(A + b^x)}{\ln b} + C$$

(c) Total sales $= \int_0^6 \frac{1100}{1 + 18(1.9)^{-t}}\, dt$

$= \left[\frac{1100 \ln(18 + (1.9)^t)}{\ln 1.9}\right]_0^6$

$\approx [1714 \ln(18 + (1.9)^t]_0^6$

$= [1714 \ln(18 + (1.9)^6]$

$\quad - [1714 \ln(18 + (1.9)^0]$

≈ 2100 thousand iPods

69. (a) Total number

$= \int_8^{14} \left[620 + \frac{900e^{0.25t}}{3 + e^{0.25t}}\right] dt$

$= [620t + 3600 \ln(3 + e^{0.25t})]_8^{14}$

(use the substitution $u = 3 + e^{0.25t}$ for the second part)

≈ 8200 wiretaps

(b) The actual number of wiretaps is the sum

1329 + 1350 + 1190 + 1491 + 1358 + 1442

= 8160,

which agrees with the answer in part (a) to two significant digits. Therefore, the integral in part (a) does give an accurate estimate.

71. $W = \int_{v_0}^{v_1} v \frac{d}{dv}(p)\, dv = \int_{v_0}^{v_1} v \frac{d}{dv}(mv)\, dv$

$= \int_{v_0}^{v_1} v \cdot m\, dv = \left[\frac{1}{2}mv^2\right]_{v_0}^{v_1}$

$= \frac{1}{2}mv_1^2 - \frac{1}{2}mv_0^2$

73. They are related by the Fundamental Theorem of Calculus, which states (summarized briefly) that the definite integral of a suitably nice function can be calculated by evaluating the indefinite integral at the two endpoints and subtracting.

75. The total sales from time a to time b are obtained from the marginal sales by taking its definite integral from a to b.

77. An example is $v(t) = t - 5$.

79. An example is $f(x) = e^{-x}$.

81. By the FTC,

$$\int_a^x f(t)\,dt = G(x) - G(a)$$

where G is an antiderivative of f. Hence,

$$F(x) = G(x) - G(a).$$

Taking derivatives of both sides,

$$F'(x) = G'(x) + 0 = f(x),$$

as required. The result gives us a formula, in terms of area, for an antiderivative of any continuous function.

Chapter 6 Review Exercises

1. $\int (x^2 - 10x + 2)\, dx = \frac{x^3}{3} - 5x^2 + 2x + C$

2. $\int (e^x + \sqrt{x})\, dx = e^x + \frac{2}{3}x^{3/2} + C$

3. $\int \left(\frac{4x^2}{5} - \frac{4}{5x^2}\right) dx = \int \left(\frac{4}{5}x^2 - \frac{4}{5}x^{-2}\right) dx$

$= \frac{4}{5}\frac{x^3}{3} - \frac{4}{5}\frac{x^{-1}}{-1} + C$

$= 4x^3/15 + 4/(5x) + C$

4. $\int \left(\frac{3x}{5} - \frac{3}{5x}\right) dx = \int \left(\frac{3}{5}x - \frac{3}{5}x^{-1}\right) dx$

$= \frac{3}{5}\frac{x^2}{2} - \frac{3}{5}\ln|x| + C$

$= 3x^2/10 - (3/5)\ln|x| + C$

5. By the shortcut,

$\int e^{-2x+11}\, dx = \frac{1}{-2}e^{-2x+11} + C$

$= -e^{-2x+11}/2 + C$

6. By the shortcut,

$\int \frac{dx}{(4x-3)^2} = \int (4x-3)^{-2}\, dx$

$= \frac{(4x-3)^{-1}}{(-1)(4)} + C = -1/[[4(4x{-}3)] + C$

7. Let $u = x^2 + 4$, $du = 2x\, dx$, $dx = \frac{1}{2x}\, du$

$\int x(x^2+4)^{10}\, dx = \int xu^{10}\frac{1}{2x}\, du$

$= \frac{1}{2}\int u^{10}\, du = \frac{1}{22}u^{11} + C$

$= \frac{1}{22}(x^2+4)^{11} + C$

8. Let $u = x^3 + 3x + 2$, $du = (3x^2 + 3)\, dx = 3(x^2+1)\, dx$, $dx = \frac{1}{3(x^2+1)}\, du$

$\int \frac{x^2+1}{(x^3+3x+2)^2}\, dx = \int \frac{x^2+1}{u^2}\cdot\frac{1}{3(x^2+1)}\, du$

$= \frac{1}{3}\int u^{-2}\, du = -\frac{1}{3}u^{-1} + C$

$= -\frac{1}{3(x^3+3x+2)} + C$

9. From the shortcut,

$\int 5e^{-2x}\, dx = -\frac{5}{2}e^{-2x} + C$

10. Let $u = -x^2/2$, $du = -x\, dx$, $dx = -\frac{1}{x}\, du$

$\int xe^{-x^2/2}\, dx = -\int xe^u\frac{1}{x}\, du$

$= -\int e^u\, du = -e^u + C = -e^{-x^2/2} + C$

11. Put $u = x + 2$. $du/dx = 1$, $du = dx$.

$\int \frac{x+1}{x+2}\, dx = \int \frac{x+1}{u}\, du = \int \frac{u-1}{u}\, du$

(Solve for x in the equation $u = x + 2$)

$= \int \left(1 - \frac{1}{u}\right) du = u - \ln|u| + C$

$= (x+2) - \ln|x+2| + C$

(or $x - \ln|x+2| + C$ if we incorporate the 2 in C).

12. Put $u = x - 1$. $du/dx = 1$, $du = dx$.

$$\int x\sqrt{x-1}\,dx = \int x\sqrt{u}\,du$$

$$= \int (u+1)\sqrt{u}\,du$$

(Solve for x in the equation $u = x - 1$)

$$= \int (u^{3/2} + u^{1/2})\,du = \frac{2u^{5/2}}{5} + \frac{2u^{3/2}}{3} + C$$

$$= \frac{2(x-1)^{5/2}}{5} + \frac{2(x-1)^{3/2}}{3} + C$$

13. $\Delta x = \frac{b-a}{n} = \frac{3-0}{6} = 0.5$

Left Sum = $[f(0) + f(0.5) + f(1) + f(1.5) + f(2) + f(2.5)]\Delta x$

$\approx [-4 + 0 + 3 + 1 + 0 + 2](0.5) = 1$

14. $\Delta x = \frac{b-a}{n} = \frac{3-1}{4} = 0.5$

Left Sum = $[f(1) + f(1.5) + f(2) + f(2.5)]\Delta x$

$\approx [-4 - 1 + 1 + 0](0.5) = -2$

15. $\Delta x = \frac{b-a}{n} = \frac{1-(-1)}{4} = 0.5$

Left Sum = $[f(-1) + f(-0.5) + f(0) + f(0.5)]\Delta x$

$= [2 + 1.25 + 1 + 1.25](0.5) = 2.75$

16. $\Delta x = \frac{b-a}{n} = \frac{4-0}{4} = 1$

Left Sum = $[f(0) + f(1) + f(2) + f(3)]\Delta x$

$= [0 - 2 - 2 + 0](1) = -4$

17. $\Delta x = \frac{b-a}{n} = \frac{1-0}{5} = 0.2$

Left Sum = $[f(0) + f(0.2) + f(0.4) + f(0.6) + f(0.8)]\Delta x$

$= [0 - 0.192 - 0.336 - 0.384 - 0.228](0.2)$

$= -0.24$

18. $\Delta x = \frac{b-a}{n} = \frac{1.5-0}{3} = 0.5$

Left Sum = $[f(0) + f(0.5) + f(1)]\Delta x$

$\approx [\frac{1}{2} + \frac{1}{3} + 0](0.5) = \frac{5}{12} \approx 0.4167$

19. Technology formulas:

TI-83/84: `e^(-x²)`

Excel: `EXP(-x^2)`

$n = 10$: 0.7778

$n = 100$: 0.7500

$n = 1000$: 0.7471

20. Technology formula: `x^(-x)`

$n = 10$: 0.7869

$n = 100$: 0.6972

$n = 1000$: 0.6885

21. Left Sum = $(0 + -0.5 + -1 + 0 + 0 + 1)(0.5) = -0.25$

Right Sum = $(-0.5 + -1 + 0 + 0 + 1 + 1)(0.5) = 0.25$

Total Change = Area under graph over $[-1, 2]$ = Average of left- and right- sums = $(-0.25 + 0.25)/2 = 0$

22. Left Sum = Left Sum = $(0 + -0.5 + 1 + 0)(0.5) = 0.25$

Right Sum =$(-0.5 + 1 + 0 + -0.5)(0.5) = 0$

Total Change = Area under graph over $[0, 2]$ = Average of left- and right- sums = $(0.25+0)/2 = 0.125$

23. $\int_0^1 (x - x^3)\,dx = \left[\frac{x^2}{2} - \frac{x^4}{4}\right]_0^1$

$= \frac{1}{2} - \frac{1}{4} - (0) = \frac{1}{4}$

24. By the shortcut,

$$\int_0^9 \frac{1}{x+1}\,dx = [\ln|x+1|]_0^9$$

$$= \ln 10 - \ln 1 = \ln 10 \approx 2.303$$

25. $\int_{-1}^{1} (1 + e^x)\,dx = [x + e^x]_{-1}^{1}$

$$= (1 + e) - (-1 + e^{-1}) = 2 + e - e^{-1}$$

26. $\int_0^4 (x + \sqrt{x})\,dx = \left[\frac{x^2}{2} + \frac{2x^{3/2}}{3}\right]_0^4$

$$= (8 + 16/3) = 40/3$$

27. Let $u = x^3 + 1$, $du = 3x^2\,dx$, $dx = \frac{1}{3x^2}$; when $x = 0$, $u = 1$; when $x = 2$, $u = 9$

$$\int_0^2 x^2\sqrt{x^3 + 1}\,dx = \int_1^9 x\sqrt{u}\,\frac{1}{3x^2}\,du$$

$$= \frac{1}{3}\int_1^9 u^{1/2}\,du = \frac{2}{9}\,[u^{3/2}]_1^9$$

$$= \frac{2}{9}(27 - 1) = \frac{52}{9}$$

28. $\int_{-1}^{1} 3^{2x-2}\,dx = \frac{1}{2\ln 3}\,[3^{2x-2}]_{-1}^{1}$

$$= \frac{1}{2\ln 3}(3^0 - 3^{-4}) = \frac{40}{81\ln 3} \approx 0.4495$$

29. $u = 1+4e^{-2x}$, $du/dx = -8e^{-2x}$, $dx = -du/(8e^{-2x})$

$x = 0 \Rightarrow u = 1+4 = 5$

$x = \ln 2 \Rightarrow u = 1+4e^{-2\ln 2} = 1+1 = 2$

$$\int_0^{\ln 2} \frac{e^{-2x}}{1 + 4e^{-2x}}\,dx = -\int_5^2 \frac{e^{-2x}}{u\,8e^{-2x}}\,du$$

$$= \int_2^5 \frac{1}{8u}\,du = \frac{1}{8}[\ln u]_2^5 = \frac{1}{8}[\ln 5 - \ln 2]$$

30. $u = -x^2$, $du/dx = -2x$, $dx = -du/(2x)$

$x = 0 \Rightarrow u = 0^2 = 0$

$x = 1 \Rightarrow u = -1^2 = -1$

$$\int_0^1 3xe^{-x^2}\,dx = -\int_0^{-1} 3xe^u\frac{du}{2x}\,du$$

$$= \int_{-1}^{0} \frac{3}{2}e^u\,du = \frac{3}{2}[e^u]_{-1}^{0} = \frac{3}{2}(1-e^{-1})$$

31. $y = 4 - x^2$ crosses the x axis when $x = \pm 2$, so we can compute the area as

$$\int_{-2}^{2} (4 - x^2)\,dx = \left[4x - \frac{x^3}{3}\right]_{-2}^{2}$$

$$= 8 - \frac{8}{3} - \left(-8 + \frac{8}{3}\right) = \frac{32}{3}.$$

32. $y = 4 - x^2$ crosses the x axis when $x = \pm 2$, so we compute two integrals:

$$\int_0^2 (4 - x^2)\,dx = \left[4x - \frac{x^3}{3}\right]_0^2$$

$$= 8 - \frac{8}{3} - (0) = \frac{16}{3}$$

and

$$\int_2^5 (4 - x^2)\,dx = \left[4x - \frac{x^3}{3}\right]_2^5$$

$$= 20 - \frac{125}{3} - \left(8 - \frac{8}{3}\right) = -27.$$

Adding the absolute values gives a total area of $\frac{16}{3}$ $+ 27 = \frac{97}{3}$.

33. $y = xe^{-x^2}$ crosses the x axis at $x = 0$, so we can compute the area as $\int_0^5 xe^{-x^2}\,dx$.

Let $u = -x^2$, $du = -2x\,dx$, $dx = -\frac{1}{2x}\,du$; when $x = 0$, $u = 0$; when $x = 5$, $u = -25$. So the area is

$$\int_0^5 xe^{-x^2}\,dx = -\int_0^{-25} xe^u \frac{1}{2x}\,du$$

$$= \frac{1}{2}\int_0^{-25} e^u\,du = -\frac{1}{2}[e^u]_0^{-25}$$

$$= \frac{1 - e^{-25}}{2}.$$

34. Since $y = -2x$ when $x < 0$ and $y = 2x$ when $x \geq 0$, we compute the area using two integrals:

$$\int_{-1}^{0} (-2x)\,dx = [-x^2]_{-1}^{0} = 0 - (-1) = 1$$

and

$$\int_0^1 2x\,dx = [x^2]_0^1 = 1 - 0 = 1.$$

Adding these together gives $1 + 1 = 2$.

35. (a) $q'(p) = -20p$

$$q = \int (-20p)\,dp = -10p^2 + C$$

$q = 100{,}000$ when $p = 0$, so

$0 + C = 100{,}000$

$C = 100{,}000$

giving $q = 100{,}000 - 10p^2$

(b) $q = 0$ when

$100{,}000 - 10p^2 = 0$

$p^2 = 10{,}000$

$p = \$100.$

36. Let $s(t)$ be the height function, $v(t)$ the velocity, and $a(t)$ the acceleration. We know that $a(t) = -32$ ft/s^2; we are told that $v(0) = 60$ ft/s and $s(0) = 100$ ft. So,

$$v(t) = \int (-32)\,dt = -32t + C$$

$60 = -32(0) + C$

$C = 60$

giving $v(t) = -32t + 60$. Thus,

$$s(t) = \int (-32t + 60)\,dt$$

$= -16t^2 + 60t + C$

$100 = -16(0) + 60(0) + C$

$C = 100$

giving $s(t) = -16t^2 + 60t + 100$.

The book will hit the ground when $s(t) = 0$

$-16t^2 + 60t + 100 = 0$

$t = 5$ seconds.

(The other solution is $t = -1.25$ s, which we reject because it is negative.)

(b) $v(5) = -100$ ft/s, so the book is traveling 100 ft/s when it hits the ground.

(c) The maximum value of $s(t)$ occurs when $v(t) = 0$, so when

$-32t + 60 = 0$

$t = 1.875$ s.

The height at that time is

$s(1.875) = 156.25$ ft above ground level.

37. $\Delta t = (5-0)/10 = 0.5$

Left Sum = (5 + 5 + 5 + 5 + 5 + 5 + 5 + 5 + 5 + 5)(0.5) = 25

Hence, sales were approximately 25,000 copies.

38. The total number of books sold is given by the definite integral of the function shown over the interval [0, 1.5]:

$\Delta t = 0.25$

Left Sum = $(2 + 1 + 0 - 1 - 1 + 0)(0.25) = 0.25$

Right Sum = $(1 + 0 - 1 - 1 + 0 + 0)(0.5) = -0.25$

$$\text{Average} = \frac{0.25 + (-0.25)}{2} = 0$$

Total net sales amounted to 0.

39. The last 10 days of the period is represented by the interval [0, 10]

$\Delta t = (10-0)/5 = 2$

Left Sum = $(n(0) + n(2) + n(4) + n(6) + n(8))(2)$
$= (0 + 1968 + 3904 + 5856 + 7872)(2) = 39{,}200$ hits

40. Total change in cost

$$= \int_2^4 (x-2)^2[8-(x-2)^3]^{3/2}\, dx.$$

Put $u = 8-(x-2)^3$; $du/dx = -3(x-2)^2$; $dx = -du/(3(x-2)^2)$

$x = 2 \Rightarrow u = 8$

$x = 4 \Rightarrow u = 2^2[8-8] = 0$

$$\int_2^4 (x-2)^2[8-(x-2)^3]^{3/2}\, dx$$

$$= -\int_8^0 (x-2)^2 u^{3/2} \frac{du}{3(x-2)^2}\, dx$$

$$= \int_0^8 \frac{u^{3/2}}{3}\, dx = \left[\frac{2}{15}u^{5/2}\right]_0^8$$

≈ 24 thousand dollars

41. If $S(t)$ is the weekly sales in week t, they were estimating that

$$S(t) = 6400(2)^{t/2}.$$

The total sales over the first five weeks would be

$$\int_0^5 S(t)\, dt = \int_0^5 6400(2)^{t/2}\, dt$$

$$= \frac{12{,}800}{\ln 2}[2^{t/2}]_0^5 = \frac{12{,}800}{\ln 2}[2^{5/2} - 2^0]$$

$\approx 86{,}000$ books.

42. Let $u = e^{0.55t} + 14.01$, $du = 0.55e^{0.55t}\, dt$, $dt = \frac{1}{0.55e^{0.55t}}\, du$; when $t = 0$, $u = 15.01$; when $t = 5$, $u \approx 29.653$

$$\int_0^5 \left(6053 + \frac{4474e^{0.55t}}{e^{0.55t} + 14.01}\right) dt$$

$$= [6053t]_0^5 + \int_{15.01}^{29.653} \frac{4474e^{0.55t}}{u} \cdot \frac{1}{0.55e^{0.55t}}\, du$$

$$= 30{,}265 + 8135 \int_{15.01}^{29.653} \frac{1}{u}\, du$$

$$= 30{,}265 + 8135[\ln u]_{15.01}^{29.653}$$

$\approx 35{,}800$ books.

7.1

1.

	D		I
$+$	$2x$	↘	e^x
$-$	2	↘	e^x
$+\int$	0	$\rightarrow$	e^x

$$\int 2xe^x\,dx = 2xe^x - 2e^x + C = 2(x-1)e^x + C$$

3.

	D		I
$+$	$3x-1$	↘	e^{-x}
$-$	3	↘	$-e^{-x}$
$+\int$	0	$\rightarrow$	e^{-x}

$$\int (3x-1)e^{-x}\,dx = -(3x-1)e^{-x} - 3e^{-x} + C$$

$$= -(3x+2)e^{-x} + C$$

5.

	D		I
$+$	x^2-1	↘	e^{2x}
$-$	$2x$	↘	$\frac{1}{2}e^{2x}$
$+$	2	↘	$\frac{1}{4}e^{2x}$
$-\int$	0	$\rightarrow$	$\frac{1}{8}e^{2x}$

$$\int (x^2-1)e^{2x}\,dx$$

$$= \frac{1}{2}(x^2-1)e^{2x} - \frac{1}{2}xe^{2x} + \frac{1}{4}e^{2x} + C$$
$$= \frac{1}{4}(2x^2-2x-1)e^{2x} + C$$

7.

	D		I
$+$	x^2+1	↘	e^{-2x+4}
$-$	$2x$	↘	$-\frac{1}{2}e^{-2x+4}$
$+$	2	↘	$\frac{1}{4}e^{-2x+4}$
$-\int$	0	$\rightarrow$	$-\frac{1}{8}e^{-2x+4}$

$$\int (x^2+1)e^{-2x+4}\,dx$$

$$= -\frac{1}{2}(x^2+1)e^{-2x+4} - \frac{1}{2}xe^{-2x+4} - \frac{1}{4}e^{-2x+4} + C$$
$$= -\frac{1}{4}(2x^2+2x+3)e^{-2x+4} + C$$

9.

	D		I
$+$	$2-x$	↘	2^x
$-$	-1	↘	$2^x/\ln 2$
$+\int$	0	$\rightarrow$	$2^x/(\ln 2)^2$

$$\int (2-x)2^x\,dx$$

$$= \frac{1}{\ln 2}(2-x)2^2 + \frac{1}{(\ln 2)^2}2^x + C$$
$$= \left[\frac{2-x}{\ln 2} + \frac{1}{(\ln 2)^2}\right]2^x + C$$

11.

	D		I
$+$	$x^2 - 1$	$\searrow$	3^{-x}
$-$	$2x$	$\searrow$	$-3^{-x}/\ln 3$
$+$	2	$\searrow$	$3^{-x}/(\ln 3)^2$
$-\int$	0	$\rightarrow$	$-3^{-x}/(\ln 3)^3$

$$\int (x^2 - 1)3^{-x}\,dx$$

$$= -\frac{1}{\ln 3}(x^2 - 1)3^{-x} - \frac{2}{(\ln 3)^2}x3^{-x} - \frac{2}{(\ln 3)^3}3^{-x} + C$$

$$= -\left[\frac{x^2 - 1}{\ln 3} + \frac{2x}{(\ln 3)^2} + \frac{2}{(\ln 3)^3}\right]3^{-x} + C$$

13.

	D		I
$+$	$x^2 - x$	$\searrow$	e^{-x}
$-$	$2x - 1$	$\searrow$	$-e^{-x}$
$+$	2	$\searrow$	e^{-x}
$-\int$	0	$\rightarrow$	$-e^{-x}$

$$\int \frac{x^2 - x}{e^x}\,dx = \int (x^2 - x)e^{-x}\,dx$$

$$= -(x^2 - x)e^{-x} - (2x - 1)e^{-x} - 2e^{-x} + C$$
$$= -(x^2 + x + 1)e^{-x} + C$$

15.

	D		I
$+$	x	$\searrow$	$(x + 2)^6$
$-$	1	$\searrow$	$(x + 2)^7/7$
$+\int$	0	$\rightarrow$	$(x + 2)^8/56$

$$\int x(x + 2)^6\,dx = \frac{1}{7}x(x + 2)^7 - \frac{1}{56}(x + 2)^8 + C$$

17.

	D		I
$+$	x	$\searrow$	$(x - 2)^{-3}$
$-$	1	$\searrow$	$-(x - 2)^{-2}/2$
$+\int$	0	$\rightarrow$	$(x - 2)^{-1}/2$

$$\int \frac{x}{(x - 2)^3}\,dx = \int x(x - 2)^{-3}\,dx$$

$$= -\frac{1}{2}x(x - 2)^{-2} - \frac{1}{2}(x - 2)^{-1} + C$$
$$= -\frac{x}{2(x - 2)^2} - \frac{1}{2(x - 2)} + C$$

19.

	D		I
$+$	$\ln x$	$\searrow$	x^3
$-\int$	$1/x$	$\rightarrow$	$x^4/4$

$$\int x^3 \ln x\,dx = \frac{1}{4}x^4 \ln x - \int \frac{1}{4}x^3\,dx$$

$$= \frac{1}{4}x^4 \ln x - \frac{1}{16}x^4 + C$$

21.

	D		I
+	$\ln(2t)$	↘	$t^2 + 1$
$-\int$	$1/t$	→	$t^3/3 + t$

$$\int (t^2 + 1)\ln(2t)\,dt = \left(\frac{1}{3}t^3 + t\right)\ln(2t) -$$

$$\int \left(\frac{1}{3}t^2 + 1\right) dt = \left(\frac{1}{3}t^3 + t\right)\ln(2t) - \frac{1}{9}t^3 - t + C$$

23.

	D		I
+	$\ln t$	↘	$t^{1/3}$
$-\int$	$1/t$	→	$3t^{4/3}/4$

$$\int t^{1/3}\ln t\,dt = \frac{3}{4}t^{4/3}\ln t - \int \frac{3}{4}t^{1/3}\,dt$$

$$= \frac{3}{4}t^{4/3}\ln t - \frac{9}{16}t^{4/3} + C = \frac{3}{4}t^{4/3}\left(\ln t - \frac{3}{4}\right) + C$$

25.

	D		I
+	$\log_3 x$	↘	1
$-\int$	$1/(x \ln 3)$	→	x

$$\int \log_3 x\,dx = x\log_3 x - \int \frac{1}{\ln 3}\,dx$$

$$= x\log_3 x - \frac{x}{\ln 3} + C$$

27. $\int (xe^{2x} - 4e^{3x})\,dx = \int xe^{2x}\,dx - \int 4e^{3x}\,dx$

$$= \int xe^{2x}\,dx - \frac{4}{3}e^{3x}$$

To evaluate the remaining integral we use integration by parts:

	D		I
+	x	↘	e^{2x}
−	1	↘	$\frac{1}{2}e^{2x}$
$+\int$	0	→	$\frac{1}{4}e^{2x}$

$$\int xe^{2x}\,dx - \frac{4}{3}e^{3x} = \frac{1}{2}xe^{2x} - \frac{1}{4}e^{2x} - \frac{4}{3}e^{3x} + C$$

$$= \left(\frac{1}{2}x - \frac{1}{4}\right)e^{2x} - \frac{4}{3}e^{3x} + C$$

29. $\int (x^2e^x - xe^{x^2})\,dx = \int x^2e^x\,dx - \int xe^{x^2}\,dx$.

To evaluate the first integral we use integration by parts:

	D		I
+	x^2	↘	e^x
−	$2x$	↘	e^x
+	2	↘	e^x
$-\int$	0	→	e^x

$$\int x^2e^x\,dx = x^2e^x - 2xe^x + 2e^x + C$$

$$= (x^2 - 2x + 2)e^x + C.$$

To evaluate the second integral we use substitution:

$u = x^2$, $du = 2x\,dx$, $dx = \frac{1}{2x}\,du$

$$\int xe^{x^2}\,dx = \int xe^u \frac{1}{2x}\,du = \frac{1}{2}\int e^u\,du = \frac{1}{2}e^u + C$$

$$= \frac{1}{2}e^{x^2} + C$$

Combining the two integrals we get

$$\int (x^2e^x - xe^{x^2})\,dx = (x^2 - 2x + 2)e^x - \frac{1}{2}e^{x^2} + C$$

31.

	D		I
+	$x + 1$	↘	e^x
−	1	↘	e^x
+∫	0	→	e^x

$$\int_0^1 (x + 1)e^x\,dx = [(x + 1)e^x - e^x]_0^1$$

$$= [xe^x]_0^1 = e - 0 = e$$

33.

	D		I
+	x^2	↘	$(x + 1)^{10}$
−	$2x$	↘	$(x + 1)^{11}/11$
+	2	↘	$(x + 1)^{12}/132$
−∫	0	→	$(x + 1)^{13}/1716$

$$\int_0^1 x^2(x + 1)^{10}\,dx =$$

$$\left[\frac{1}{11}x^2(x + 1)^{11} - \frac{1}{66}x(x + 1)^{12} + \frac{1}{858}(x + 1)^{13}\right]_0^1$$

$$= \frac{1}{11}2^{11} - \frac{1}{66}2^{12} + \frac{1}{858}2^{13} - \frac{1}{858} = \frac{38{,}229}{286}$$

35.

	D		I
+	$\ln(2x)$	↘	x
−∫	$1/x$	→	$x^2/2$

$$\int_1^2 x\ln(2x)\,dx = \left[\frac{1}{2}x^2\ln(2x)\right]_1^2 - \int_1^2 \frac{1}{2}x\,dx$$

$$= \left[\frac{1}{2}x^2\ln(2x)\right]_1^2 - \left[\frac{1}{4}x^2\right]_1^2$$

$$= 2\ln 4 - \frac{1}{2}\ln 2 - \left(1 - \frac{1}{4}\right)$$

$$= 4\ln 2 - \frac{1}{2}\ln 2 - \frac{3}{4} = \frac{7}{2}\ln 2 - \frac{3}{4}$$

37.

	D		I
+	$\ln(x + 1)$	↘	x
−∫	$1/(x + 1)$	→	$x^2/2$

$$\int_0^1 x\ln(x + 1)\,dx$$

$$= \left[\frac{1}{2}x^2\ln(x + 1)\right]_0^1 - \int_0^1 \frac{x^2}{2(x + 1)}\,dx$$

$$= \frac{1}{2}\ln 2 - \int_0^1 \frac{x^2}{2(x + 1)}\,dx$$

We evaluate this integral using a substitution: $u = x + 1$, $du = dx$; $x = u - 1$; when $x = 0$, $u = 1$; when $x = 1$, $u = 2$.

$$\int_0^1 \frac{x^2}{2(x + 1)}\,dx = \int_1^2 \frac{x^2}{2u}\,du = \int_1^2 \frac{(u - 1)^2}{2u}\,du$$

$$= \frac{1}{2}\int_1^2 \frac{u^2 - 2u + 1}{u}\,du = \frac{1}{2}\int_1^2 \left(u - 2 + \frac{1}{u}\right) du$$

$$= \left[\frac{1}{4}u^2 - u + \frac{1}{2}\ln u\right]_1^2$$

$$= 1 - 2 + \frac{1}{2}\ln 2 - \left(\frac{1}{4} - 1 + 0\right)$$

$$= -\frac{1}{4} + \frac{1}{2}\ln 2$$

Combining with our earlier calculation we get

$$\int_0^1 x\ln(x+1)\,dx = \frac{1}{2}\ln 2 - \left(-\frac{1}{4} + \frac{1}{2}\ln 2\right) = \frac{1}{4}$$

39. We calculate the area using $\int_0^{10} xe^{-x}\,dx$. To evaluate this integral we use integration by parts:

	D		I
+	x	↘	e^{-x}
−	1	↘	$-e^{-x}$
$+\int$	0	→	e^{-x}

$$\int_0^{10} xe^{-x}\,dx = [-xe^{-x} - e^{-x}]_0^{10}$$

$$= -10e^{-10} - e^{-10} - (0 - e^0) = 1 - 11e^{-10}$$

41. We calculate the area using $\int_1^2 (x+1)\ln x\,dx$.

To evaluate this integral we use integration by parts:

	D		I
+	$\ln x$	↘	$x + 1$
$-\int$	$1/x$	→	$x^2/2 + x$

$$\int_1^2 (x+1)\ln x\,dx$$

$$= \left[\left(\frac{1}{2}x^2 + x\right)\ln x\right]_1^2 - \int_1^2 \left(\frac{1}{2}x + 1\right) dx$$

$$= \left[\left(\frac{1}{2}x^2 + x\right)\ln x\right]_1^2 - \left[\frac{1}{4}x^2 + x\right]_1^2$$

$$= 4\ln 2 - 0 - \left[1 + 2 - \left(\frac{1}{4} + 1\right)\right] = 4\ln 2 - \frac{7}{4}$$

43. We compute the displacement by integrating the velocity over the first two minutes, or 120 seconds: $\int_0^{120} 2000te^{-t/120}\,dt$. We evaluate this integral using integration by parts:

	D		I
+	t	↘	$e^{-t/120}$
−	1	↘	$-120e^{-t/120}$
$+\int$	0	→	$14{,}400e^{-t/120}$

$$\int_0^{120} 2000te^{-t/120}\,dt$$

$$= 2000[-120te^{-t/120} - 14{,}400e^{-t/120}]_0^{120}$$

$$= 2000[-14{,}400e^{-1} - 14{,}400e^{-1} - (-14{,}400e^0)]$$

$$= 28{,}800{,}000(1 - 2e^{-1})\text{ ft} \approx 7{,}610{,}000\text{ ft}$$

45. We are given $C'(x) = 10 + \dfrac{\ln(x+1)}{(x+1)^2}$ and $C(0) = 5000$. So

$$C(x) = \int \left[10 + \frac{\ln(x+1)}{(x+1)^2}\right] dx$$

$$= 10x + \int (x+1)^{-2} \ln(x+1)\, dx$$

To evaluate this integral we use integration by parts:

	D		I
+	$\ln(x+1)$	↘	$(x+1)^{-2}$
$-\int$	$1/(x+1)$	→	$-(x+1)^{-1}$

$$C(x) = 10x - (x+1)^{-1}\ln(x+1) + \int (x+1)^{-2}\, dx$$

$$= 10x - (x+1)^{-1}\ln(x+1) - (x+1)^{-1} + K$$

To determine K we substitute $C(0) = 5000$

$5000 = -\ln 1 - 1 + K$

$K = 5001$

So,

$$C(x) = 10x - \frac{\ln(x+1)}{x+1} - \frac{1}{x+1} + 5001$$

47. If $p(t)$ is the price in week t, we are told that $p(t) = 10 + 0.5t$. If $q(t)$ is the weekly sales, we are told that $q(t) = 50e^{-0.02t}$. The weekly revenue is therefore

$R(t) = p(t)q(t) = 50(10 + 0.5t)e^{-0.02t}$

and the total revenue over the coming year is

$$\int_0^{52} 50(10 + 0.5t)e^{-0.02t}\, dt$$

$$= \int_0^{52} 500e^{-0.02t}\, dt + 25\int_0^{52} te^{-0.02t}\, dt$$

We evaluate each integral:

$$\int_0^{52} 500e^{-0.02t}\, dt = [-25{,}000e^{-0.02t}]_0^{52}$$

$= -25{,}000e^{-1.04} - (-25{,}000e^0)$

$= 25{,}000 - 25{,}000e^{-1.04}$

We evaluate the second integral using integration by parts:

	D		I
+	t	↘	$e^{-0.02t}$
−	1	↘	$-50e^{-0.02t}$
$+\int$	0	→	$2500e^{-0.02t}$

$$\int_0^{52} te^{-0.02t}\, dt = [-50te^{-0.02t} - 2500e^{-0.02t}]_0^{52}$$

$= -2600e^{-1.04} - 2500e^{-1.04} - (-2500e^0)$

$= 2500 - 5100e^{-1.04}$

So, the total revenue is

$25{,}000 - 25{,}000e^{-1.04} + 25(2500 - 5100e^{-1.04})$

$\approx \$33{,}598$.

49. If $p(t)$ is the price t years after 1990, we are told that $p(t) = 3e^{0.03t}$. The annual revenue is therefore

$R(t) = p(t)q(t) = 3(17t^2 + 100t + 2300)e^{0.03t}$,

so the total revenue from 1993 to 2003 is

$$\int_3^{13} 3(17t^2 + 100t + 2300)e^{0.03t}\, dt$$

$$= \int_3^{13} (51t^2 + 300t + 6900)e^{0.03t}\, dt$$

We evaluate using integration by parts:

	D		I
+	$51t^2 + 300t + 6900$	↘	$e^{0.03t}$
−	$102t + 300$	↘	$(100/3)e^{0.03t}$
+	102	↘	$(10,000/9)e^{0.03t}$
−∫	0	→	$(1,000,000/27)e^{0.03t}$

$$\int_3^{13} (51t^2 + 300t + 6900)e^{0.03t}\,dt$$

$= [(100/3)(51t^2 + 300t + 6900)e^{0.03t}$
$- (10,000/9)(102t + 300)e^{0.03t}$
$+ (1,000,000/27)(102)e^{0.03t}]_3^{13}$
$\approx \$170,000$ million.

51. Answers will vary. Examples are xe^{x^2} and $e^{x^2} = 1\cdot e^{x^2}$

53. $n+1$ times

55. If $f(x)$ is a polynomial of degree n, then $f^{(n+1)}(x) = 0$. Using integration by parts to evaluate $\int_0^b f(x)e^{-x}\,dx$ we get the following table:

	D		I
+	$f(x)$	↘	e^{-x}
−	$f'(x)$	↘	$-e^{-x}$
+	$f''(x)$	↘	e^{-x}
		…	
±	$f^{(n)}(x)$	↘	$\pm e^{-x}$
∓∫	0	→	$\mp e^{-x}$

So, $\int_0^b f(x)e^{-x}\,dx$

$= [-f(x)e^{-x} - f'(x)e^{-x} - \dots - f^{(n)}(x)e^{-x}]_0^b$
$= -[f(b) + f'(b) + \dots$
$+ f^{(n)}(b)]e^{-b} + [f(0) + f'(0) + \dots + f^{(n)}(0)]e^0$
$= F(0) - F(b)e^{-b}$

7.2

In these solutions we always take the integral of $f(x) - g(x)$ with $f(x) \ge g(x)$. Remember that, if you reverse the order, you will simply get the negative of that integral and should then take the absolute value.

1. We have $x^2 \ge 0 > -1$ for all x, so the two graphs do not cross. The area is $\int_{-1}^{1} [x^2 - (-1)]\,dx = \int_{-1}^{1} (x^2 + 1)\,dx = \left[\frac{1}{3}x^3 + x\right]_{-1}^{1}$ $= \frac{1}{3} + 1 - \left(-\frac{1}{3} - 1\right) = \frac{8}{3}$.

3. $-x = x$ when $x = 0$. The area is $\int_0^2 [x - (-x)]\,dx = \int_0^2 2x\,dx = [x^2]_0^2 = 4 - 0 = 4$.

5. $x = x^2$ when $x^2 - x = 0$, $x(x - 1) = 0$, $x = 0$ or $x = 1$. We calculate the area using two integrals: $\int_{-1}^{0} (x^2 - x)\,dx = \left[\frac{1}{3}x^3 - \frac{1}{2}x^2\right]_{-1}^{0} =$ $0 - \left(-\frac{1}{3} - \frac{1}{2}\right) = \frac{5}{6}$ and $\int_0^1 (x - x^2)\,dx =$ $\left[\frac{1}{2}x^2 - \frac{1}{3}x^3\right]_0^1 = \frac{1}{2} - \frac{1}{3} - (0) = \frac{1}{6}$. The total area is therefore $\frac{5}{6} + \frac{1}{6} = 1$.

7. $e^x > x$ for all x. (Examine the graphs, or consider the fact that $e^x - x$ has its minimum value when its derivative, $e^x - 1$, is 0, which occurs when $x = 0$.) The area is $\int_0^1 (e^x - x)\,dx =$ $\left[e^x - \frac{1}{2}x^2\right]_0^1 = e - \frac{1}{2} - (1) = e - \frac{3}{2} \approx 1.218$.

9. $(x - 1)^2 \ge 0 \ge -(x - 1)^2$, so the area is $\int_0^1 [(x - 1)^2 + (x - 1)^2]\,dx = \int_0^1 2(x - 1)^2\,dx =$ $\left[\frac{2}{3}(x - 1)^3\right]_0^1 = 0 - \left(-\frac{2}{3}\right) = \frac{2}{3}$.

11. $x = x^4$ when $x^4 - x = 0$, $x(x^3 - 1) = 0$, $x = 0$ or $x = 1$. So, the area is $\int_0^1 (x - x^4)\,dx =$ $\left[\frac{1}{2}x^2 - \frac{1}{5}x^5\right]_0^1 = \frac{1}{2} - \frac{1}{5} = \frac{3}{10}$.

13. $x^3 = x^4$ when $x^4 - x^3 = 0$, $x^3(x - 1) = 0$, $x = 0$ or $x = 1$. So, the area is $\int_0^1 (x^3 - x^4)\,dx =$ $\left[\frac{1}{4}x^4 - \frac{1}{5}x^5\right]_0^1 = \frac{1}{4} - \frac{1}{5} - (0) = \frac{1}{20}$.

15. $x^2 = x^4$ when $x^4 - x^2 = 0$, $x^2(x^2 - 1) = 0$, $x = 0$ or $x = \pm 1$. So, the area is $\int_{-1}^{0} (x^2 - x^4)\,dx +$ $\int_0^1 (x^2 - x^4)\,dx = \left[\frac{1}{3}x^3 - \frac{1}{5}x^5\right]_{-1}^{0} + \left[\frac{1}{3}x^3 - \frac{1}{5}x^5\right]_0^1$ $= 0 - \left(-\frac{1}{3} + \frac{1}{5}\right) + \left(\frac{1}{3} - \frac{1}{5}\right) = \frac{4}{15}$. (In fact, since $x^2 \ge x^4$ on all of $[-1, 1]$, we could have used the single integral $\int_{-1}^{1} (x^2 - x^4)\,dx$ to calculate this area.)

17. Here are the graphs of $y = e^x$ and $y = 2$:

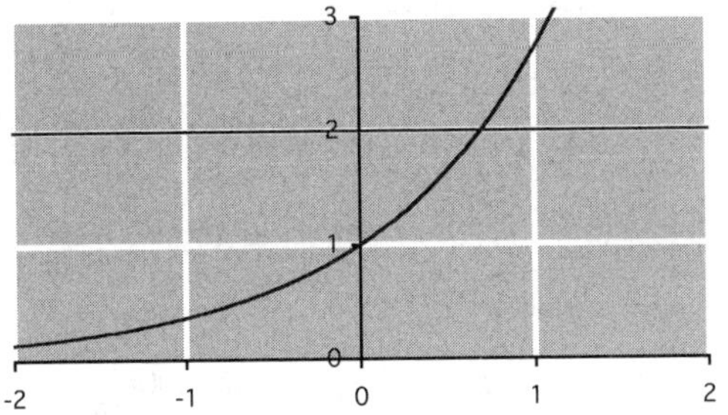

The two graphs intersect where $e^x = 2$, $x = \ln 2$. From the graph we can see that the area we want is the area between these two graphs for $0 \le x \le \ln 2$. So we compute $\int_0^{\ln 2} (2 - e^x)\, dx =$

$[2x - e^x]_0^{\ln 2} = 2 \ln 2 - e^{\ln 2} - (0 - 1) =$
$2 \ln 2 - 2 + 1 = 2 \ln 2 - 1.$

19. Here are the graphs of $y = \ln x$ and $y = 2 - \ln x$:

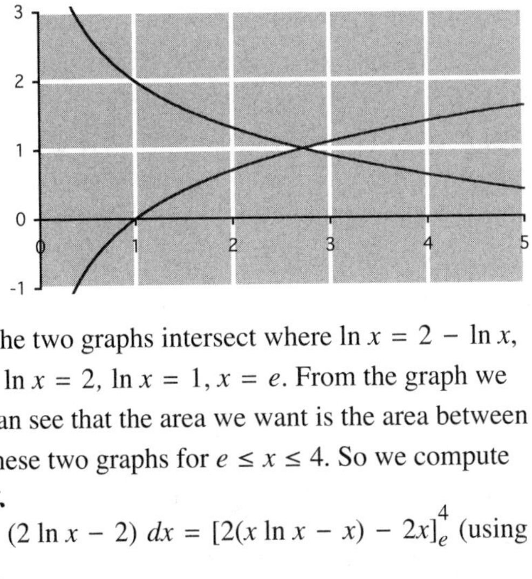

The two graphs intersect where $\ln x = 2 - \ln x$, $2 \ln x = 2$, $\ln x = 1$, $x = e$. From the graph we can see that the area we want is the area between these two graphs for $e \le x \le 4$. So we compute $\int_e^4 (2 \ln x - 2)\, dx = [2(x \ln x - x) - 2x]_e^4$ (using the antiderivative of $\ln x$ we derived in Section 7.1) $= [2x \ln x - 4x]_e^4 = 8 \ln 4 - 16 -$
$(2e \ln e - 4e) = 8 \ln 4 + 2e - 16 \approx 0.5269.$

21. Formula for
Online Utilities
→ Numerical Integration Utility:

```
abs(e^x-(2x+1))
```

Left End-Point: –1, Right End-Point: 1 (Use "Adaptive Quadrature")
Formula for TI-83:

```
fnInt(abs(e^x-(2x+1)),X,-1,1)
```

Answer: 0.9138

23. Here are the graphs of $y = \ln x$ and $y = \frac{1}{2}x - \frac{1}{2}$:

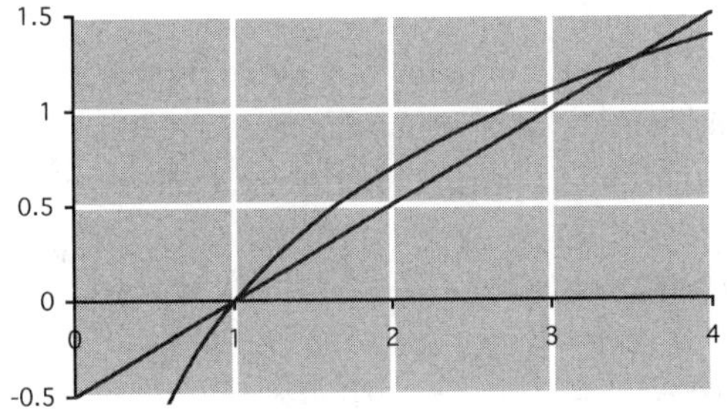

The two graphs intersect at $x = 1$ and at a point somewhere between $x = 3$ and $x = 4$. We cannot solve the equation $\ln x = \frac{1}{2}x - \frac{1}{2}$ algebraically, but we can use technology to estimate the second intersection point. To four decimal places it is $x \approx 3.5129$. Therefore, the area is approximately

$$\int_1^{3.5129} \left(\ln x - \frac{1}{2}x + \frac{1}{2}\right) dx =$$

$$\left[x \ln x - x - \frac{1}{4}x^2 + \frac{1}{2}x\right]_1^{3.5129} =$$

$$3.5129 \ln(3.5129) - 3.5129 - \frac{1}{4}(3.5129)^2 + \frac{1}{2}3.5129 - \left(0 - 1 - \frac{1}{4} + \frac{1}{2}\right) \approx 0.3222.$$

25. Area $= \int_0^5 [R(t) - C(t)]\, dt$

$$= \int_0^5 [(100 + 10t) - (90 + 5t)]\, dt$$

$$= \int_0^5 (10 + 5t)\, dt = \left[10t + \frac{5}{2}t^2\right]_0^5 = 112.5.$$

Since area under a curve represents total change, this area represents your total profit for the week, \$112.50.

27. (a) Since area under a curve represents total change, the area represents the accumulated U.S. trade deficit with China (total excess value of imports over exports) for the 8-year period 1996–2004.

(b) Area $= \int_1^9 [I(t) - E(t)]\, dt$

$$= \int_1^9 [(t^2 + 3.5t + 50) - (0.4t^2 - 1.6t + 14)]\, dt$$

$$= \int_1^9 (0.6t^2 + 5.1t + 36)\, dt$$

$$= \left[0.2t^3 + \frac{5.1}{2}t^2 + 36t\right]_1^9$$

$$= 637.6 \approx 640.$$

The U.S. accumulated a \$640 billion trade deficit with China over the period 1996–2004.

29. (a) Since 2000 corresponds to $t = 10$ and 2010 corresponds to $t = 20$, we compute

$$\int_{10}^{20} [P(t) - I(t)]\, dt$$

$$= \int_{10}^{20} [(20t^3 + 1000t^2 + 28{,}000t + 360{,}000)$$

$$- (15t^3 + 800t^2 + 19{,}000 + 200{,}000)]\, dt$$

$$= \int_{10}^{20} (5t^3 + 200t^2 + 9000t + 160{,}000)\, dt$$

$$= \left[\frac{5}{4}t^4 + \frac{200}{3}t^3 + 4500t^2 + 160{,}000t\right]_{10}^{20}$$

$$\approx 3{,}600{,}000.$$

(b) This is the area of the region between the graphs of $P(t)$ and $I(t)$ for $10 \le t \le 20$.

31. The area between the export and import curves represents Canada's accumulated trade surplus (that is, the total excess of exports over imports) from January 1997 to January 2001.

33. (A)

35. The claim is wrong because the area under a curve can only represent income if the curve is a graph of income *per unit time*. The value of a stock price is not income per unit time—the income can only be realized when the stock is sold and it amounts to the current market price. The total net income (per share) from the given investment would be the stock price on the date of sale minus the purchase price of \$40.

7.3

1. Average $= \frac{1}{2}\int_0^2 x^3\,dx = \frac{1}{2}\left[\frac{1}{4}x^4\right]_0^2$

$= \frac{1}{2}(4 - 0) = 2$

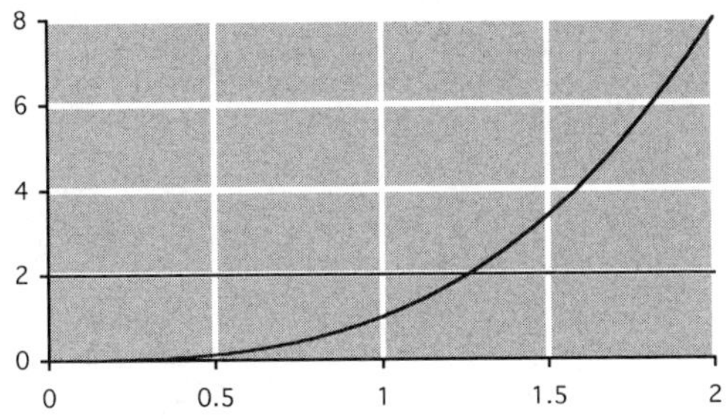

3. Average $= \frac{1}{2}\int_0^2 (x^3 - x)\,dx = \frac{1}{2}\left[\frac{1}{4}x^4 - \frac{1}{2}x^2\right]_0^2$

$= \frac{1}{2}(4 - 2 - 0) = 1$

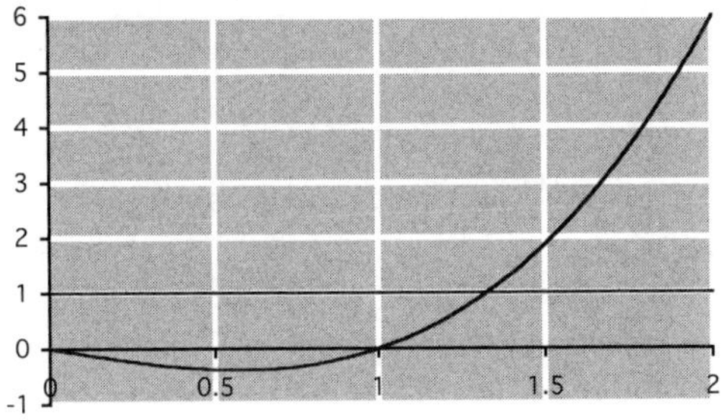

5. Average $= \frac{1}{2}\int_0^2 e^{-x}\,dx = \frac{1}{2}[-e^{-x}]_0^2$

$= \frac{1}{2}(1 - e^{-2}) \approx 0.43$

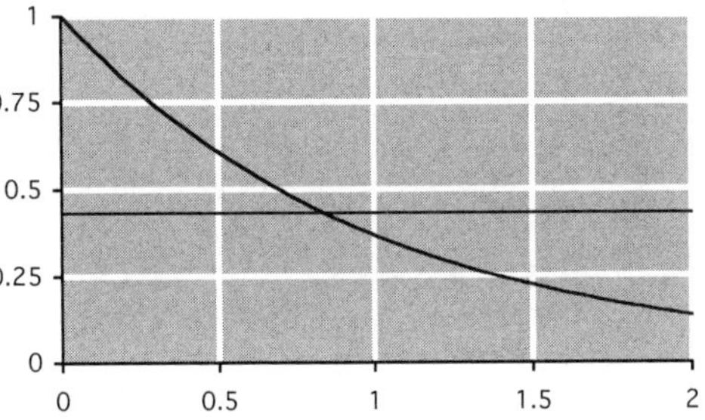

7. $\bar{r}(2) = (3 + 5 + 10)/3 = 6$, and so on.

x	0	1	2	3	4	5	6	7
$r(x)$	3	5	10	3	2	5	6	7
$\bar{r}(x)$			6	6	5	10/3	13/3	6

9. We must have $(1 + 2 + r(2))/3 = \bar{r}(2) = 3$, so $r(2) = 6$. Working from left to right we fill in the other missing values similarly.

x	0	1	2	3	4	5	6	7
$r(x)$	1	2	6	7	11	15	10	2
$\bar{r}(x)$			3	5	8	11	12	9

11. Moving average: $\bar{f}(x) = \frac{1}{5}\int_{x-5}^{x} t^3\,dt$

$= \frac{1}{5}\left[\frac{1}{4}t^4\right]_{x-5}^{x} = \frac{1}{20}[x^4 - (x-5)^4]$

$= \frac{1}{20}(20x^3 - 150x^2 + 500x - 625)$

$= x^3 - \frac{15}{2}x^2 + 25x - \frac{125}{4}$

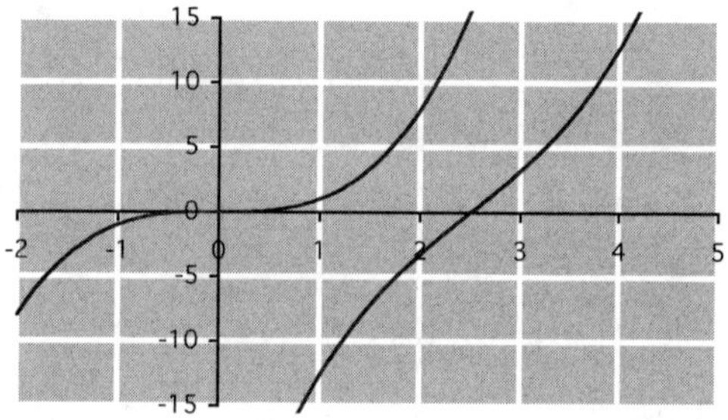

13. Moving average: $\bar{f}(x) = \frac{1}{5}\int_{x-5}^{x} t^{2/3}\,dt$

$= \frac{1}{5}\left[\frac{3}{5}t^{5/3}\right]_{x-5}^{x} = \frac{3}{25}[x^{5/3} - (x-5)^{5/3}]$

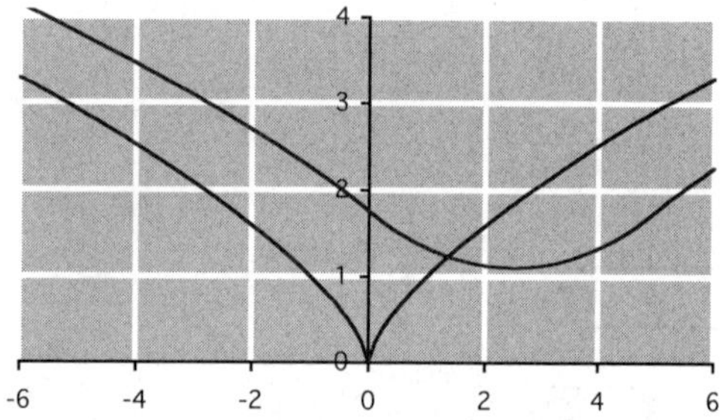

15. Moving average: $\bar{f}(x) = \frac{1}{5}\int_{x-5}^{x} e^{0.5t}\,dt$

$= \frac{1}{5}[2e^{0.5t}]_{x-5}^{x} = \frac{2}{5}[e^{0.5x} - e^{0.5(x-5)}]$

$= \frac{2}{5}(1 - e^{-2.5})e^{0.5x} \approx 0.367e^{0.5x}$

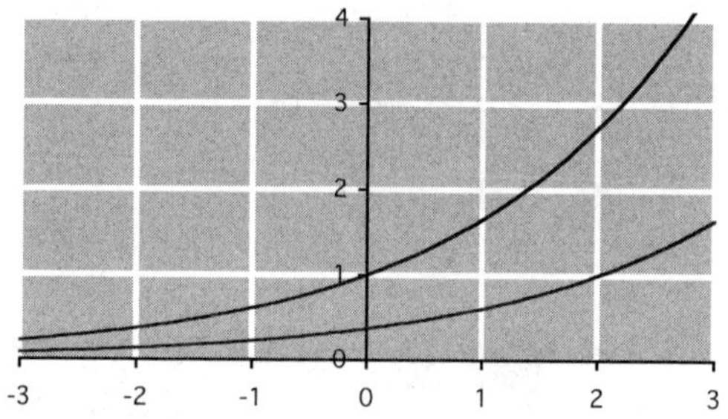

17. Moving average: $\bar{f}(x) = \frac{1}{5}\int_{x-5}^{x} t^{1/2}\,dt$

$= \frac{1}{5}\left[\frac{2}{3}t^{3/2}\right]_{x-5}^{x} = \frac{2}{15}[x^{3/2} - (x-5)^{3/2}]$

(Note that the domain is $x \geq 5$.)

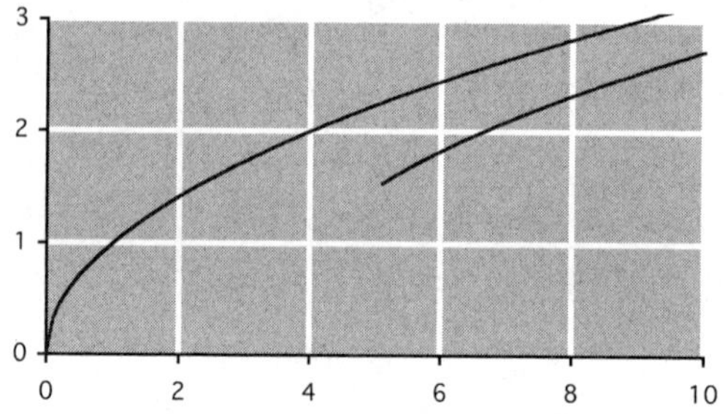

19. Ploting the moving average on a TI-83/84:

```
Y1 = 10x/(1+5*abs(x))
Y2 = (1/3)fnInt(Y1(T),T,X-3,X)
```

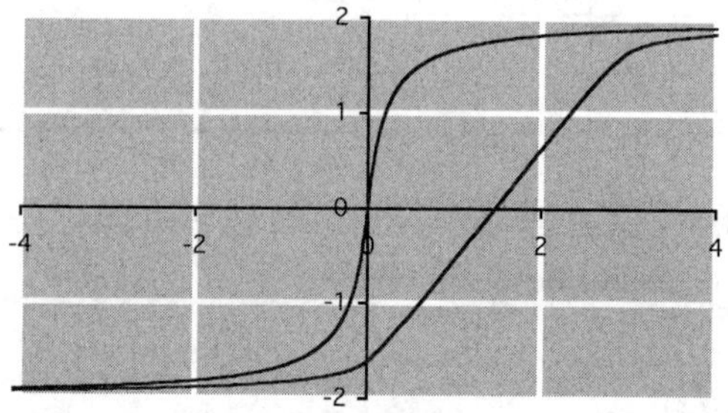

21. Ploting the moving average on a TI-83/84:

```
Y1 = ln(1+x^2)
Y2 = (1/3)fnInt(Y1(T),T,X-3,X)
```

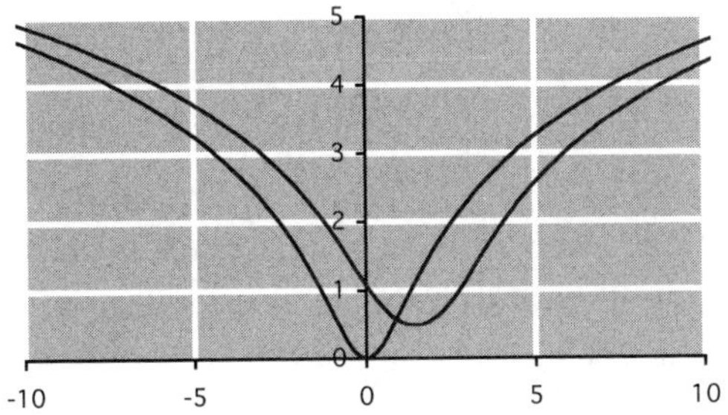

23. (117+120+123+126+129+132+132+130 +130+131)/10 = 1270/10 = 127

25. $\frac{1}{3}\int_{8}^{11}(0.355t - 1.6)\,dt = \frac{1}{3}[0.1755t^2 - 1.6t]_8^{11}$

= \$1.7345 million

27. The amount you have in the account at time t is $A(t) = 10{,}000e^{0.08t}$, $0 \le t \le 1$. The average amount over the first year is

$$\int_0^1 10{,}000e^{0.08t}\,dt = [125{,}000e^{0.08t}]_0^1$$

= \$10,410.88.

29. The amount in the account begins at \$3000 at the beginning of the month and then declines linearly to 0 by the end of the month. So, the amount in the account during the month is

$A(t) = 3000 - 3000t$, $0 \le t \le 1$.

The average over one month is therefore

$$\int_0^1 (3000 - 3000t)\,dt = [3000t - 1500t^2]_0^1$$

= \$1500.

Since the average over each month is \$1500, the average over several months is also \$1500.

31.

Year t	1995	1996	1997	1998	1999	2000	2001	2002	2003	2004
Employment (millions)	117	120	123	126	129	132	132	130	130	131
Moving average (millions)				122	125	128	130	131	131	131

Each 4-year moving average is computed by averaging that year's figure with that of the preceding three years:

1998: (117+120+123+126)/4 = 121.5 ≈ 122

1999: (120+123+126+129)/4 = 124.5 ≈ 125

and so on.

Some changes are larger (for example, 2000 to 2001) and others are smaller (for example, 2003 to 2004).

33. (a)

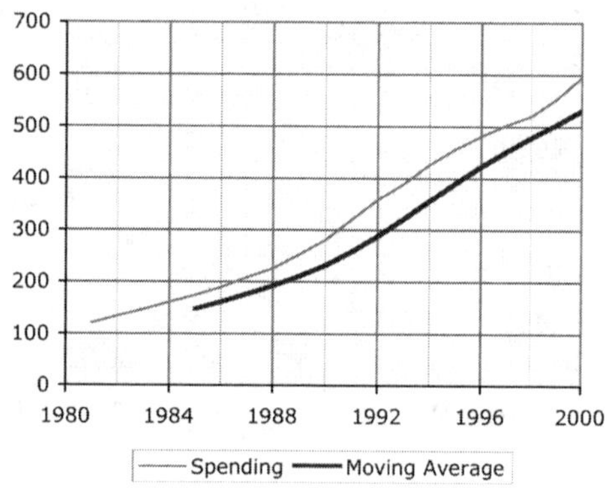

(b) The moving average figures in 1991 and 1997 are 321 and 504 respectively. The average rate of change of the moving average over the interval [1991, 1997] is therefore

$$\frac{504-321}{1997-1991} = \frac{183}{6} = 30.5 \approx \$31 \text{ billion/year}$$

Public spending on health care in the U.S. was increasing at a rate of approximately \$31 billion per year during the given period.

35. (a) Average $= \frac{1}{10}\int_3^{13}(17t^2 + 100t + 2300)\,dt$

$= \frac{1}{10}\left[\frac{17}{3}t^3 + 50t^2 + 2300t\right]_3^{13} \approx 4300$

(b) Moving average $= \frac{1}{2}\int_{t-2}^{t}(17t^2 + 100t + 2300)dt$

$= \frac{1}{2}\left[\frac{17}{3}t^3 + 50t^2 + 2300t\right]_{t-2}^{t}$

=

$\frac{1}{2}\left[\frac{17}{3}[t^3 - (t-2)^3] + 50[t^2 - (t-2)^2] + 4600\right]$

(c) The function is quadratic because the t^3 terms cancel.

37. (a) The line through $(t, s) = (0, 240)$ and $(25, 600)$ is

$s = 14.4t + 240.$

(b) $\bar{s}(t) = \frac{1}{4}\int_{t-4}^{t}(14.4x + 240)\,dx$

$= \frac{1}{4}[7.2x^2 + 240x]_{t-4}^{t}$

$= \frac{1}{4}\{7.2[t^2 - (t-4)^2] + 960\}$

$= \frac{1}{4}(57.6t + 844.4) = 14.4t + 211.2$

(c) The slope of the moving average is the same as the slope of the original function (because the original is linear).

39. $\bar{f}(x) = \frac{1}{a}\int_{x-a}^{x}(mt + b)\,dt = \frac{1}{a}\left[\frac{m}{2}t^2 + bt\right]_{x-a}^{x}$

$= \frac{1}{a}\left[\frac{m}{2}[x^2 - (x-a)^2] + bx - b(x-a)\right]$

$= \frac{1}{a}\left[max - \frac{ma^2}{2} + ab\right] = mx + b - \frac{ma}{2}$

41. (a)

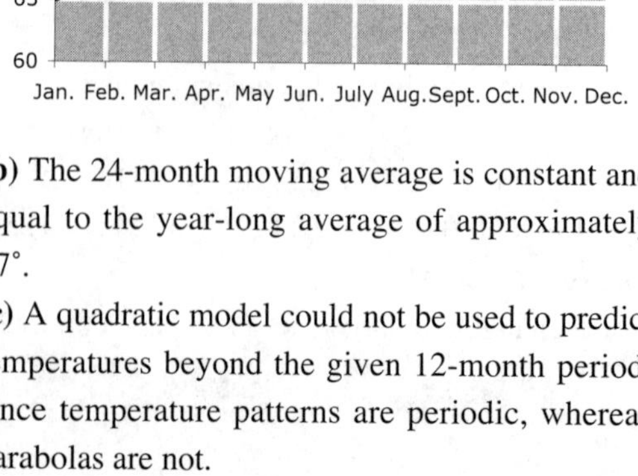

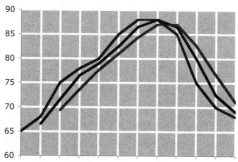

(b) The 24-month moving average is constant and equal to the year-long average of approximately 77°.

(c) A quadratic model could not be used to predict temperatures beyond the given 12-month period, since temperature patterns are periodic, whereas parabolas are not.

43. The moving average "blurs" the effects of short-term oscillations in the price, and shows the longer-term trend of the stock price.

45. The area above the x-axis equals the area below the x-axis. Example: $y = x$ on $[-1,1]$.

47. This need not be the case; for instance, the function $f(x) = x^2$ on $[0, 1]$ has average value 1/3, whereas the value midway between the maximum and minimum is 1/2.

49. (C): A shorter term moving average most closely approximates the original function, since it averages the function over a shorter period, and continuous functions change by only a small amount over a small period.

7.4

1. $q = 5 - p/2$, so $\bar{q} = 5 - 5/2 = 5/2$. The consumers' surplus is

$$\int_0^{5/2} (10 - 2q - 5)\, dq = \int_0^{5/2} (5 - 2q)\, dq$$

$$= [5q - q^2]_0^{5/2} = 5\left(\frac{5}{2}\right) - \left(\frac{5}{2}\right)^2 - (0) = \$6.25.$$

3. $q = (100 - p)^2/9$, so $\bar{q} = (100 - 76)^2/9 = 64$. The consumers' surplus is

$$\int_0^{64} (100 - 3\sqrt{q} - 76)\, dq = \int_0^{64} (24 - 3\sqrt{q})\, dq$$

$$= [24q - 2q^{3/2}]_0^{64} = 24(64) - 2(64)^{3/2} - (0)$$

$$= \$512.$$

5. $q = -\frac{1}{2}\ln(p/500)$, so

$$\bar{q} = -\frac{1}{2}\ln(1/5) = \frac{1}{2}\ln 5.$$

The consumers' surplus is

$$\int_0^{(\ln 5)/2} (500e^{-2q} - 100)\, dq$$

$$= [-250e^{-2q} - 100q]_0^{(\ln 5)/2}$$

$$= -250(1/5) - 50\ln 5 - (-250) = \$119.53.$$

7. $\bar{q} = 100 - 2(20) = 60$; $p = 50 - q/2$. The consumers' surplus is

$$\int_0^{60} \left(50 - \frac{1}{2}q - 20\right) dq = \int_0^{60} \left(30 - \frac{1}{2}q\right) dq$$

$$= \left[30q - \frac{1}{4}q^2\right]_0^{60} = 30(60) - \frac{1}{4}(60)^2 - (0)$$

$$= \$900.$$

9. $\bar{q} = 100 - 0.25(10)^2 = 75$; $p = 2\sqrt{100 - q}$. The consumers' surplus is

$$\int_0^{75} (2\sqrt{100 - q} - 10)\, dq$$

$$= \left[-\frac{4}{3}(100 - q)^{3/2} - 10q\right]_0^{75}$$

$$= -\frac{4}{3}(100 - 75)^{3/2} - 10(75)$$

$$- \left(-\frac{4}{3}(100)^{3/2} - 0\right)$$

$$= \$416.67.$$

11. $\bar{q} = 500e^{-0.5} - 50$; $p = -2\ln\left(\frac{1}{500}q + \frac{1}{10}\right)$.

The consumers' surplus is

$$\int_0^{500e^{-0.5}-50} \left[-2\ln\left(\frac{1}{500}q + \frac{1}{10}\right) - 1\right] dq$$

$$= -2\int_0^{500e^{-0.5}-50} \ln\left(\frac{1}{500}q + \frac{1}{10}\right) dq - [q]_0^{500e^{-0.5}-50}$$

$$= -2\int_0^{500e^{-0.5}-50} \ln\left(\frac{1}{500}q + \frac{1}{10}\right) dq$$

$$- 500e^{-0.5} + 50.$$

We evaluate the remaining integral using substitution (and the integral of ln u we found by integration by parts):

Let $u = \frac{1}{500}q + \frac{1}{10}$; $du = \frac{1}{500}dq$, $dq = 500\, du$.

When $q = 0$, $u = 0.1$; when $q = 500e^{-0.5} - 50$, $u = e^{-0.5}$.

$$\int_0^{500e^{-0.5}-50} \ln\left(\frac{1}{500}q + \frac{1}{10}\right) dq = \int_{0.1}^{e^{-0.5}} 500\ln u\, du$$

$$= 500[u\ln u - u]_{0.1}^{e^{-0.5}}$$

$$= 500[e^{-0.5}\ln(e^{-0.5}) - e^{-0.5} - (0.1\ln(0.1) - 0.1)]$$

$$\approx -289.769.$$

The consumers' surplus is therefore

$-2(-289.769) - 500e^{-0.5} + 50 = \$326.27.$

13. $q = p/2 - 5$, so $\bar{q} = 20/2 - 5 = 5$. The producers' surplus is

$$\int_0^5 [20 - (10 + 2q)]\, dq = \int_0^5 (10 - 2q)\, dq$$

$$= [10q - q^2]_0^5 = 50 - 25 - (0) = \$25.$$

15. $q = (p/2 - 5)^3$, so $\bar{q} = (12/2 - 5)^3 = 1$. The producers' surplus is

$$\int_0^1 [12 - (10 + 2q^{1/3})]\, dq = \int_0^1 (2 - 2q^{1/3})\, dq$$

$$= \left[2q - \frac{3}{2} q^{4/3}\right]_0^1 = 2 - \frac{3}{2} - (0) = \$0.50.$$

17. $q = 2 \ln(p/500)$, so

$\bar{q} = 2 \ln(1000/500) = 2 \ln 2.$

The producers' surplus is

$$\int_0^{2 \ln 2} (1000 - 500e^{0.5q})\, dq$$

$$= [1000q - 1000e^{0.5q}]_0^{2 \ln 2}$$

$$= 2000 \ln 2 - 2000 - (-1000)$$

$$= 2000 \ln 2 - 1000 = \$386.29.$$

19. $\bar{q} = 2(40) - 50 = 30$; $p = q/2 + 25$. The producers' surplus is

$$\int_0^{30} \left[40 - \left(\frac{1}{2} q + 25\right)\right] dq$$

$$= \int_0^{30} \left(15 - \frac{1}{2} q\right) dq = \left[15q - \frac{1}{4} q^2\right]_0^{30}$$

$$= 450 - 225 - (0) = \$225.$$

21. $\bar{q} = 0.25(10)^2 - 10 = 15$

$p = \sqrt{4q + 40} = 2\sqrt{q + 10}\,.$

The producers' surplus is

$$\int_0^{15} (10 - 2\sqrt{q + 10})\, dq$$

$$= \left[10q - \frac{4}{3}(q + 10)^{3/2}\right]_0^{15}$$

$$= 150 - \frac{500}{3} - \left(-\frac{4}{3} \cdot 10^{3/2}\right) = \$25.50.$$

23. $\bar{q} = 500e^{0.05(10)} - 50 = 500e^{0.5} - 50$

$p = 20 \ln\left(\frac{1}{500} q + \frac{1}{10}\right).$

The producers' surplus is

$$\int_0^{500e^{0.5}-50} \left[10 - 20 \ln\left(\frac{1}{500} q + \frac{1}{10}\right)\right] dq$$

$$= \left[10q - (20q + 1000) \ln\left(\frac{1}{500} q + \frac{1}{10}\right) + (20q + 1000)\right]_0^{500e^{0.5}-50}$$

(using the substitution $u = \frac{1}{500} q + \frac{1}{10}$ as in Exercise 11)

$$= 5000e^{0.5} - 500 - 10{,}000e^{0.5} \ln e^{0.5} + 10{,}000e^{0.5} - \left(-1000 \ln\left(\frac{1}{10}\right) + 1000\right)$$

$$= \$12{,}684.63.$$

25. $TV = \int_0^{10} 30{,}000\, dt = [30{,}000t]_0^{10} = \$300{,}000;$

$FV = \int_0^{10} 30{,}000e^{0.07(10-t)}\, dt =$

$$\left[-\frac{30{,}000}{0.07} e^{0.07(10-t)}\right]_0^{10} = \$434{,}465.45$$

27. $TV = \int_0^{10} (30{,}000 + 1000t)\, dt$

$= [30{,}000t + 500t^2]_0^{10} = \$350{,}000;$

$FV = \int_0^{10} (30{,}000 + 1000t)e^{0.07(10-t)}\, dt =$

$\left[-\frac{1}{0.07}(30{,}000 + 1000t)\, e^{0.07(10-t)} - \frac{1000}{0.07^2}\, e^{0.07(10-t)}\right]_0^{10}$

(using integration by parts) $= \$498{,}496.61$

29. $TV = \int_0^{10} 30{,}000e^{0.05t}\, dt = [600{,}000e^{0.05t}]_0^{10}$

$= \$389{,}232.76;$

$FV = \int_0^{10} 30{,}000e^{0.05t}e^{0.07(10-t)}\, dt$

$= \int_0^{10} 30{,}000e^{0.7}e^{-0.02t}\, dt$

$= [-1{,}500{,}000e^{0.7}e^{-0.02t}]_0^{10} = \$547{,}547.16$

31. $TV = \int_0^{5} 20{,}000\, dt = [20{,}000t]_0^5 = \$100{,}000;$

$PV = \int_0^{5} 20{,}000e^{-0.08t}\, dt = [-250{,}000e^{-0.08t}]_0^5$

$= \$82{,}419.99$

33. $TV = \int_0^{5} (20{,}000 + 1000t)\, dt$

$= [20{,}000t + 500t^2]_0^5 = \$112{,}500;$

$PV = \int_0^{5} (20{,}000 + 1000t)e^{-0.08t}\, dt$

$= [-(250{,}000 + 12{,}500t)e^{-0.08t} - 156{,}250e^{-0.08t}]_0^5$

(by integration by parts)

$= \$92{,}037.48$

35. $TV = \int_0^{5} 20{,}000e^{0.03t}\, dt = \left[\frac{20{,}000}{0.03}\, e^{0.03t}\right]_0^5$

$= \$107{,}889.50;$

$PV = \int_0^{5} 20{,}000e^{0.03t}e^{-0.08t}\, dt$

$= \int_0^{5} 20{,}000e^{-0.05t}\, dt = [-400{,}000e^{-0.05t}]_0^5$

$= \$88{,}479.69$

37. To find the equilibrium tuition set demand equal to supply:

$20{,}000 - 2p = 7500 + 0.5p$

so $\bar{p} = \$5000$.

The equilibrium supply is thus

$\bar{q} = 20{,}000 - 2(5{,}000) = 10{,}000.$

To find the consumers' surplus we solve for

$p = 10{,}000 - q/2$ and compute

$CS = \int_0^{10{,}000} \left(10{,}000 - \frac{q}{2} - 5000\right) dq$

$= \int_0^{10{,}000} \left(5000 - \frac{q}{2}\right) dq = \left[5000q - \frac{q^2}{4}\right]_0^{10{,}000}$

$= \$25{,}000{,}000.$

To find the producers' surplus we solve for

$p = 2q - 15{,}000$

and compute

$$PS = \int_0^{10,000} [5000 - (2q - 15,000)]\, dq$$
$$= \int_0^{10,000} (20,0000 - 2q)\, dq$$
$$= [20,000q - q^2]_0^{10,000} = \$100,000,000.$$

The total social gain is \$125 million.

39. $\bar{q} = b - m\bar{p},\ p = \frac{1}{m}(b - q)$

$$CS = \int_0^{b-m\bar{p}} \left[\frac{1}{m}(b - q) - \bar{p}\right] dq$$
$$= \left[\frac{b}{m}q - \frac{1}{2m}q^2 - \bar{p}q\right]_0^{b-m\bar{p}}$$
$$= \frac{b}{m}(b - m\bar{p}) - \frac{1}{2m}(b - m\bar{p})^2$$
$$- \bar{p}(b - m\bar{p}) - (0)$$
$$= \frac{1}{2m}(b - m\bar{p})[2b - (b - m\bar{p}) - 2m\bar{p}]$$
$$= \frac{1}{2m}(b - m\bar{p})^2$$

41. Total revenue $= \int_{-1}^{4} (-1.7t^2 + 5t + 28)\, dt$

$$= \left[-\frac{1.7}{3}t^3 + 2.5t^2 + 28t\right]_{-1}^{4} \approx 140$$

43. Total revenue $= \int_{-1}^{4} 150e^{0.14t}\, dt$

$$= \left[\frac{150}{0.14}e^{0.14t}\right]_{-1}^{4} \approx 940$$

45. Total revenue $= \int_{-1}^{4} (-1.7t^2 + 5t + 28)e^{0.04(4-t)}\, dt$

$$= \left[-\frac{(-1.7t^2 + 5t + 28)e^{0.04(4-t)}}{0.04} - \frac{(-3.4t + 5)e^{0.04(4-t)}}{0.04^2} - \frac{-3.4e^{0.04(4-t)}}{0.04^3}\right]_{-1}^{4}$$
$$\approx 160$$

47. Total revenue $= \int_{-1}^{4} 150e^{0.14t}e^{-0.05(4-t)}\, dt$

$$= \int_{-1}^{4} 150e^{-0.2}e^{0.19t}\, dt = \left[\frac{150}{0.19}e^{-0.2}e^{0.19t}\right]_{-1}^{4}$$
$$\approx 850$$

49. $R(t) = 12 \times 700 = \$8400\text{/year}.$

$$FV = \int_0^{45} 8400e^{0.06(45-t)}\, dt$$
$$= [-140,000e^{0.06(45-t)}]_0^{45} = \$1,943,162.44$$

51. $R(t) = 12 \times 700e^{0.03t} = 8400e^{0.03t}$

$$FV = \int_0^{45} 8400e^{0.03t}e^{0.06(45-t)}\, dt$$
$$= \int_0^{45} 8400e^{2.7}e^{-0.03t}\, dt = [-280,000e^{2.7}e^{-0.03t}]_0^{45}$$
$$= \$3,086,245.73$$

53. $R(t) = 3375.\ PV = \int_0^{30} 3375e^{-0.04t}\, dt$

$$= [-84,375e^{-0.04t}]_0^{30} = \$58,961.74$$

55. $R(t) = 100{,}000 + 5000t$

$$PV = \int_0^{20} (100{,}000 + 5000t)e^{-0.05t}\, dt$$

$$= [-20(100{,}000 + 5000t)e^{-0.05t} - 400(5000)e^{-0.05t}]_0^{20}$$

$$= \$1{,}792{,}723.35$$

57. Total

59. She is correct, provided there is a positive rate of return, in which case the future value (which includes interest) is greater than the total value (which does not).

61. $PV < TV < FV$

7.5

1. $\displaystyle\int_1^{+\infty} x\,dx = \lim_{M\to+\infty}\int_1^M x\,dx = \lim_{M\to+\infty}\left[\frac{1}{2}x^2\right]_1^M$

$\displaystyle = \lim_{M\to+\infty}\left(\frac{1}{2}M^2 - \frac{1}{2}\right) = +\infty$; diverges

3. $\displaystyle\int_{-2}^{+\infty} e^{-0.5x}\,dx = \lim_{M\to+\infty}\int_{-2}^{M} e^{-0.5x}\,dx$

$\displaystyle = \lim_{M\to+\infty}[-2e^{-0.5x}]_{-2}^{M}$

$\displaystyle = \lim_{M\to+\infty}(-2e^{-0.5M} + 2e) = 2e$; converges

5. $\displaystyle\int_{-\infty}^{2} e^x\,dx = \lim_{M\to-\infty}\int_M^2 e^x\,dx = \lim_{M\to-\infty}[e^x]_M^2$

$\displaystyle = \lim_{M\to-\infty}(e^2 - e^M) = e^2$; converges

7. $\displaystyle\int_{-\infty}^{-2}\frac{1}{x^2}\,dx = \lim_{M\to-\infty}\int_M^{-2}\frac{1}{x^2}\,dx = \lim_{M\to-\infty}[-x^{-1}]_M^{-2}$

$\displaystyle = \lim_{M\to-\infty}\left(\frac{1}{2} + \frac{1}{M}\right) = \frac{1}{2}$; converges

9. $\displaystyle\int_0^{+\infty} x^2e^{-6x}\,dx = \lim_{M\to+\infty}\int_0^M x^2e^{-6x}\,dx$

$\displaystyle = \lim_{M\to+\infty}\left[-\frac{1}{6}x^2e^{-6x} - \frac{2}{36}xe^{-6x} - \frac{2}{216}e^{-6x}\right]_0^M$

$\displaystyle = \lim_{M\to+\infty}\left(-\frac{1}{6}M^2e^{-6M} - \frac{2}{36}Me^{-6M} - \frac{2}{216}e^{-6M} + \frac{2}{216}\right)$

$\displaystyle = \frac{2}{216} = \frac{1}{108}$; converges

11. $\displaystyle\int_0^5\frac{2}{x^{1/3}}\,dx = \lim_{r\to0^+}\int_r^5\frac{2}{x^{1/3}}\,dx = \lim_{r\to0^+}[3x^{2/3}]_r^5$

$\displaystyle = \lim_{r\to0^+}(3\times5^{2/3} - 3r^{2/3}) = 3\times5^{2/3}$; converges

13. $\displaystyle\int_{-1}^{2}\frac{3}{(x+1)^2}\,dx = \lim_{r\to-1^+}\int_r^2\frac{3}{(x+1)^2}\,dx$

$\displaystyle = \lim_{r\to-1^+}[-3(x+1)^{-1}]_r^2$

$\displaystyle = \lim_{r\to-1^+}\left(-1 + \frac{3}{r+1}\right) = +\infty$; diverges

15. $\displaystyle\int_{-1}^{2}\frac{3x}{x^2-1}\,dx = \lim_{r\to-1^+}\int_r^0\frac{3x}{x^2-1}\,dx$

$\displaystyle + \lim_{r\to1^-}\int_0^r\frac{3x}{x^2-1}\,dx + \lim_{r\to1^+}\int_r^2\frac{3x}{x^2-1}\,dx.$

Now, $\displaystyle\int\frac{3x}{x^2-1}\,dx = \frac{3}{2}\ln|x^2-1| + C$

by substitution, so

$\displaystyle\lim_{r\to-1^+}\int_r^0\frac{3x}{x^2-1}\,dx = \lim_{r\to-1^+}\left[\frac{3}{2}\ln|x^2-1|\right]_r^0$

$\displaystyle = \lim_{r\to-1^+}\left(-\frac{3}{2}\ln|r^2-1|\right) = +\infty$

Since this one part diverges, the whole integral diverges. (In fact, all three parts diverge.)

17. $\displaystyle\int_{-2}^{2} \frac{1}{(x+1)^{1/5}}\,dx$

$$= \lim_{r\to -1^-} \int_{-2}^{r} \frac{1}{(x+1)^{1/5}}\,dx + \lim_{r\to -1^+} \int_{r}^{2} \frac{1}{(x+1)^{1/5}}\,dx$$

$$= \lim_{r\to -1^-} \left[\frac{5}{4}(x+1)^{4/5}\right]_{-2}^{r} + \lim_{r\to -1^+} \left[\frac{5}{4}(x+1)^{4/5}\right]_{r}^{2}$$

$$= \lim_{r\to -1^-} \left[\frac{5}{4}(r+1)^{4/5} - \frac{5}{4}\right] + \lim_{r\to -1^+} \left[\frac{5}{4}3^{4/5} - \frac{5}{4}(r+1)^{4/5}\right]$$

$= \frac{5}{4}(3^{4/5} - 1)$; converges

19. $\displaystyle\int_{-1}^{1} \frac{2x}{x^2-1}\,dx$

$$= \lim_{r\to -1^+} \int_{r}^{0} \frac{2x}{x^2-1}\,dx + \lim_{r\to 1^-} \int_{0}^{r} \frac{2x}{x^2-1}\,dx$$

$$= \lim_{r\to -1^+} [\ln|x^2-1|]_r^0 + \lim_{r\to 1^-} [\ln|x^2-1|]_0^r$$

(use the substitution $u = x^2 - 1$)

$$= \lim_{r\to -1^+} (-\ln|r^2-1|) + \lim_{r\to 1^-} \ln|r^2-1|$$

$= +\infty - \infty$; diverges

(Note that the infinities don't cancel. For convergence we need each part of the integral to converge on its own.)

21. $\displaystyle\int_{-\infty}^{+\infty} xe^{-x^2}\,dx$

$$= \lim_{M\to -\infty} \int_{M}^{0} xe^{-x^2}\,dx + \lim_{M\to +\infty} \int_{0}^{M} xe^{-x^2}\,dx$$

$$= \lim_{M\to -\infty} \left[-\frac{1}{2}e^{-x^2}\right]_M^0 + \lim_{M\to +\infty} \left[-\frac{1}{2}e^{-x^2}\right]_0^M$$

(use the substitution $u = -x^2$)

$$= \lim_{M\to -\infty} \left(-\frac{1}{2} + \frac{1}{2}e^{-M^2}\right) + \lim_{M\to +\infty} \left(-\frac{1}{2}e^{-M^2} + \frac{1}{2}\right)$$

$= -\frac{1}{2} + \frac{1}{2} = 0$; converges

23. $\displaystyle\int_{0}^{+\infty} \frac{1}{x\ln x}\,dx$

$$= \lim_{r\to 0^+} \int_{r}^{1/2} \frac{1}{x\ln x}\,dx + \lim_{r\to 1^-} \int_{1/2}^{r} \frac{1}{x\ln x}\,dx + \lim_{r\to 1^+} \int_{r}^{2} \frac{1}{x\ln x}\,dx + \lim_{M\to +\infty} \int_{2}^{M} \frac{1}{x\ln x}\,dx$$

$$\lim_{r\to 0^+} \int_{r}^{1/2} \frac{1}{x\ln x}\,dx = \lim_{r\to 0^+} [\ln|\ln x|]_r^{1/2}$$

(use the substitution $u = \ln x$)

$$= \lim_{r\to 0^+} [\ln|\ln(1/2)| - \ln|\ln r|] = -\infty$$

Without checking the remaining parts of the integral we can say that the whole integral diverges.

25. $\displaystyle\int_0^{+\infty} \frac{2x}{x^2-1}\,dx$

$\displaystyle = \lim_{r\to 1^-}\int_0^r \frac{2x}{x^2-1}\,dx + \lim_{r\to 1^+}\int_r^2 \frac{2x}{x^2-1}\,dx$

$\displaystyle + \lim_{M\to +\infty}\int_2^M \frac{2x}{x^2-1}\,dx$

$\displaystyle = \lim_{r\to 1^-}[\ln|x^2-1|]_0^r + \lim_{r\to 1^+}[\ln|x^2-1|]_r^2$

$\displaystyle + \lim_{M\to +\infty}[\ln|x^2-1|]_2^M$

$\displaystyle = \lim_{r\to 1^-}\ln|r^2-1| + \lim_{r\to 1^+}(\ln 3 - \ln|r^2-1|)$

$\displaystyle + \lim_{M\to +\infty}(\ln|M^2-1| - \ln 3)$

$= -\infty + \infty + \infty$; diverges

27. Total revenue $\displaystyle = \int_0^{+\infty} 91.7(0.90)^t\,dt$

$\displaystyle = \lim_{M\to +\infty}\int_0^M 91.7(0.90)^t\,dt$

$\displaystyle = \lim_{M\to +\infty}\left[\frac{91.7}{\ln 0.90}(0.90)^t\right]_0^M$

$\displaystyle = \lim_{M\to +\infty}\left[\frac{91.7}{\ln 0.90}(0.90)^M - \frac{91.7}{\ln 0.90}\right]$

$\displaystyle = -\frac{91.7}{\ln 0.90} \approx \$870$ million

29. Annual sales $= S(t) = 400(0.945)^t$ billion cigarettes per year.

Total sales $\displaystyle = \int_0^{+\infty} 400(0.945)^t\,dt$

$\displaystyle = \lim_{M\to +\infty}\int_0^M 400(0.945)^t\,dt$

$\displaystyle = \lim_{M\to +\infty}\left[\frac{400}{\ln 0.945}(0.945)^t\right]_0^M$

$\displaystyle = \lim_{M\to +\infty}\left[\frac{400}{\ln 0.945}(0.945)^M - \frac{400}{\ln 0.945}\right]$

$\displaystyle = -\frac{400}{\ln 0.945} \approx 7100$ billion cigarettes

31. Annual sales $= S(t) = 200(0.90)^t$.

Total sales $\displaystyle = \int_0^{+\infty} 200(0.90)^t\,dt$

$\displaystyle = \lim_{M\to +\infty}\int_0^M 200(0.90)^t\,dt$

$\displaystyle = \lim_{M\to +\infty}\left[\frac{200}{\ln 0.90}(0.90)^t\right]_0^M$

$\displaystyle = \lim_{M\to +\infty}\left[\frac{200}{\ln 0.90}(0.90)^M - \frac{200}{\ln 0.90}\right]$

$\displaystyle = -\frac{200}{\ln 0.90} \approx 1900.$

No, you will not sell more than about 2000 of them.

33. $\displaystyle \lim_{t\to +\infty} N(t) = 2.5 + \int_1^{+\infty} 0.214t^{-0.91}\,dt$

$\displaystyle = 2.5 + \lim_{M\to +\infty}\int_1^M 0.214t^{-0.91}\,dt$

$\displaystyle = 2.5 + \lim_{M\to +\infty}[2.378t^{0.09}]_1^M$

$\displaystyle = 2.5 + \lim_{M\to +\infty}[2.378M^{0.09} - 2.378] = +\infty.$

The integral diverges, so the number of graduates each year will rise without bound.

35. (a) The revenue per cell phone user is $P(t) = 350e^{-0.1t}$; multiplying by the number of users gives the annual revenue as $R(t) = 350e^{-0.1t}(39t + 68)$ million dollars per year.

(b) Total revenue $= \int_0^{+\infty} 350e^{-0.1t}(39t + 68)\,dt$

$$= \lim_{M\to+\infty} \int_0^M 350e^{-0.1t}(39t + 68)\,dt$$

$$= \lim_{M\to+\infty} [-3500e^{-0.1t}(39t + 68) - 1{,}365{,}000e^{-0.1t}]_0^M$$

$$= \lim_{M\to+\infty} [-3500e^{-0.1M}(39M + 68) - 1{,}365{,}000e^{-0.1M} + 3500(68) + 1{,}365{,}000]$$

$$= 3500(68) + 1{,}365{,}000$$

$$= \$1{,}603{,}000 \text{ million.}$$

37. The annual investment, in billion of constant dollars per year, is

$$Q(t) = (1.7t^2 - 0.5t + 8)e^{-0.05t}$$

The total investment is

$$\int_0^{+\infty} (1.7t^2 - 0.5t + 8)e^{-0.05t}\,dt$$

$$= \lim_{M\to+\infty} \int_0^M (1.7t^2 - 0.5t + 8)e^{-0.05t}\,dt$$

$$= \lim_{M\to+\infty} [-20(1.7t^2 - 0.5t + 8)e^{-0.05t} - 400(3.4t - 0.5)e^{-0.05t} - 27{,}200e^{-0.05t}]_0^M$$

$$= \lim_{M\to+\infty} [-20(1.7M^2 - 0.5M + 8)e^{-0.05M} - 400(3.4M - 0.5)e^{-0.05M} - 27{,}200e^{-0.05M} + 160 - 200 + 27{,}200]$$

$$\approx \$27{,}000 \text{ billion}$$

39. $\int_0^{+\infty} N(t)\,dt = \int_0^{+\infty} \frac{82.8(7.14)^t}{21.8 + (7.14)^t}\,dt$

$$= \lim_{M\to+\infty} \int_0^M \frac{82.8(7.14)^t}{21.8 + (7.14)^t}\,dt$$

$$= \lim_{M\to+\infty} \left[\frac{82.8}{\ln 7.14} \ln(21.8 + 7.14^t)\right]_0^M$$

(use the substitution $u = 21.8 + 7.14^t$)

$$= \lim_{M\to+\infty} \left[\frac{82.8}{\ln 7.14} \ln(21.8 + 7.14^M) - \frac{82.8}{\ln 7.14} \ln(22.8)\right]$$

$$= +\infty.$$

$\int_0^{+\infty} N(t)\,dt$ diverges, indicating that there is no bound to the expected total future online sales of books.

$$\int_{-\infty}^0 N(t)\,dt = \int_{-\infty}^0 \frac{82.8(7.14)^t}{21.8 + (7.14)^t}\,dt$$

$$= \lim_{M\to-\infty} \int_M^0 \frac{82.8(7.14)^t}{21.8 + (7.14)^t}\,dt$$

$$= \lim_{M\to-\infty} \left[\frac{82.8}{\ln 7.14} \ln(21.8 + 7.14^t)\right]_M^0$$

$$= \lim_{M\to-\infty} \left[\frac{82.8}{\ln 7.14} \ln(22.8) - \frac{82.8}{\ln 7.14} \ln(21.8 + 7.14^M)\right]$$

$$= \frac{82.8}{\ln 7.14} \ln(22.8) - \frac{82.8}{\ln 7.14} \ln(21.8))$$

$$\approx 1.889.$$

$\int_{-\infty}^0 N(t)\,dt$ converges to approximately 1.889, indicating that total online sales of books prior to 1997 amounted to approximately 1.889 million books.

41. 1

43. 0.1587

45. The value per bottle is $P(t) = 85e^{0.4t}$. The annual sales rate is $Q(t) = 500e^{-t}$. The annual income is

$$R(t) = P(t)Q(t) = 42{,}500e^{-0.6t}$$

The total income is

$$\int_0^{+\infty} 42{,}500e^{-0.6t}\,dt = \lim_{M\to+\infty}\int_0^M 42{,}500e^{-0.6t}\,dt$$

$$= \lim_{M\to+\infty}[-70{,}833e^{-0.6t}]_0^M$$

$$= \lim_{M\to+\infty}[-70{,}833e^{-0.6M} + 70{,}833]$$

$$= \$70{,}833$$

47. (a) $$\int_{0.2}^{+\infty}\frac{1}{5.6997k^{1.081}}\,dk$$

$$= \lim_{M\to+\infty}\int_{0.2}^{M}\frac{1}{5.6997k^{1.081}}\,dk$$

$$= \lim_{M\to+\infty}[-2.166k^{-0.081}]_{0.2}^M$$

$$= \lim_{M\to+\infty}[-2.166M^{-0.081} + 2.166(0.2)^{-0.081}]$$

≈ 2.468 meteors on average

(b) $$\int_0^1\frac{1}{5.6997k^{1.081}}\,dk = \lim_{r\to0}\int_r^1\frac{1}{5.6997k^{1.081}}\,dk$$

$$= \lim_{r\to0}[-2.166k^{-0.081}]_r^1$$

$$= \lim_{r\to0}[-2.166 + 2.166r^{-0.081}]$$

$= +\infty$; the integral diverges.

We can interpret this as saying that the number of impacts by meteors smaller than 1 megaton is very large. (This makes sense because, for example, this number includes meteors no larger than a grain of dust.)

49. (a) $$\Gamma(1) = \int_0^{+\infty}e^{-t}\,dt = \lim_{M\to+\infty}\int_0^M e^{-t}\,dt$$

$$= \lim_{M\to+\infty}[-e^{-t}]_0^M = \lim_{M\to+\infty}[-e^{-M} + 1]$$

$$= 1$$

$$\Gamma(2) = \int_0^{+\infty}te^{-t}\,dt = \lim_{M\to+\infty}\int_0^M te^{-t}\,dt$$

$$= \lim_{M\to+\infty}[-te^{-t} - e^{-t}]_0^M$$

$$= \lim_{M\to+\infty}[-Me^{-M} - e^{-M} + 1]$$

$$= 1$$

(b) $$\Gamma(n+1) = \int_0^{+\infty}t^ne^{-t}\,dt = \lim_{M\to+\infty}\int_0^M t^ne^{-t}\,dt$$

$$= \lim_{M\to+\infty}\left([-t^ne^{-t}]_0^M + \int_0^M nt^{n-1}e^{-t}\,dt\right)$$

$$= \lim_{M\to+\infty}[-M^ne^{-M} + 0] + n\int_0^{+\infty}t^{n-1}e^{-t}\,dt$$

$$= n\Gamma(n)$$

(c) If n is a positive integer, then applying part (b) several times we get

$$\Gamma(n) = (n-1)\Gamma(n-1)$$
$$= (n-1)(n-2)\Gamma(n-2)$$
$$= \dots$$
$$= (n-1)(n-2)\cdots1\Gamma(1)$$
$$= (n-1)!$$

by part (a).

51. The integral does not converge, so the number given by the FTC is meaningless.

53. Yes; the integrals converge to 0, and the FTC also gives 0.

55. In all cases, you need to rewrite the improper integral as a limit and use technology to evaluate the integral of which you are taking the limit. Evaluate for several values of the endpoint approaching the limit. In the case of an integral in which one of the limits of integration is infinite, you may have to instruct the calculator or computer to use more subdivisions as you approach $+\infty$.

57. Answers will vary.

7.6

1. $y = \int (x^2 + \sqrt{x})\, dx = \frac{x^3}{3} + \frac{2x^{3/2}}{3} + C$

3. $y\, dy = x\, dx$; $\int y\, dy = \int x\, dx$; $\frac{y^2}{2} = \frac{x^2}{2} + C$

5. $\frac{1}{y}\, dy = x\, dx$; $\int \frac{1}{y}\, dy = \int x\, dx$; $\ln |y| = \frac{x^2}{2} + C$; $|y| = e^{x^2/2 + C} = e^C e^{x^2/2}$; $y = Ae^{x^2/2}$ (where $A = \pm e^C$)

7. $\frac{1}{y^2}\, dy = (x + 1)\, dx$; $\int \frac{1}{y^2}\, dy = \int (x + 1)\, dx$;

$-\frac{1}{y} = \frac{1}{2}(x + 1)^2 + K = \frac{(x + 1)^2 + C}{2}$;

$y = -\frac{2}{(x + 1)^2 + C}$

9. $y\, dy = \frac{\ln x}{x}\, dx$; $\int y\, dy = \int \frac{\ln x}{x}\, dx$;

$\frac{y^2}{2} = (\ln x)^2 + C$ (substitute $u = \ln x$);

$y = \pm\sqrt{(\ln x)^2 + C}$

11. $y = \int (x^3 - 2x)\, dx = \frac{x^4}{4} - x^2 + C$;

$1 = 0 - 0 + C$; $C = 1$; $y = \frac{x^4}{4} - x^2 + 1$

13. $y^2\, dy = x^2\, dx$; $\int y^2\, dy = \int x^2\, dx$;

$\frac{y^3}{3} = \frac{x^3}{3} + C$; $y^3 = x^3 + K$; $8 = 0 + K$; $K = 8$;

$y^3 = x^3 + 8$; $y = (x^3 + 8)^{1/3}$

15. $\frac{1}{y}\, dy = \frac{1}{x}\, dx$; $\int \frac{1}{y}\, dy = \int \frac{1}{x}\, dx$;

$\ln |y| = \ln |x| + C$; $|y| = e^{\ln|x| + C} = e^C e^{\ln|x|} = e^C|x|$;

$y = Ax$; $2 = A \times 1$; $A = 2$; $y = 2x$

17. $\frac{1}{y + 1}\, dy = x\, dx$; $\int \frac{1}{y + 1}\, dy = \int x\, dx$;

$\ln |y + 1| = \frac{x^2}{2} + C$; $\ln 1 = 0 + C$; $C = 0$;

$\ln |y + 1| = \frac{x^2}{2}$; $|y + 1| = e^{x^2/2}$; $y + 1 = e^{x^2/2}$

(note that $y(0) + 1 = 1 > 0$); $y = e^{x^2/2} - 1$

19. $\frac{1}{y^2}\, dy = \frac{x}{x^2 + 1}\, dx$; $\int \frac{1}{y^2}\, dy = \int \frac{x}{x^2 + 1}\, dx$;

$-\frac{1}{y} = \frac{1}{2}\ln(x^2 + 1) + C$; $1 = 0 + C$; $C = 1$;

$-\frac{1}{y} = \frac{1}{2}\ln(x^2 + 1) + 1$; $y = -\frac{2}{\ln(x^2 + 1) + 2}$

21. With $s(t)$ = monthly sales after t months, $\frac{ds}{dt} = -0.05s$; $s = 1000$ when $t = 0$. $\frac{1}{s}\, ds = -0.05\, dt$; $\int \frac{1}{s}\, ds = \int (-0.05)\, dt$; $\ln |s| = -0.05t + C$; $|s| = e^{-0.05t + C} = e^C e^{-0.05t}$; $s = Ae^{-0.05t}$; $1000 = A \times 1$; $A = 1000$; $s = 1000e^{-0.05t}$ quarts per month.

23. $\frac{dH}{dt} = -k(H - 75)$; $\frac{1}{H - 75}\, dH = -k\, dt$;

$\int \frac{1}{H - 75}\, dH = \int (-k)\, dt$; $\ln(H - 75) = -kt + C$; $H(t) = 75 + Ae^{-kt}$. $H(0) = 190$, so $190 = 75 + A$; $A = 115$. $H(10) = 150$, so

$150 = 75 + 115e^{-10k}$; $k = -\dfrac{1}{10}\ln\left(\dfrac{150-75}{115}\right) \approx 0.04274$. $H(t) = 75 + 115e^{-0.04274t}$ degrees Fahrenheit after t minutes.

25. With $S(t)$ = total sales after t months, $\dfrac{dS}{dt} = 0.1(100{,}000 - S)$; $S(0) = 0$. $\dfrac{1}{100{,}000 - S}\,dS = 0.1\,dt$; $\displaystyle\int \frac{1}{100{,}000 - S}\,dS = \int 0.1\,dt$;

$-\ln(100{,}000 - S) = 0.1t + C$; $S(t) = 100{,}000 - Ae^{-0.1t}$. $0 = 100{,}000 - A$; $A = 100{,}000$. $S(t) = 100{,}000 - 100{,}000e^{-0.1t} = 100{,}000(1 - e^{-0.1t})$ monitors after t months.

27. (a) $\dfrac{dp}{dt} = k[D(p) - S(p)]$

$= k(20{,}000 - 1000p)$

(b) $\dfrac{1}{20{,}000 - 1000p}\,dp = k\,dt$;

$\displaystyle\int \frac{1}{20{,}000 - 1000p}\,dp = \int k\,dt$;

$-\ln(20{,}000 - 1000p) = kt + C$;
$p(t) = 20 - Ae^{-kt}$.
(c) $p(0) = 10$ and $p(1) = 12$, so $10 = 20 - A$; $A = 10$; $12 = 20 - 10e^{-k}$; $k = -\ln\left(\dfrac{20-12}{10}\right) \approx 0.2231$; $p(t) = 20 - 10e^{-0.2231t}$ dollars after t months.

29. $-\dfrac{p}{q}\dfrac{dq}{dp} = 0.05p - 1.5$; $\dfrac{1}{q}\,dq = \left(-0.05 + \dfrac{1.5}{p}\right)dp$; $\displaystyle\int \frac{1}{q}\,dq = \int\left(-0.05 + \frac{1.5}{p}\right)dp$

; $\ln q = -0.05p + 1.5\ln p + C$; $q = Ae^{-0.05p}p^{1.5}$. $q(20) = 20$, so $20 = Ae^{-1}(20)^{1.5}$; $A = e(20)^{-0.5} \approx 0.6078$. $q = 0.6078e^{-0.05p}p^{1.5}$.

31. If $y = \dfrac{CL}{e^{-aLt} + C}$ then $\dfrac{dy}{dt} = \dfrac{-aCL^2e^{-aLt}}{(e^{-aLt} + C)^2}$ and $ay(L - y) = a\left(\dfrac{CL}{e^{-aLt} + C}\right)\left(\dfrac{-Le^{-aLt}}{e^{-aLt} + C}\right) = \dfrac{-aCL^2e^{-aLt}}{(e^{-aLt} + C)^2}$ also, so this y satisfies the differential equation.

33. $a = 1/4$ and $L = 2$, so $S = \dfrac{2C}{e^{-0.5t} + C}$ for some C. $S = 0.001$ when $t = 0$, so $0.001 = \dfrac{2C}{1 + C}$; $C = \dfrac{0.001}{2 - 0.001} = \dfrac{1}{1999}$. $S = \dfrac{2/1999}{e^{-0.5t} + 1/1999}$. Graph:

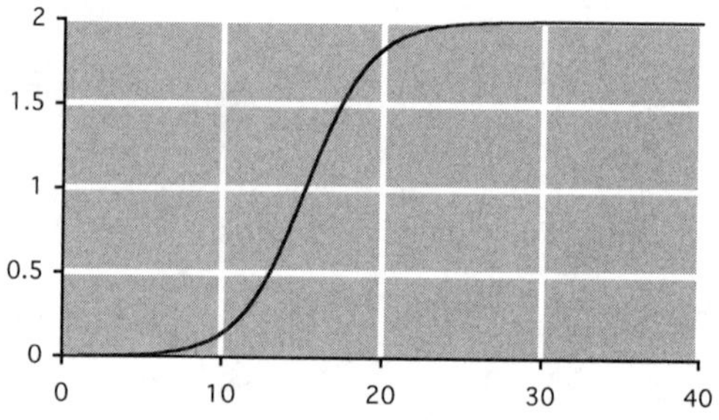

It will take about 27 months to saturate the market.

35. (a) $\dfrac{1}{y\ln(y/b)}\,dy = -a\,dt$; $\displaystyle\int \frac{1}{y\ln(y/b)}\,dy = \int(-a)\,dt$; $\ln[\ln(y/b)] = -at + C$ [use the substitution $u = \ln(y/b)$]; $y = be^{Ae^{-at}}$, for some constant A.

(b) $5 = 10e^A$, $A = \ln 0.5 \approx -0.69315$;
$y = 10e^{-0.69315e^{-t}}$. Graph:

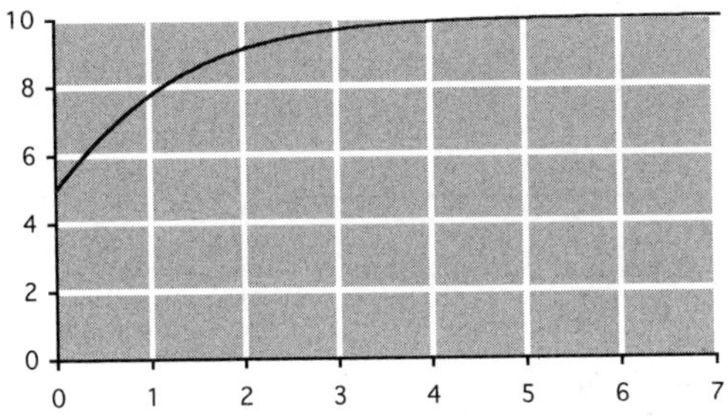

37. A general solution gives all possible solutions to the equation, using at least one arbitrary constant. A particular solution is one specific function that satisfies the equation. We obtain a particular solution by substituting specific values for any arbitrary constants in the general solution.

39. Example: $\frac{d^2y}{dx^2} = 1$ has general solution
$y = \frac{1}{2}x^2 + Cx + D$ (integrate twice).

41. Differentiate to get the differential equation
$\frac{dy}{dx} = -4e^{-x} + 3$.

Chapter 7 Review Exercises

1.

	D		I
+	$x^2 + 2$		e^x
		↘	
−	$2x$		e^x
		↘	
+	2		e^x
		↘	
$-\int$	0	→	e^x

$$\int (x^2 + 2)e^x\,dx = (x^2 + 2)e^x - 2xe^x + 2e^x + C =$$

$$(x^2 - 2x + 4)e^x + C$$

2.

	D		I
+	$x^2 - x$		e^{-3x+1}
		↘	
−	$2x-1$		$-\frac{1}{3}e^{-3x+1}$
		↘	
+	2		$\frac{1}{9}e^{-3x+1}$
		↘	
$-\int$	0	→	$-\frac{1}{27}e^{-3x+1}$

$$\int (x^2-x)e^{-3x+1}\,dx = -\frac{1}{3}(x^2 - x)\,e^{-3x+1} - \frac{1}{9}(2x-1)$$

$$e^{-3x+1} - \frac{1}{27}2e^{-3x+1} + C$$

$$= \frac{1}{27}(-9x^2 + 9x - 6x + 3 - 2)\,e^{-3x+1} + C$$

$$= \frac{1}{27}(-9x^2 + 3x + 1)\,e^{-3x+1} + C$$

3.

	D		I
+	$\ln(2x)$		x^2
		↘	
$-\int$	$1/x$	→	$x^3/3$

$$\int x^2\ln(2x)\,dx = \frac{1}{3}x^3\ln(2x) - \int \frac{1}{3}x^2\,dx$$

$$= \frac{1}{3}x^3\ln(2x) - \frac{1}{9}x^3 + C$$

4.

	D		I
+	$\log_5 x$		1
		↘	
$-\int$	$1/(x\ln 5)$	→	x

$$\int \log_5 x\,dx = x\log_5 x - \int \frac{1}{\ln 5}\,dx$$

$$= x\log_5 x - \frac{x}{\ln 5} + C$$

5.

	D		I
+	$x^3 + 1$		e^{-x}
		↘	
−	$3x^2$		$-e^{-x}$
		↘	
+	$6x$		e^{-x}
		↘	
−	6		$-e^{-x}$
		↘	
$+\int$	0	→	e^{-x}

$$\int_{-2}^{2} (x^3 + 1)e^{-x}\,dx$$

$$= [-(x^3 + 1)e^{-x} - 3x^2e^{-x} - 6xe^{-x} - 6e^{-x}]_{-2}^{2}$$

$$= [-(x^3 + 3x^2 + 6x + 7)e^{-x}]_{-2}^{2}$$

$$= -39e^{-2} - e^2 \approx -12.67$$

6.

	D		I
+	$\ln x$	↘	x^2
$-\int$	$1/x$	→	$x^3/3$

$$\int_1^e x^2 \ln x\,dx = \left[\frac{1}{3}x^3 \ln x\right]_1^e - \int_1^e \frac{1}{3}x^2\,dx$$

$$= \left[\frac{1}{3}x^3 \ln x - \frac{1}{9}x^3\right]_1^e$$

$$= \frac{1}{3}e^3 - \frac{1}{9}e^3 + \frac{1}{9} = \frac{2e^3 + 1}{9}$$

7. $$\int_1^{+\infty} \frac{1}{x^5}\,dx = \lim_{M \to +\infty} \int_1^M \frac{1}{x^5}\,dx$$

$$= \lim_{M \to +\infty} \left[-\frac{1}{4}x^{-4}\right]_1^M$$

$$= \lim_{M \to +\infty} \left[-\frac{1}{4}M^{-4} + \frac{1}{4}\right] = \frac{1}{4}$$

8. $$\int_0^1 \frac{1}{x^5}\,dx = \lim_{r \to 0^+} \int_r^1 \frac{1}{x^5}\,dx$$

$$= \lim_{r \to 0^+} [-x^{-4}/4]_r^1$$

$$= \lim_{r \to 0^+} \left(-\frac{1}{4} + \frac{1}{4r^4}\right) = +\infty;\ \text{diverges}$$

9. $$\int_{-2}^{2} \frac{1}{(x + 1)^{1/3}}\,dx$$

$$= \lim_{r \to -1^-} \int_{-2}^{r} \frac{1}{(x + 1)^{1/3}}\,dx + \lim_{r \to -1^+} \int_{r}^{2} \frac{1}{(x + 1)^{1/3}}\,dx$$

$$= \lim_{r \to -1^-} \left[\frac{3}{2}(x + 1)^{2/3}\right]_{-2}^{r} + \lim_{r \to -1^+} \left[\frac{3}{2}(x + 1)^{2/3}\right]_{r}^{2}$$

$$= \lim_{r \to -1^-} \left[\frac{3}{2}(r + 1)^{2/3} - \frac{3}{2}\right] + \lim_{r \to -1^+} \left[\frac{3}{2}3^{2/3} - \frac{3}{2}(r + 1)^{2/3}\right]$$

$$= \frac{3}{2}(3^{2/3} - 1)$$

10. $$\int_0^1 \frac{1}{\sqrt{1 - x}}\,dx = \lim_{r \to 1^-} \int_0^r \frac{1}{\sqrt{1 - x}}\,dx$$

$$= \lim_{r \to 1^-} [-2\sqrt{1 - x}\,]_0^r$$

$$= \lim_{r \to 1^-} (-2\sqrt{1 - r} + 2) = 2$$

11. $x^3 = 1 - x^3$ when $x^3 = 1/2$, $x = 1/2^{1/3}$.

$$\text{Area} = \int_0^{1/2^{1/3}} [(1 - x^3) - x^3]\,dx + \int_{1/2^{1/3}}^{1} [x^3 - (1 - x^3)]\,dx$$

$$= \int_0^{1/2^{1/3}} (1 - 2x^3)\,dx + \int_{1/2^{1/3}}^{1} (2x^3 - 1)\,dx$$

$$= \left[x - \frac{1}{2}x^4\right]_0^{1/2^{1/3}} + \left[\frac{1}{2}x^4 - x\right]_{1/2^{1/3}}^{1}$$

$$= \frac{1}{2^{1/3}} - \frac{1}{4\cdot 2^{1/3}} + \frac{1}{2} - 1 - \frac{1}{4\cdot 2^{1/3}} + \frac{1}{2^{1/3}}$$

$$= \frac{3}{2\cdot 2^{1/3}} - \frac{1}{2} \approx 0.6906$$

12. $e^x \geq 1 \geq e^{-x}$ for x in [0, 2], so

$$\text{Area} = \int_0^2 (e^x - e^{-x})\,dx = [e^x + e^{-x}]_0^2$$

$$= e^2 + e^{-2} - 2$$

13. $1 - x^2 = x^2$ when $x = \pm 1/\sqrt{2}$, so

$$\text{Area} = \int_{-1/\sqrt{2}}^{1/\sqrt{2}} [(1 - x^2) - x^2]\,dx$$

$$= \int_{-1/\sqrt{2}}^{1/\sqrt{2}} (1 - 2x^2)\,dx$$

$$= \left[x - \frac{2}{3}x^3\right]_{-1/\sqrt{2}}^{1/\sqrt{2}}$$

$$= \frac{1}{\sqrt{2}} - \frac{2}{6\sqrt{2}} + \frac{1}{\sqrt{2}} - \frac{2}{6\sqrt{2}}$$

$$= \frac{4}{3\sqrt{2}} = \frac{2\sqrt{2}}{3}$$

14. $x \geq xe^{-x}$ for x in [0, 2] because $e^{-x} \leq 1$ in that range.

$$\text{Area} = \int_0^2 (x - xe^{-x})\,dx$$

$$= \left[\frac{1}{2}x^2 + xe^{-x} + e^{-x}\right]_0^2$$

(using integration by parts)

$$= 2 + 2e^{-2} + e^{-2} - 1 = 1 + 3e^{-2}$$

15. $\bar{f} = \frac{1}{2-(-2)} \int_{-2}^{2} x^3 - 1\,dx = \frac{1}{4}\left[\frac{x^4}{4} - x\right]_{-2}^{2}$

$$= \frac{1}{4}[2 - 6] = -1$$

16. $\bar{f} = \frac{1}{1-0} \int_0^1 \frac{x}{x^2+1}\,dx = \left[\frac{1}{2}\ln(x^2+1)\right]_0^1$

(using $u = x^2+1$)

$$= \frac{1}{2}\ln 2$$

17. Average $= \frac{1}{1-0}\int_0^1 x^2e^x\,dx$

$$= [x^2e^x - 2xe^x + 2e^x]_0^1$$

(using integration by parts)

$$= e - 2$$

18. Average $= \frac{1}{2e-1}\int_1^{2e} (x+1)\ln x\,dx$

$$= \frac{1}{2e-1}\left[\left(\frac{1}{2}x^2 + x\right)\ln x\right]_1^{2e} - \frac{1}{2e-1}\int_1^{2e}\left(\frac{1}{2}x + 1\right)dx$$

(using integration by parts)

$$= \frac{1}{2e-1}\left[\left(\frac{1}{2}x^2 + x\right)\ln x - \frac{1}{4}x^2 - x\right]_1^{2e}$$

$$= \frac{1}{2e-1}\left[(2e^2 + 2e)\ln(2e) - e^2 - 2e + \frac{1}{4} + 1\right]$$

$$= \frac{2(e^2+e)\ln 2 + e^2 + 5/4}{2e-1} \approx 5.10548$$

19. $\bar{f} = \frac{1}{2}\int_{x-2}^{x} (3t+1)\,dt = \frac{1}{2}\left[\frac{3t^2}{2} + t\right]_{x-2}^{x}$

$$= \frac{1}{2}\left[\frac{3x^2}{2} + x - \left(\frac{3(x-2)^2}{2} + x - 2\right)\right]$$

$$= 3x - 2$$

20. $\bar{f} = \frac{1}{2}\int_{x-2}^{x}(6t^2 + 12)\,dt = \frac{1}{2}[2t^3+12t]_{x-2}^{x}$

$= \frac{1}{2}[2x^3 + 12x - (2(x-2)^3 + 12(x-2))]$

$= 6x^2 - 12x + 20$

21. $\frac{1}{2}\int_{x-2}^{x} t^{4/3}\,dt = \left[\frac{3}{14}t^{7/3}\right]_{x-2}^{x}$

$= \frac{3}{14}[x^{7/3} - (x-2)^{7/3}]$

22. $\frac{1}{2}\int_{x-2}^{x}\ln t\,dt = \frac{1}{2}[t\ln t - t]_{x-2}^{x}$

$= \frac{1}{2}[x\ln x - x - (x-2)\ln(x-2) + x - 2]$

$= \frac{1}{2}[x\ln x - (x-2)\ln(x-2) - 2]$

23. $q = 100 - 2p$, so $\bar{q} = 100 - 20 = 80$.
The consumers' surplus is

$CS = \int_{0}^{80}(50 - \frac{1}{2}q - 10)\,dq$

$= \int_{0}^{80}(40 - \frac{1}{2}q)\,dq$

$= [40q - \frac{1}{4}q^2]_0^{80} = 3200 - 1600 = \$1600.$

24. $q = (10-p)^2$, so $\bar{q} = (10-4)^2 = 36$.
The consumers' surplus is

$CS = \int_{0}^{36}(10 - q^{1/2} - 4)\,dq$

$= \int_{0}^{36}(6 - q^{1/2})\,dq$

$= [6q - \frac{2}{3}q^{3/2}]_0^{36} = \$72.$

25. $q = 2p - 100$, so $\bar{q} = 200 - 100 = 100$.
The producers' surplus is

$PS = \int_{0}^{100}(100 - 50 - \frac{1}{2}q)\,dq$

$= \int_{0}^{100}(50 - \frac{1}{2}q)\,dq$

$= [50q - \frac{1}{4}q^2]_0^{100} = 5000 - 2500 = \$2500.$

26. $q = (p-10)^2$, so $\bar{q} = (40-10)^2 = 900$.
The producers' surplus is

$PS = \int_{0}^{900}(40 - (10 + q^{1/2}))\,dq$

$= \int_{0}^{900}(30 - q^{1/2})\,dq$

$= [30q - \frac{2}{3}q^{3/2}]_0^{900} = \$9000$

27. $\frac{1}{y^2}\,dy = x^2\,dx$; $\int\frac{1}{y^2}\,dy = \int x^2\,dx$;

$-\frac{1}{y} = \frac{x^3}{3} + K = \frac{x^3 + C}{3}$; $y = -\frac{3}{x^3 + C}$

28. $\frac{1}{y+2}\,dy = x\,dx$; $\int\frac{1}{y+2}\,dy = \int x\,dx$;

$\ln|y+2| = \frac{x^2}{2} + C$; $y + 2 = Ae^{x^2/2}$;
$y = Ae^{x^2/2} - 2$

29. $y\,dy = \frac{1}{x}\,dx$; $\int y\,dy = \int\frac{1}{x}\,dx$; $\frac{y^2}{2} = \ln|x| +$

C; $\frac{1}{2} = 0 + C$; $\frac{y^2}{2} = \ln|x| + \frac{1}{2}$; $y = \sqrt{2\ln|x| + 1}$

(note that $y(1) > 0$)

30. $\frac{1}{y}\,dy = \frac{x}{x^2+1}\,dx$; $\int \frac{1}{y}\,dy = \int \frac{x}{x^2+1}\,dx$;
$\ln|y| = \frac{1}{2}\ln(x^2+1) + C$ (substitute $u = x^2+1$);
$\ln 2 = 0 + C$; $\ln y = \frac{1}{2}\ln(x^2+1) + \ln 2 =$
$\ln(2\sqrt{x^2+1})$; $y = 2\sqrt{x^2+1}$

31. (a) The amount in the account at any given time is $1{,}000{,}000e^{0.06t}$ dollars after t years, so the average amount over two years is

$$\frac{1{,}000{,}000}{2-0}\int_0^2 e^{0.06t}\,dt = 500{,}000\left[\frac{e^{0.06t}}{0.06}\right]_0^2$$

$$= 500{,}000\,\frac{e^{0.12}-1}{0.06} \approx \$1{,}062{,}500$$

(b) $$\frac{1{,}000{,}000}{1/12-0}\int_{t-1/12}^{t} e^{0.06s}\,ds$$

$$= 200{,}000{,}000[e^{0.06s}]_{t-1/12}^{t}$$
$$= 200{,}000{,}000(e^{0.06t} - e^{0.06(t-1/12)})$$
$$= 200{,}000{,}000e^{0.06t}(1 - e^{-0.06/12})$$
$$\approx 997{,}500e^{0.06t}$$

32. (a) $1000\sqrt{200-2p} = 1000\sqrt{10p-400}$;
$200 - 2p = 10p - 400$; $\bar{p} = 50$; $\bar{q} =$
$1000\sqrt{200-2(50)} = 10{,}000$

(b) Solve the demand equation for p:

$$p = 100 - q^2/2{,}000{,}000.$$

$$CS = \int_0^{10{,}000}\left(100 - \frac{q^2}{2{,}000{,}000} - 50\right)dq$$

$$= \int_0^{10{,}000}\left(50 - \frac{q^2}{2{,}000{,}000}\right)dq$$

$$= \left[50q - \frac{q^3}{6{,}000{,}000}\right]_0^{10{,}000}$$

$$\approx \$333{,}000$$

Solve the supply equation for p:

$$p = 40 + q^2/10{,}000{,}000$$

$$PS = \int_0^{10{,}000}\left(50 - 40 - \frac{q^2}{10{,}000{,}000}\right)dq$$

$$= \int_0^{10{,}000}\left(10 - \frac{q^2}{10{,}000{,}000}\right)dq$$

$$= \left[10q - \frac{q^3}{30{,}000{,}000}\right]_0^{10{,}000}$$

$$\approx \$66{,}700$$

33. The price follows the function $p(t) = 40 - 2t$ while the quantity sold per week follows $q(t) = 5000e^{-0.1t}$, where t is measured in weeks. The revenue per week is therefore $R(t) = p(t)q(t)$ and the total revenue over the next 8 weeks is

$$\int_0^8 5000(40 - 2t)e^{-0.1t}\,dt$$

$$= 5000[-(400 - 20t)e^{-0.1t} + 200e^{-0.1t}]_0^8$$

(using integration by parts)

$$= 5000[(20t - 200)e^{-0.1t}]_0^8$$
$$= 5000[-40e^{-0.8} + 200] \approx \$910{,}000$$

34. (a) The rate at which money is deposited is $100{,}000 + 10{,}000(12t)$ dollars per month after t years; converting to dollars per year gives $R(t) = 1{,}200{,}000 + 1{,}440{,}000t$ dollars per year after t years. The total deposited over two years is

$$\int_0^2 (1{,}200{,}000 + 1{,}440{,}000t)e^{0.06(2-t)}\,dt$$

$$= \left[-\frac{1{,}200{,}000 + 1{,}440{,}000t}{0.06}e^{0.06(2-t)} - \frac{1{,}440{,}000}{0.06^2}e^{0.06(2-t)}\right]_0^2$$

$$\approx \$5{,}549{,}000$$

(b) The principal is given by

$$\int_0^2 (1{,}200{,}000 + 1{,}440{,}000t)\, dt$$

$$= [1{,}200{,}000t + 720{,}000t^2]_0^2 = \$5{,}280{,}000,$$

so the interest is the remaining \$269,000.

35. The revenue stream is $R(t) = 50e^{0.1t}$ million dollars per year. The present value if the next year's revenue is

$$\int_0^1 50e^{0.1t}e^{-0.06t}\, dt$$

$$= \int_0^1 50e^{0.04t}\, dt = [1250e^{0.04t}]_0^1$$

$\approx$ \$51 million

36. The money $y(t)$ in the account satisfies the differential equation $\frac{dy}{dt} = 0.0001y^2$. We solve this equation:

$$\frac{1}{y^2}\, dy = 0.0001\, dt$$

$$\int \frac{1}{y^2}\, dy = \int 0.0001\, dt$$

$$-\frac{1}{y} = 0.0001t + C$$

$$y(0) = 10{,}000$$

so $C = -1/10{,}000 = -0.0001$

$$y = \frac{1}{0.0001 - 0.0001t} = \frac{10{,}000}{1 - t}$$

The amount in the account would approach infinity one year after the deposit.

Chapter 8
8.1

1. (a) $f(0, 0) = 0^2 + 0^2 - 0 + 1 = 1$
(b) $f(1, 0) = 1^2 + 0^2 - 1 + 1 = 1$
(c) $f(0, -1) = 0^2 + (-1)^2 - 0 + 1 = 2$
(d) $f(a, 2) = a^2 + 2^2 - a + 1 = a^2 - a + 5$
(e) $f(y, x) = y^2 + x^2 - y + 1$
(f) $f(x + h, y + k) =$
$(x + h)^2 + (y + k)^2 - (x + h) + 1$

3. (a) $f(0, 0) = 0 + 0 - 0 = 0$
(b) $f(1, 0) = 0.2 + 0 - 0 = 0.2$
(c) $f(0, -1) = 0 + 0.1(-1) - 0 = -0.1$
(d) $f(a, 2) = 0.2a + 0.1(2) - 0.01a(2) =$
$0.18a + 0.2$
(e) $f(y, x) = 0.2y + 0.1x - 0.01xy =$
$0.1x + 0.2y - 0.01xy$
(f) $f(x + h, y + k) =$
$0.2(x + h) + 0.1(y + k) - 0.01(x + h)(y + k)$

5. (a) $g(0, 0, 0) = e^{0+0+0} = 1$
(b) $g(1, 0, 0) = e^{1+0+0} = e$
(c) $g(0, 1, 0) = e^{0+1+0} = e$
(d) $g(z, x, y) = e^{z+x+y} = e^{x+y+z}$
(e) $g(x + h, y + k, z + l) = e^{x+h+y+k+z+l}$

7. (a) $g(0, 0, 0) = \dfrac{0}{0 + 0 + 0}$ does not exist
(b) $g(1, 0, 0) = \dfrac{0}{1 + 0 + 0} = 0$
(c) $g(0, 1, 0) = \dfrac{0}{0 + 1 + 0} = 0$
(d) $g(z, x, y) = \dfrac{zxy}{z^2 + x^2 + y^2} = \dfrac{xyz}{x^2 + y^2 + z^2}$
(e) $g(x + h, y + k, z + l) =$
$\dfrac{(x + h)(y + k)(z + l)}{(x + h)^2 + (y + k)^2 + (z + l)^2}$

9. (a) f increases by 2.3 units for every 1 unit of increase in x. **(b)** f decreases by 1.4 units for every 1 unit of increase in y. **(c)** f decreases by 2.5 units for every 1 unit increase in z.

11. Neither, because of the y^2 term

13. Linear

15. Linear

17. Interaction, because $g(x, y, z) = \frac{1}{4}x - \frac{3}{4}y + \frac{1}{4}yz$

19. (a) $f(20, 10) = 107$
(b) $f(40, 20) = -14$
(c) $f(10, 20) - f(20, 10) = -6 - 107 = -113$

21. $x \rightarrow$, $y \downarrow$

	10	20	30	40
10	52	107	162	217
20	94	194	294	394
30	136	281	426	571
40	178	368	558	748

25.

18
4
0.0965
47,040

27.

6.9078
1.5193
5.4366
0

29. Let z = annual sales of Z (in millions of dollars), x = annual sales of X, and y = annual sales of Y. A linear model has the form $z = ax + by + c$. We are told that $a = -2.1$ and $b = 0.4$, and that $z(6, 6) = 6$. Thus, $6 = -2.1(6) + 0.4(6) + c$, so $c = 16.2$. The model is $z = -2.1x + 0.4y + 16.2$.

31. $\sqrt{(2-1)^2 + (-2+1)^2} = \sqrt{2}$

33. $\sqrt{(0-a)^2 + (b-0)^2} = \sqrt{a^2 + b^2}$

35. Set the two distances equal and solve: $\sqrt{(1-0)^2 + (k-0)^2} = \sqrt{(1-2)^2 + (k-1)^2}$; $\sqrt{1+k^2} = \sqrt{2-2k+k^2}$;
$1 + k^2 = 2 - 2k + k^2$; $2k = 1$; $k = \frac{1}{2}$

37. Rewrite the equation as $\sqrt{(x-2)^2 + (y+1)^2} = 3$. The set is the set of points at a distance of 3 from $(2, -1)$, that is, the circle with center $(2, -1)$ and radius 3.

39. The marginal cost of cars is \$6000 per car, the coefficient of x. The marginal cost of trucks is \$4000 per truck, the coefficient of y.

41. $C(x, y) = 10 + 0.03x + 0.04y$ where C is the cost in dollars, x is the number of video clips sold per month, and y is the number of audio clips sold per month.

43. (a) 2003 is represented by $t = 20$. We are told that $x = 38$. Therefore,

$y = 82 - 0.78t - 1.02x$
$= 82 - 0.78(20) - 1.02(38) = 27.64 \approx 28\%$

in 2003, 28% of the articles were written by researchers in the U.S.

(b) 1983 is represented by $t = 0$, We are told that $y = 61$. Therefore

$$y = 82 - 0.78t - 1.02x$$
$$61 = 82 - 0.78(0) - 1.02x$$
$$1.02x = 82 - 61 = 21$$
$$x = 21/1.02 \approx 21\%$$

In 1983, 21% of the articles were written by researchers in Europe.

(c) Since y is measured in percentage points and t in years, the coefficient -0.78 is measured in percentage points per year.

45. (a) $c(18, 12, 12) =$
$48.4 - 0.06(18) - 0.40(12) - 1.3(12) \approx 29\%$

(b) $32 = c(x, 16, 8) =$
$48.4 + 0.06x - 0.40(16) - 1.3(8)$; $x \approx 7\%$

(c) 0.4-point drop (the coefficient of y)

47. (a) $R(12{,}000, 5000, 5000) = \$9980$

(b) $R(z) = R(5000, 5000, z) =$
$10{,}000 - 0.01(5000) - 0.02(5000) - 0.01z + 0.00001(5000)z = 9850 + 0.04z$

49. $s(c, t) = Ac + Bt + C$. Substituting the given data:

1997 ($t = 0$): $60 = 50A + C$
2002 ($t = 5$): $110 = 100A + 5B + C$
2005 ($t = 8$): $200 = 180A + 8B + C$

This is a system of 3 linear equations in 3 unknowns. We can eliminate C by subtracting the first equation from the second and the second from the third:

$$50 = 50A + 5B$$
$$90 = 80A + 3B$$

from which we get $A = 1.2$, $B = -2$. Substituting this into the first original equation gives $C = 0$. So, our model is $s(c, t) = 1.2c - 2t$

51. $U(11, 10) - U(10, 10) \approx 5.75$. This means that, if your company now has 10 copies of Macro Publish and 10 copies of Turbo Publish, then the purchase of one additional copy of Macro Publish will result in a productivity increase of approximately 5.75 pages per day.

53. (a) $(a, b, c) = (3, 1/4, 1/\pi)$ and $(a, b, c) = (1/\pi, 3, 1/4)$ both work. In fact, if we take any positive values for a and b we can take $c = \dfrac{3}{4\pi ab}$.
(b) $V(a, a, a) = \frac{4}{3}\pi a^3 = 1$ gives $a = \left(\dfrac{3}{4\pi}\right)^{1/3}$. The resulting ellipsoid is a sphere with radius a.

55. $P(100, 500{,}000) = 1000(100^{0.5})(500{,}000^{0.5}) \approx 7{,}000{,}000$

57. (a) $100 = K(1000)^a(1{,}000{,}000)^{1-a}$, $10 = K(1000)^a(10{,}000)^{1-a}$
(b) Taking logs of both sides of the first equation we get
$\log 100$
$= \log K + a \log 1000 + (1 - a) \log 1{,}000{,}000;$
$2 = \log K + 3a + 6(1 - a);$
$\log K - 3a = -4.$
From the second equation we get $\log K - a = -3$ similarly.
(c) Solving we get $\log K = -2.5$ and $a = 0.5$, so $K = 10^{-2.5} \approx 0.003162$
(d) $P(500, 1{,}000{,}000) = 0.003162(500^{0.5})(1{,}000{,}000^{0.5}) = 71$ pianos (to the nearest piano)

59. (a) We first need to convert n into years: 5 days = 5/365 years. So, $B(1.5\times10^{14}, 5/365) = \dfrac{1.5\times10^{14}\times5}{5.1\times10^{14}\times365} \approx 4 \times 10^{-3}$ grams per square meter.
(b) The total weight of sulfates in the Earth's atmosphere.

61. (a) The value of N would be doubled. **(b)** $N(R, f_p, n_e, f_l, f_i, L) = R f_p n_e f_l f_i L$, where here L is the average lifetime of an intelligent civilization. **(c)** Take the logarithm of both sides, since this would yield the linear function $\ln N = \ln R + \ln f_p + \ln n_e + \ln f_l + \ln f_i + \ln f_c + \ln L$.

63. (a)

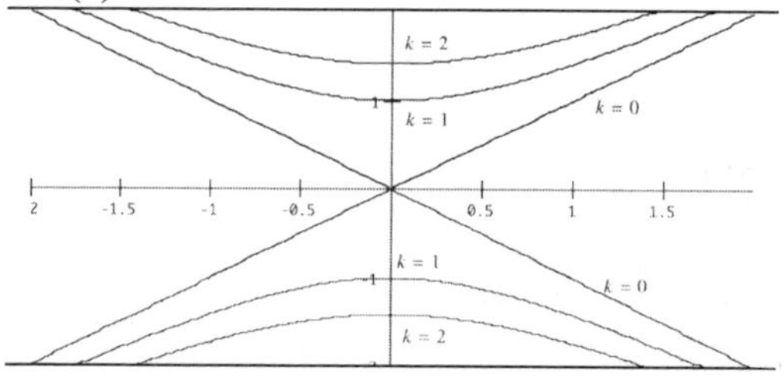

(b) The level curve $f(x, y) = 3$ is similar to the curves for 1 and 2, just further away from the origin.

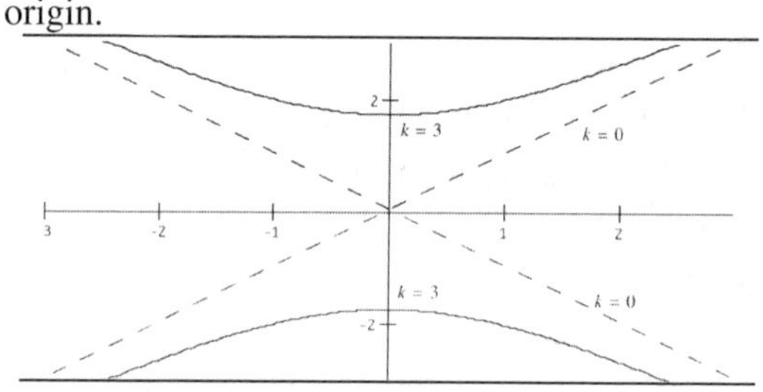

(c) These level curves would be similar to those in (a), but with the roles of x and y reversed.

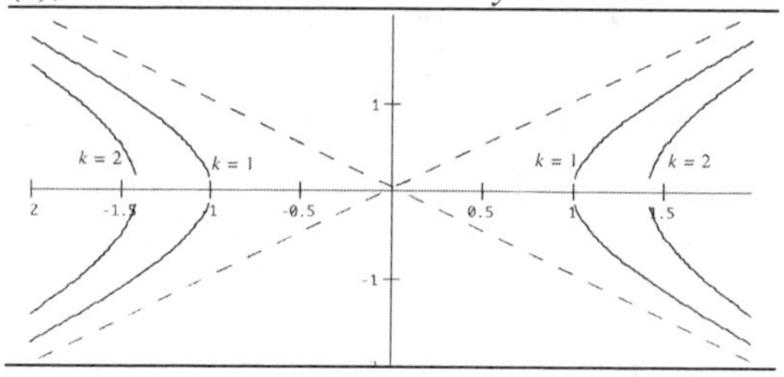

65. They are reciprocals of each other.

67. For example, $f(x, y) = x^2 + y^2$.

69. For example, $f(x, y, z) = xyz$

71. For example, take $f(x, y) = x + y$. Then setting $y = 3$ gives $f(x, 3) = x + 3$. This can be viewed as a function of the single variable x. Choosing other values for y gives other functions of x.

73. If $f = ax + by + c$, then fixing $y = k$ gives $f = ax + (bk + c)$, a linear function with slope a and intercept $bk + c$. The slope is independent of the choice of k.

75. That CDs cost more than cassettes.

8.2

1.

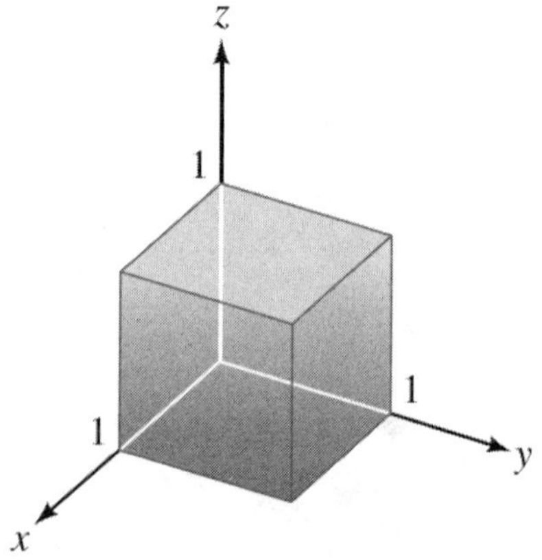

3.

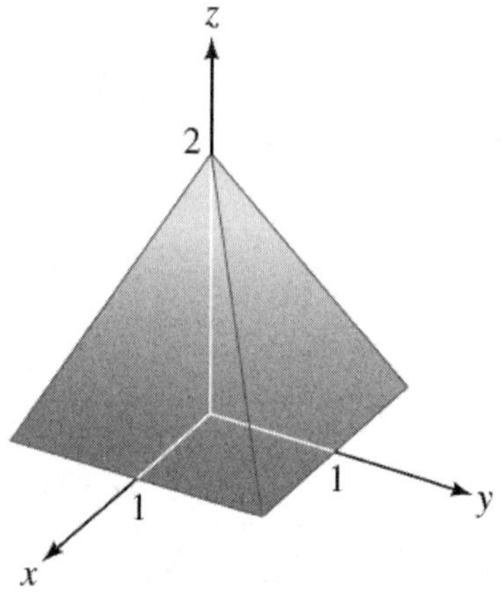

5.

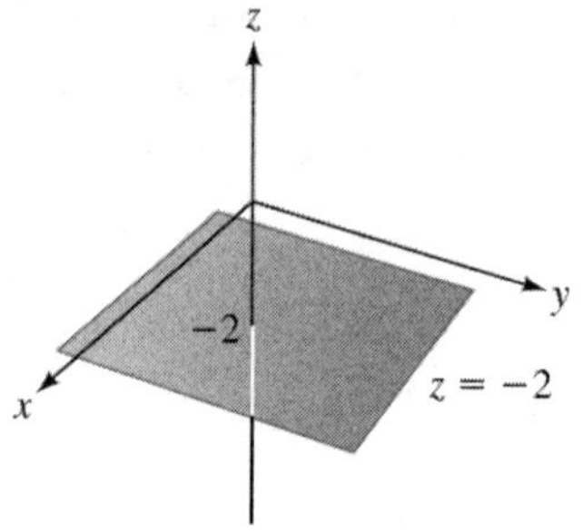

7.

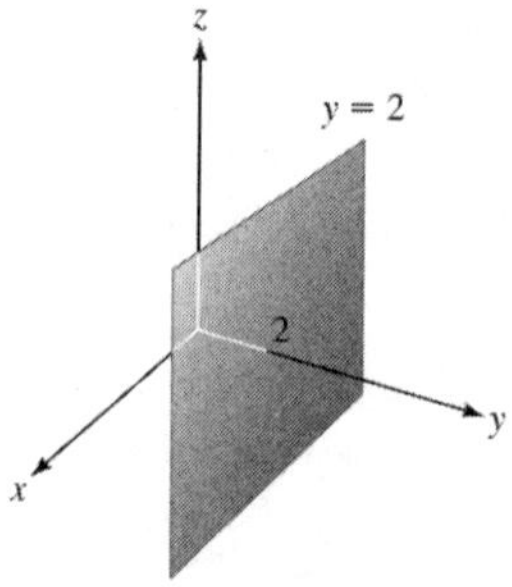

9.

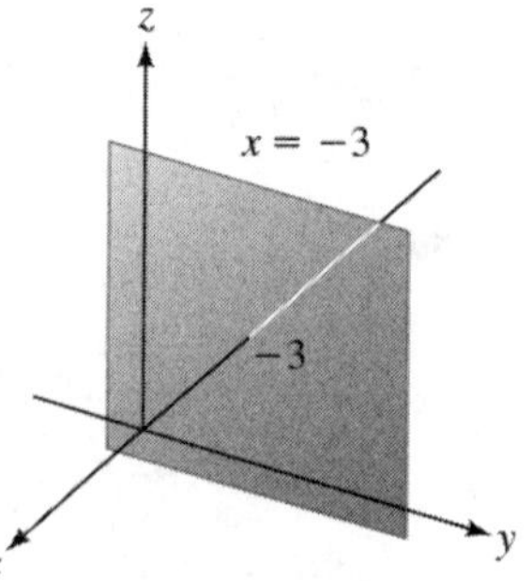

11. (H): The plane with z intercept 1, x intercept 1/3, and y intercept $-1/2$.

13. (B): An inverted paraboloid with apex at 1 on the z axis.

15. (F): A hemisphere lying below the xy plane.

17. (C): A surface of revolution having the z axis as a vertical asymptote.

19. The plane with x, y, and z intercepts all 1.

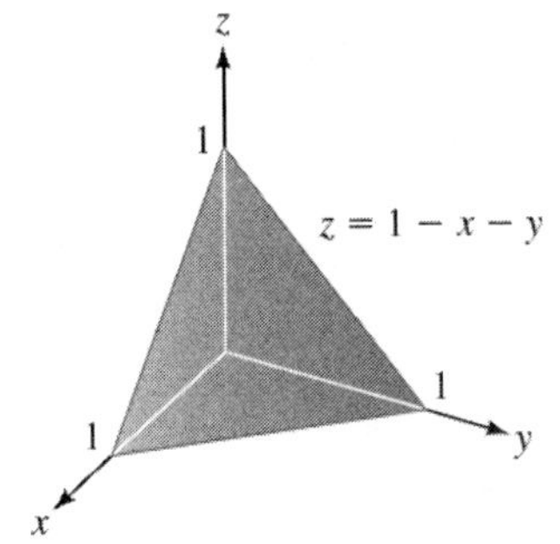

21. The plane with x intercept 1, y intercept 2, and z intercept -2.

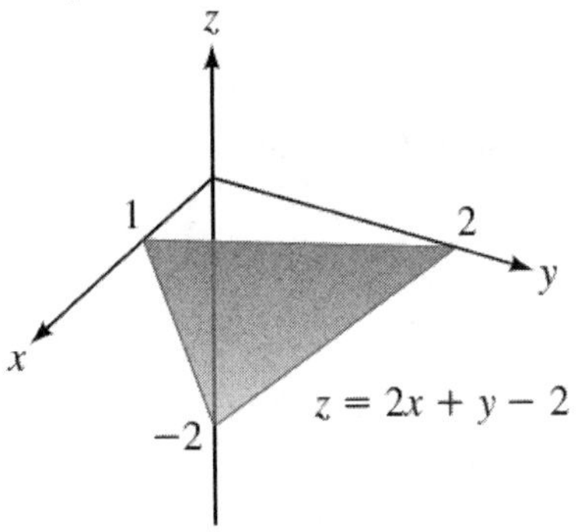

23. The plane with x intercept -2 and z intercept 2, parallel to the y axis.

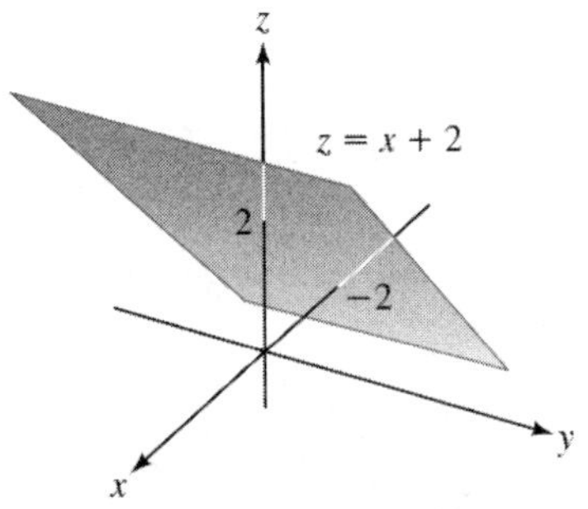

25. The plane passing through the origin, the line $z = x$ in the xz plane, and the line $z = y$ in the yz plane.

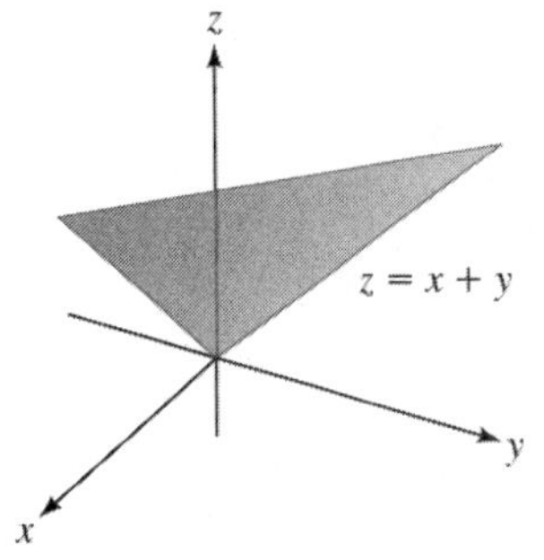

27. The cross section at $z = 1$ is the circle $x^2 + y^2 = 1/2$ and the cross section at $z = 2$ is the circle $x^2 + y^2 = 1$.

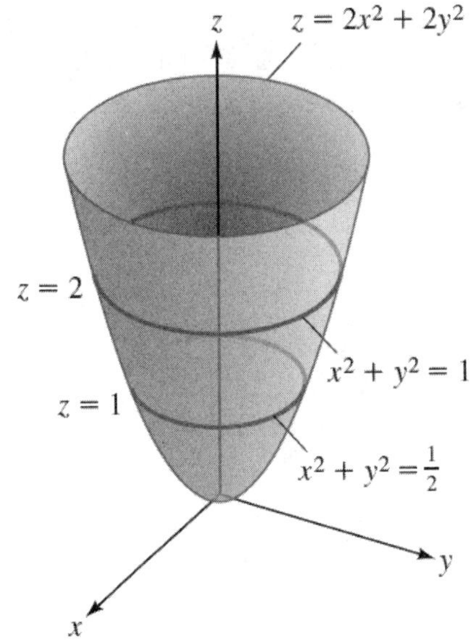

29. The cross section at $x = 0$ is the parabola $z = 2y^2$ and the cross section at $z = 1$ is the ellipse $x^2 + 2y^2 = 1$.

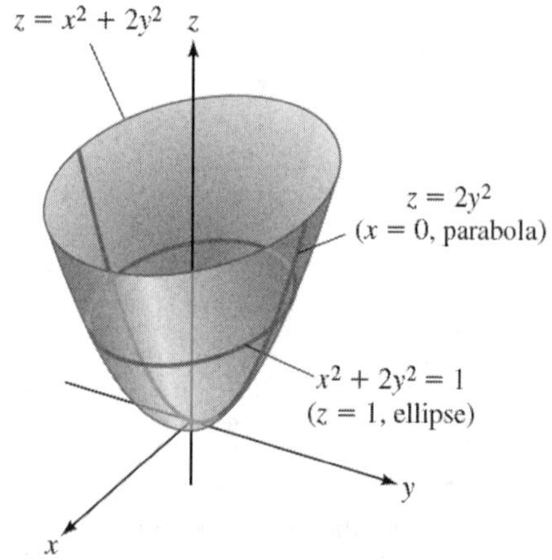

31. The graph is a cone with vertex at 2 on the z axis. The cross section at $y = 0$ is $z = 2 + |x|$ and the cross section at $z = 3$ is the circle $x^2 + y^2 = 1$.

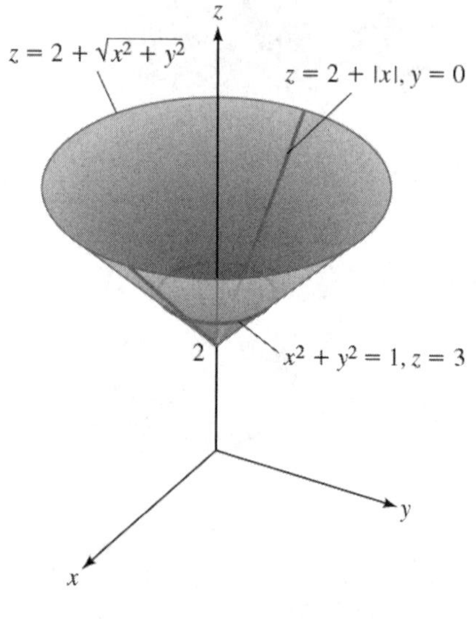

33. The graph is an inverted cone with apex at the origin. The cross section at $z = -4$ is the circle $x^2 + y^2 = 4$ and the cross section at $y = 1$ is a part of the hyperbola $z^2 - 4x^2 = 4$.

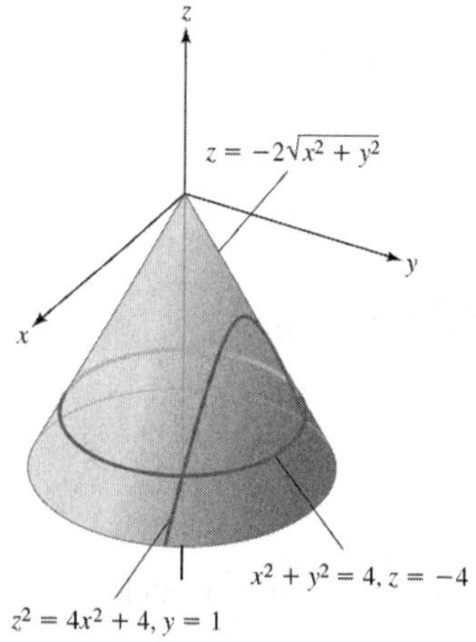

35. The cross sections $x = c$ are all copies of the parabola $z = y^2$.

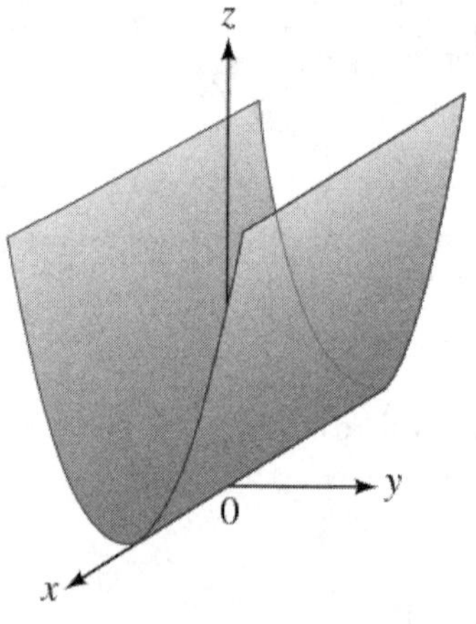

37. The cross sections $x = c$ are all copies of the hyperbola $z = 1/y$.

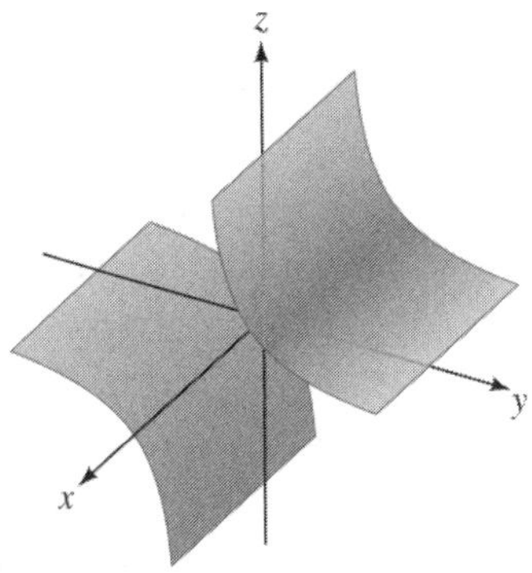

39. The cross sections $z = c$ are circles. The cross section $y = 0$ is the bell-shaped curve $z = e^{-x^2}$.

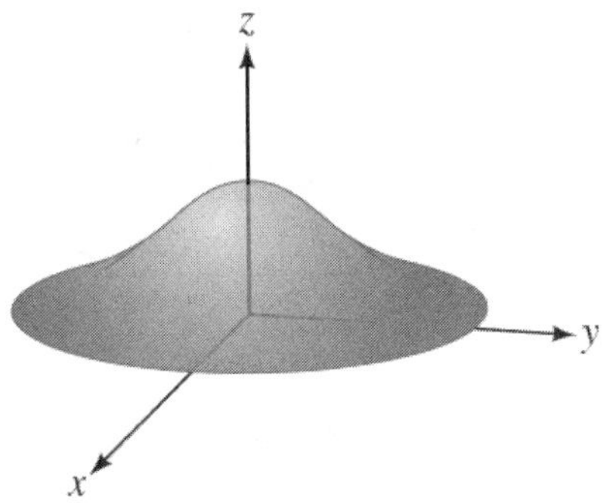

41. **(a)** The graph is a plane with x intercept -40, y intercept -60, and z intercept 240,000. **(b)** The slice $x = 10$ is the straight line with equation $z = 300{,}000 + 4000y$. It describes the cost function for the manufacture of trucks if car production is held fixed at 10 cars per week. **(c)** The level curve $z = 480{,}000$ is the straight line $6000x + 4000y = 240{,}000$. It describes the number of cars and trucks you can manufacture to maintain weekly costs at \$480,000.

43. The graph is a plane with x_1 intercept 0.3, x_2 intercept 33, and x_3 intercept 0.66. The slices by $x_1 = constant$ are straight lines that are parallel to each other. Thus, the rate of change of General Motors' share as a function of Ford's share does not depend on Chrysler's share. Specifically, GM's share decreases by 0.02 percentage points per 1 percentage-point increase in Ford's market share, regardless of Chrysler's share.

45. **(a)** The slices $x = constant$ and $y = constant$ are straight lines. **(b)** No. Even though the slices $x = constant$ and $y = constant$ are straight lines, the level curves are not, and so the surface is not a plane. **(c)** The slice $x = 10$ has a slope of 3800. The slice $x = 20$ has a slope of 3600. Manufacturing more cars lowers the marginal cost of manufacturing trucks.

47. Both level curves are quarter-circles. (We see only the portion in the first quadrant because $e \geq 0$ and $k \geq 0$.) The level curve $C = 30{,}000$ represents the relationship between the number of electricians and the number of carpenters used in building a home that costs \$30,000. Similarly for the level curve $C = 40{,}000$.

49. The following figure shows several level curves together with several lines of the form $h + w = c$. (The horizontal axis is the h axis and the vertical axis is the w axis.)

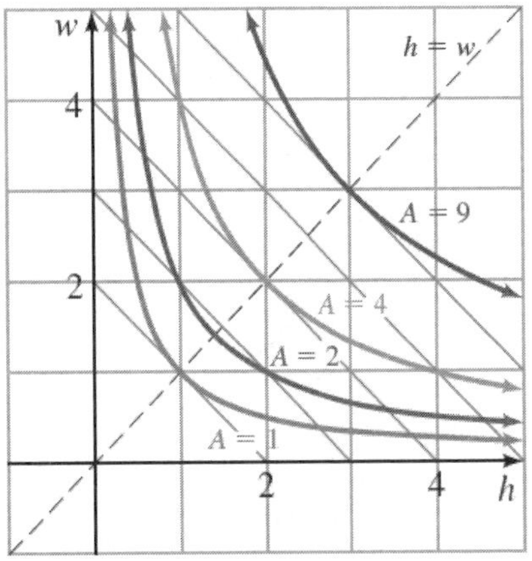

From the figure, thinking of the curves as contours on a map, we see that the largest value of A anywhere along any of the lines $h + w = c$ occurs midway along the line, when $h = w$. Thus, the largest area rectangle with a fixed perimeter occurs when $h = w$ (that is, when the rectangle is a square).

51. The level curve at $z = 3$ has the form $3 = x^{0.5}y^{0.5}$, or $y = 9/x$, and shows the relationship between the number of workers and the operating budget at a production level of 3 units.

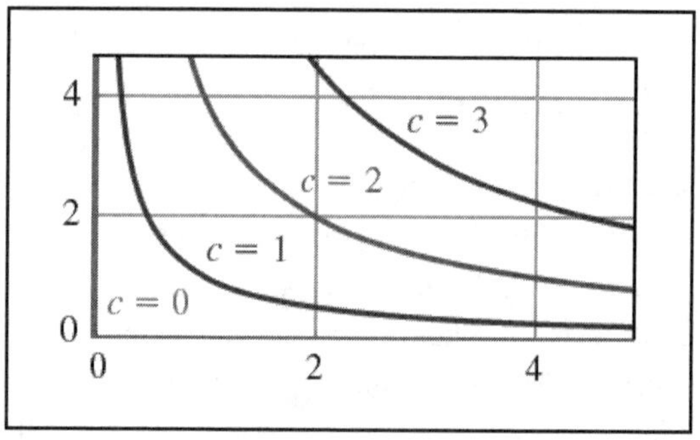

53. The level curve at $z = 0$ consists of the nonnegative y axis ($x = 0$) and tells us that zero utility corresponds to zero copies of Macro Publish, regardless of the number of copies of Turbo Publish. (Zero copies of Turbo Publish does not necessarily result in zero utility, according to the formula.)

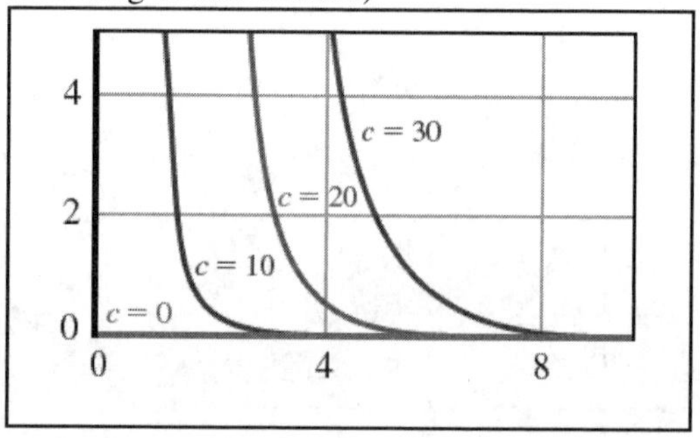

55. plane

57. Agree: Any slice through a plane is a straight line.

59. The graph of a function of three or more variables lives in four-dimensional (or higher) space, which makes it difficult to draw and visualize.

61. Referring to the diagram, we first compute the distance s. Since (a, b, z) and (x, y, z) lie in the same plane, the distance between them is the same as the distance between (a, b) and (x, y) in two dimensions, so $s = \sqrt{(x - a)^2 + (y - b)^2}$. The distance t is just the difference in height between (a, b, c) and (a, b, z), so $t = z - c$. Now d is the length of the hypotenuse of a right triangle in which the other two sides are s and t, so we have

$$d = \sqrt{s^2 + t^2} = \sqrt{(x - a)^2 + (y - b)^2 + (z - c)^2}$$

as claimed.

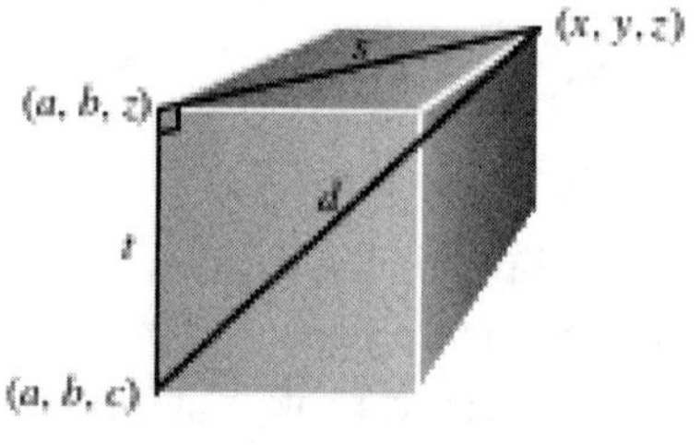

63. We need one dimension for each of the variables plus one dimension for the value of the function.

8.3

1. $f_x(x, y) = -40$; $f_y(x, y) = 20$; $f_x(1, -1) = -40$; $f_y(1, -1) = 20$

3. $f_x(x, y) = 6x + 1$; $f_y(x, y) = -3y^2$; $f_x(1, -1) = 7$; $f_y(1, -1) = -3$

5. $f_x(x, y) = -40 + 10y$; $f_y(x, y) = 20 + 10x$; $f_x(1, -1) = -50$; $f_y(1, -1) = 30$

7. $f_x(x, y) = 6xy$; $f_y(x, y) = 3x^2$; $f_x(1, -1) = -6$; $f_y(1, -1) = 3$

9. $f_x(x, y) = 2xy^3 - 3x^2y^2 - y$; $f_y(x, y) = 3x^2y^2 - 2x^3y - x$; $f_x(1, -1) = -4$; $f_y(1, -1) = 4$

11. $f_x(x, y) = 6y(2xy + 1)^2$; $f_y(x, y) = 6x(2xy + 1)^2$; $f_x(1, -1) = -6$; $f_y(1, -1) = 6$

13. $f_x(x, y) = e^{x+y}$; $f_y(x, y) = e^{x+y}$; $f_x(1, -1) = 1$; $f_y(1, -1) = 1$

15. $f_x(x, y) = 3x^{-0.4}y^{0.4}$; $f_y(x, y) = 2x^{0.6}y^{-0.6}$; $f_x(1, -1)$ is undefined; $f_y(1, -1)$ is undefined

17. $f_x(x, y) = 0.2ye^{0.2xy}$; $f_y(x, y) = 0.2xe^{0.2xy}$; $f_x(1, -1) = -0.2e^{-0.2}$; $f_y(1, -1) = 0.2e^{-0.2}$

19. $f_{xx}(x, y) = 0$; $f_{yy}(x, y) = 0$; $f_{xy}(x, y) = f_{yx}(x, y) = 0$; $f_{xx}(1, -1) = 0$; $f_{yy}(1, -1) = 0$; $f_{xy}(1, -1) = f_{yx}(1, -1) = 0$

21. $f_{xx}(x, y) = 0$; $f_{yy}(x, y) = 0$; $f_{xy}(x, y) = f_{yx}(x, y) = 10$; $f_{xx}(1, -1) = 0$; $f_{yy}(1, -1) = 0$; $f_{xy}(1, -1) = f_{yx}(1, -1) = 10$

23. $f_{xx}(x, y) = 6y$; $f_{yy}(x, y) = 0$; $f_{xy}(x, y) = f_{yx}(x, y) = 6x$; $f_{xx}(1, -1) = -6$; $f_{yy}(1, -1) = 0$; $f_{xy}(1, -1) = f_{yx}(1, -1) = 6$

25. $f_{xx}(x, y) = e^{x+y}$; $f_{yy}(x, y) = e^{x+y}$; $f_{xy}(x, y) = f_{yx}(x, y) = e^{x+y}$; $f_{xx}(1, -1) = 1$; $f_{yy}(1, -1) = 1$; $f_{xy}(1, -1) = f_{yx}(1, -1) = 1$

27. $f_{xx}(x, y) = -1.2x^{-1.4}y^{0.4}$; $f_{yy}(x, y) = -1.2x^{0.6}y^{-1.6}$; $f_{xy}(x, y) = f_{yx}(x, y) = 1.2x^{-0.4}y^{-0.6}$; $f_{xx}(1, -1)$ is undefined; $f_{yy}(1, -1)$ is undefined; $f_{xy}(1, -1)$ and $f_{yx}(1, -1)$ are undefined

29. $f_x(x, y, z) = yz$; $f_y(x, y, z) = xz$; $f_z(x, y, z) = xy$; $f_x(0, -1, 1) = -1$; $f_y(0, -1, 1) = 0$; $f_z(0, -1, 1) = 0$

31. $f_x(x, y, z) = \dfrac{4}{(x + y + z^2)^2}$; $f_y(x, y, z) = \dfrac{4}{(x + y + z^2)^2}$; $f_z(x, y, z) = \dfrac{8z}{(x + y + z^2)^2}$; $f_x(0, -1, 1)$ is undefined; $f_y(0, -1, 1)$ is undefined; $f_z(0, -1, 1)$ is undefined

33. $f_x(x, y, z) = e^{yz} + yze^{xz}$; $f_y(x, y, z) = xze^{yz} + e^{xz}$; $f_z(x, y, z) = xy(e^{yz} + e^{xz})$; $f_x(0, -1, 1) = e^{-1} - 1$; $f_y(0, -1, 1) = 1$; $f_z(0, -1, 1) = 0$

35. $f_x(x, y, z) = 0.1x^{-0.9}y^{0.4}z^{0.5}$; $f_y(x, y, z) = 0.4x^{0.1}y^{-0.6}z^{0.5}$; $f_z(x, y, z) = 0.5x^{0.1}y^{0.4}z^{-0.5}$; $f_x(0, -1, 1)$ is undefined; $f_y(0, -1, 1)$ is undefined, $f_z(0, -1, 1)$ is undefined

37. $f_x(x, y, z) = yze^{xyz}$, $f_y(x, y, z) = xze^{xyz}$, $f_z(x, y, z) = xye^{xyz}$; $f_x(0, -1, 1) = -1$; $f_y(0, -1, 1) = f_z(0, -1, 1) = 0$

39. $f_x(x, y, z) = 0$; $f_y(x, y, z) = -\dfrac{600z}{y^{0.7}(1 + y^{0.3})^2}$; $f_z(x, y, z) = \dfrac{2000}{1 + y^{0.3}}$; $f_x(0, -1, 1)$ is undefined (because $f(0, -1, 1)$ is); $f_y(0, -1, 1)$ is undefined; $f_z(0, -1, 1)$ is undefined

41. $\partial C/\partial x = 6000$: The marginal cost to manufacture each car is \$6000. $\partial C/\partial y = 4000$: The marginal cost to manufacture each truck is \$4000.

43. $y = 82 - 0.78t - 1.02x$; $\frac{\partial y}{\partial t} = -0.78$. Since units of $\partial y/\partial t$ are units of y (percentage of articles written by researchers in the U.S.) per unit of t (years), we conclude that the number of articles written by researchers in the U.S. was decreasing at a rate of 0.78 percentage points per year. $\frac{\partial y}{\partial x} = -1.02$. Since units of $\partial y/\partial x$ are units of y (percentage of articles written by researchers in the U.S.) per unit of y (percentage of articles written by researchers in Europe), we conclude that the number of articles written by researchers in the U.S. was decreasing at a rate of 1.02 percentage points per one percentage point increase in articles written in Europe.

45. $C_x(x, y) = 6000 - 20y$; $C_x(10, 20) = \$5600$ per car

47. **(a)** $\partial M/\partial c = -3.8$, $\partial M/\partial f = 2.2$. For every 1 point increase in the percentage of Chrysler owners who remain loyal, the percentage of Mazda owners who remain loyal decreases by 3.8 points. For every 1 point increase in the percentage of Ford owners who remain loyal, the percentage of Mazda owners who remain loyal increases by 2.2 points.
(b) $M(0.56, 0.56, 0.72, 0.50, 0.43) \approx 16\%$

49. **(a)** Writing $z(t, x)$, $z(10, 0) = \$16{,}500$. **(b)** $z(10, 1) = \$28{,}600$ **(c)** $z_t(t, x) = 350 + 220x$; $z_t(10, 0) = \$350$ per year **(d)** $z_t(10, 1) = \$570$ per year **(e)** The gap was widening: The median income of a white family was rising faster.

51. The marginal cost of a car is $C_x(x, y) =$ I\$$6000 + 1000e^{-0.01(x+y)}$ per car. The marginal cost of a truck is $C_y(x, y) = \$4000 + 1000e^{-0.01(x+y)}$ per truck. Both marginal costs decrease as production rises.

53. $\bar{C}(x, y) = \frac{200{,}000 + 6000x + 4000y - 100{,}000e^{-0.01(x+y)}}{x + y}$;

$$\bar{C}_x(x, y) = \frac{(6000 + 1000e^{-0.1(x+y)})(x + y) - (200{,}000 + 6000x + 4000y - 100{,}000e^{-0.01(x+y)})}{(x + y)^2} = \frac{-200{,}000 + 2000y + (1000x + 1000y + 100{,}000)e^{-0.01(x+y)}}{(x + y)^2};$$

$\bar{C}_x(50, 50) = -\$2.64$ per car. This means that at a production level of 50 cars and 50 trucks per week, the average cost per vehicle is decreasing by \$2.64 for each additional car manufactured.

$$\bar{C}_y(x, y) = \frac{(4000 + 1000e^{-0.01(x+y)})(x + y) - (200{,}000 + 6000x + 4000y - 100{,}000e^{-0.01(x+y)})}{(x + y)^2} = \frac{-200{,}000 - 2000x + (1000x + 1000y + 100{,}000)e^{-0.01(x+y)}}{(x + y)^2};$$

$\bar{C}_y(50, 50) = -\$22.64$ per truck. This means that at a production level of 50 cars and 50 trucks per week, the average cost per vehicle is decreasing by \$22.64 for each additional truck manufactured.

55. No. Your revenue function is $R(x, y) = 15{,}000x + 10{,}000y - 5000\sqrt{x+y}$, so your marginal revenue from the sale of cars is $R_x(x, y) = \$15{,}000 - \frac{2500}{\sqrt{x+y}}$ per car and your marginal revenue from the sale of trucks is $R_y(x, y) = \$10{,}000 - \frac{2500}{\sqrt{x+y}}$ per truck. These increase with increasing x and y. In other words, you will earn more revenue per vehicle with increasing sales, and so the rental company will pay more for each additional vehicle it buys.

57. $P_z(x, y, z) = 0.016x^{0.4}y^{0.2}z^{-0.6}$; $P_z(10, 100{,}000, 1{,}000{,}000) \approx 0.0001010$ papers/\$, or approximately 1 paper per \$10,000 increase in the subsidy.

59. (a) $U_x(x, y) = 4.8x^{-0.2}y^{0.2} + 1$; $U_y(x, y) = 1.2x^{0.8}y^{-0.8}$; $U_x(10, 5) = 5.18$, $U_y(10, 5) = 2.09$. This means that, if 10 copies of Macro Publish and 5 copies of Turbo Publish are purchased, the company's daily productivity is increasing at a rate of 5.18 pages per day for each additional copy of Macro purchased and by 2.09 pages per day for each additional copy of Turbo purchased.
(b) $\frac{U_x(10, 5)}{U_y(10, 5)} \approx 2.48$ is the ratio of the usefulness of one additional copy of Macro to one of Turbo. Thus, with 10 copies of Macro and 5 copies of Turbo, the company can expect approximately 2.48 times the productivity per additional copy of Macro compared to Turbo.

61. $F_y(x, y, z) = \frac{2KQq(y - b)}{[(x - a)^2 + (y - b)^2 + (z - c)^2]^2}$. With $(a, b, c) = (0, 0, 0)$, $K = 9\times 10^9$, $Q = 10$, and $q = 5$, $F_y(2, 3, 3) \approx -6\times 10^9$ N/sec.

63. (a) $A_P(P, r, t) = (1 + r)^t$; $A_r(P, r, t) = tP(1 + r)^{t-1}$; $A_t(P, r, t) = P(1 + r)^t \ln(1 + r)$; $A_P(100, 0.1, 10) = 2.59$; $A_r(100, 0.1, 10) = 2{,}357.95$; $A_t(100, 0.1, 10) = 24.72$. Thus, for a \$100 investment at 10% interest, after 10 years the accumulated amount is increasing at a rate of \$2.59 per \$1 of principal, at a rate of \$2,357.95 per increase of 1 in r (note that this would correspond to an increase in the interest rate of 100%), and at a rate of \$24.72 per year.
(b) $A_P(100, 0.1, t)$ tells you the rate at which the accumulated amount in an account bearing 10% interest with a principal of \$100 is growing per \$1 increase in the principal, t years after the investment.

65. (a) $P_x = Ka\left(\frac{y}{x}\right)^b$ and $P_y = Kb\left(\frac{x}{y}\right)^a$. They are equal precisely when $\frac{a}{b} = \left(\frac{x}{y}\right)^b\left(\frac{x}{y}\right)^a$. Substituting $b = 1-a$ now gives $\frac{a}{b} = \frac{x}{y}$.
(b) The given information implies that $P_x(100, 200) = P_y(100, 200)$. By part (a), this occurs precisely when $a/b = x/y = 100/200 = 1/2$. But $b = 1 - a$, so $a/(1 - a) = 1/2$, giving $a = 1/3$ and $b = 2/3$.

67. $u_t(r, t) = -\frac{1}{4\pi Dt^2}e^{-r^2/(4Dt)} + \frac{r^2}{16\pi D^2t^3}e^{-r^2/(4Dt)}$.
Taking $D = 1$, $u_t(1, 3) \approx -0.0075$, so the concentration is decreasing at 0.0075 parts of nutrient per part of water/sec.

69. f is increasing at a rate of s units per unit of x, f is increasing at a rate of t units per unit of y, and the value of f is r when $x = a$ and $y = b$

71. The marginal cost of building an additional orbicus; zonars per unit.

73. Answers will vary. One example is $f(x, y) = -2x + 3y$. Others are $f(x, y) = -2x + 3y + 9$ and $f(x, y) = xy - 3x + 2y + 10$.

75. **(a)** b is the z-intercept of the plane, m is the slope of the intersection of the plane with the xz-plane, n is the slope of the intersection of the plane with the yz-plane. **(b)** Write $z = b + rx + sy$. We are told that $\partial z/\partial x = m$, so $r = m$. Similarly, $s = n$. Thus, $z = b + mx + ny$. We are also told that the plane passes through (h, k, l). Substituting gives $l = b + mh + nk$. This gives b as $l - mh - nk$. Substituting in the equation for z therefore gives $z = l - mh - nk + mx + ny = l + m(x - h) + n(y - k)$, as required.

8.4

1. P: relative minimum; Q: none of the above; R: relative maximum

3. P: saddle point; Q: relative maximum; R: none of the above

5. relative minimum

7. Neither

9. Saddle point

11. $f_x = 2x$; $f_y = 2y$; $f_{xx} = 2$; $f_{yy} = 2$; $f_{xy} = 0$. $f_x = 0$ when $x = 0$; $f_y = 0$ when $y = 0$, so $(0, 0)$ is the only critical point. $H = 4$ and $f_{xx} > 0$, so f has a relative minimum at $(0, 0, 1)$.

13. $g_x = -2x - 1$; $g_y = -2y + 1$; $g_{xx} = -2$; $g_{yy} = -2$; $g_{xy} = 0$. $g_x = 0$ when $x = -1/2$; $g_y = 0$ when $y = 1/2$, so $(-1/2, 1/2)$ is the only critical point.
$H = 4$ and $g_{xx} < 0$, so g has a relative maximum at $(-1/2, 1/2, 3/2)$.

15. $h_x = 2xy - 4x$; $h_y = x^2 - 8y$; $h_{xx} = 2y - 4$; $h_{yy} = -8$; $h_{xy} = 2x$. $h_x = 0$ when $x = 0$ or $y = 2$; $h_y = 0$ when $x^2 = 8y$. The two possibilities are $x = 0$, so $y = 0$, or $y = 2$, so $x^2 = 16$ or $x = \pm 4$. This gives three critical points: $(0, 0)$, $(-4, 2)$, and $(4, 2)$. $H(x, y) = -8(2y - 4) - 4x^2 = 32 - 16y - 4x^2$; $H(0, 0) = 32$ and $h_{xx}(0, 0) = -4 < 0$; $H(-4, 2) = -64 = H(4, 2)$. Hence h has a relative maximum at $(0, 0, 0)$ and saddle points at $(\pm 4, 2, -16)$.

17. $s_x = 2xe^{x^2+y^2}$; $s_y = 2ye^{x^2+y^2}$; $s_{xx} = (2 + 4x^2)e^{x^2+y^2}$; $s_{yy} = (2 + 4y^2)e^{x^2+y^2}$; $s_{xy} = 4xye^{x^2+y^2}$. $s_x = 0$ when $x = 0$; $s_y = 0$ when $y = 0$; so the only critical point is $(0, 0)$. $H(0, 0) = 4 - 0 = 4$ and $s_{xx}(0, 0) = 2 > 0$, so s has a relative minimum at $(0, 0, 1)$.

19. $t_x = 4x^3 + 8y^2$; $t_y = 16xy + 8y^3$; $t_{xx} = 12x^2$; $t_{yy} = 16x + 24y^2$; $t_{xy} = 16y$. $t_x = 0$ when $x^3 = -2y^2$ (notice that $x \le 0$ in this case); $t_y = 0$ when $y = 0$ or $y^2 = -2x$; if $y = 0$ then $x = 0$; if $y^2 = -2x$ then $x^3 = 4x$, so $x = 0$ (and $y = 0$) or $x = -2$ and $y = \pm 2$. This gives three critical points: $(0, 0)$ and $(-2, \pm 2)$. $H(-2, \pm 2) = 3072 \pm 32 > 0$ and $t_{xx}(-2, \pm 2) = 48 > 0$. $H(0, 0) = 0$ so the second derivative test is inconclusive. We can see from the graph of t that the origin is not a max, min, or saddle point. Or we can look at the slice along $x = y$ (suggested by the graph) where $t = 3x^4 + 8x^3$; this function increases as x approaches 0 and then increases again as x becomes larger than 0. So, t has two relative minima at $(-2, \pm 2, -16)$ and $(0, 0)$ is a critical point that is not a relative extremum or saddle point.

21. $f_x = 2x$; $f_y = 1 - e^y$; $f_{xx} = 2$; $f_{yy} = -e^y$; $f_{xy} = 0$. $f_x = 0$ when $x = 0$; $f_y = 0$ when $y = 0$. This gives one critical point, $(0, 0)$. $H(0, 0) = -2$, so f has a saddle point at $(0, 0, -1)$ and no other critical points.

23. $f_x = -(2x + 2)e^{-(x^2+y^2+2x)}$;
$f_y = -2ye^{-(x^2+y^2+2x)}$;
$f_{xx} = (4x^2 + 8x + 2)e^{-(x^2+y^2+2x)}$;
$f_{yy} = (4y^2 - 2)e^{-(x^2+y^2+2x)}$;
$f_{xy} = 2y(2x + 2)e^{-(x^2+y^2+2x)}$.
$f_x = 0$ when $x = -1$; $f_y = 0$ when $y = 0$. This gives one critical point, $(-1, 0)$. $H(-1, 0) = 4e^2 > 0$ and $f_{xx}(-1, 0) = -2e < 0$, so f has a relative maximum at $(-1, 0, e)$.

25. $f_x = y - \frac{2}{x^2}$; $f_y = x - \frac{2}{y^2}$; $f_{xx} = \frac{4}{x^3}$; $f_{yy} = \frac{4}{y^3}$; $f_{xy} = 1$. $f_x = 0$ when $y = \frac{2}{x^2}$; $f_y = 0$ when $x = \frac{2}{y^2} = \frac{1}{2}x^4$, $x = 2^{1/3}$ ($x = 0$ is excluded because it is not in the domain of f). The corresponding value of y is $y = 2^{1/3}$. This gives one critical point, $(2^{1/3}, 2^{1/3})$. $H(2^{1/3}, 2^{1/3}) = 3$ and $f_{xx}(2^{1/3}, 2^{1/3}) = 2 > 0$, so f has a relative minimum at

$(2^{1/3}, 2^{1/3}, 3(2^{2/3}))$.

27. $g_x = 2x - \frac{2}{x^2y}$; $g_y = 2y - \frac{2}{xy^2}$; $g_{xx} = 2 + \frac{4}{x^3y}$; $g_{yy} = 2 + \frac{4}{xy^3}$; $g_{xy} = \frac{2}{x^2y^2}$. $g_x = 0$ when $y = \frac{1}{x^3}$; $g_y = 0$ when $x = \frac{1}{y^3} = x^9$, $x = \pm 1$ ($x = 0$ is excluded because it is not in the domain of f). This gives two critical points, $(1, 1)$ and $(-1, -1)$. $H(1, 1) = 32$ and $g_{xx}(1, 1) = 6 > 0$; $H(-1, -1) = 32$ and $g_{xx}(-1, -1) = 6 > 0$. So, g has relative minima at $(1, 1, 4)$ and $(-1, -1, 4)$.

29. f has an absolute minimum at $(0, 0, 1)$: $x^2 + y^2 + 1 \ge 1$ for all x and y because $x^2 + y^2 \ge 0$.

31. The relative maximum at $(0, 0, 0)$ is not absolute. For example,

$h(10, 10) = 400 > h(0, 0)$.

33. $M_c = 8 + 8c - 20f$; $M_f = -40 - 20c + 80f$; $M_{cc} = 8$; $M_{ff} = 80$; $M_{cf} = -20$. $M_c = 0$ and $M_f = 0$ when $(c, f) = (2/3, 2/3)$; $H(2/3, 2/3) = 240$ and $M_{cc}(2/3, 2/3) = 8 > 0$, so M has a minimum at $(2/3, 2/3, 1/3)$. Thus, at least 1/3 of all Mazda owners would choose another new Mazda, and this lowest loyalty occurs when 2/3 of Chrysler and Ford owners remain loyal to their brands.

35. The subsidy is $S(x, y) = 500x + 100y$, so the net cost is $N(x, y) = C(x, y) - S(x, y) = 4000 + 100x^2 + 50y^2 - 500x - 100y$. $N_x = 200x - 500$; $N_y = 100y - 100$; $N_{xx} = 200$; $N_{yy} = 100$; $N_{xy} = 0$. $N_x = 0$ when $x = 2.5$; $N_y = 0$ when $y = 1$. $H = 20{,}000$ and $N_{xx} = 100 > 0$, so N has a maximum at $(2.5, 1)$. The firm should remove 2.5 pounds of sulfur and 1 pound of lead per day.

37. The total revenue is $R = p_1q_1 + p_2q_2 = 100{,}000p_1 - 100p_1^2 + 10p_1p_2 + 150{,}000p_2 + 10p_1p_2 - 100p_2^2 = 100{,}000p_1 + 150{,}000p_2 - 100p_1^2 + 20p_1p_2 - 100p_2^2$. For convenience, write R_1 for $\partial R/\partial p_1$ and R_2 for $\partial R/\partial p_2$. Then $R_1 = 100{,}000 - 200p_1 + 20p_2$; $R_2 = 150{,}000 + 20p_1 - 200p_2$; $R_{11} = -200$; $R_{22} = -200$; $R_{12} = 20$. $R_1 = 0$ and $R_2 = 0$ when $(p_1, p_2) = (580.81, 808.08)$. $H = 39{,}600$ and $R_{11} = -200 < 0$, so R has a maximum at this critical point. You should charge \$580.81 for the Ultra Mini and \$808.08 for the Big Stack.

39. Let l = length, w = width, and h = height. We are told that $l + w + h \le 62$; for the largest possible volume we will want $l + w + h = 62$, so $l = 62 - w - h$. The volume is $V = lwh = (62 - w - h)wh = 62wh - w^2h - wh^2$. $V_w = 62h - 2wh - h^2$; $V_h = 62w - w^2 - 2wh$; $V_{ww} = -2h$; $V_{hh} = -2w$; $V_{wh} = 62 - 2w - 2h$. $V_w = 0$ when $62 - 2w - h = 0$ ($h = 0$ is not in the domain); $V_h = 0$ when $62 - w - 2h = 0$; this occurs when $w = h = 62/3 \approx 20.67$. $H \approx 1280$ and $V_{ww} < 0$, so V has a maximum at this critical point. The largest volume bag has dimensions $l = w = h \approx 20.67$ in and volume ≈ 8827 cubic inches.

41. Let l = length, w = width, and h = height. We are told that $l + 2(w + h) \le 108$; for the largest possible volume we will want $l + 2(w + h) = 108$, so $l = 108 - 2(w + h)$. The volume is $V = lwh = (108 - 2w - 2h)wh = 108wh - 2w^2h - 2wh^2$. $V_w = 108h - 4wh - 2h^2$; $V_h = 108w - 2w^2 - 4wh$; $V_{ww} = -4h$; $V_{hh} = -4w$; $V_{wh} = 108 - 4w - 4h$. $V_w = 0$ when $108 - 4w - 2h = 0$; $V_h = 0$ when $108 - 2w - 4h = 0$; this occurs when $w = h = 18$. $H = 3888$ and $V_{ww} < 0$, so V has a maximum at this critical point. The corresponding length is $l = 108 - 2(w + h) = 36$. So, the largest volume package has dimensions 18 in $\times$18 in $\times$ 36 in and volume = 11,664 cubic inches.

43.

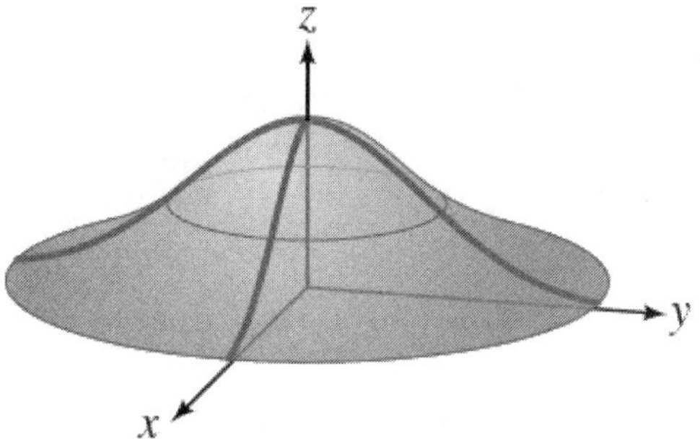

45. The function below has a relative max at $(0, 0, -1)$, has the unit circle $x^2 + y^2 = 1$ as a vertical asymptote, and takes on values larger than -1 as $x^2 + y^2$ gets large. The particular function graphed is $f(x, y) = -\dfrac{1}{|x^2 + y^2 - 1|}$.

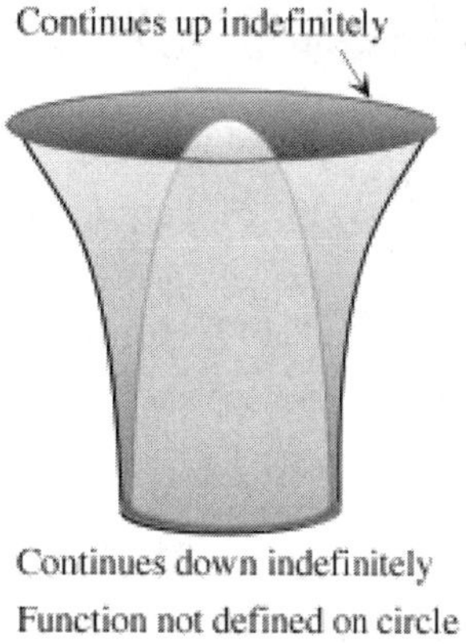

47. H must be positive.

49. No. In order for there to be a relative maximum at (a, b), *all* vertical planes through (a, b) should yield a curve with a relative maximum at (a, b). It could happen that a slice by another vertical plane through (a, b) (such as $x - a = y - b$) does not yield a curve with a relative maximum at (a, b). An example is $f(x, y) = x^2 + y^2 - \sqrt{xy}$, at the point (0, 0). Look at the slices through $x = 0$, $y = 0$ and $y = x$.

51. $\bar{C}_x = \dfrac{\partial}{\partial x}\left(\dfrac{C}{x + y}\right) = \dfrac{(x + y)C_x - C}{(x + y)^2}$. If this is zero, then $(x + y)C_x = C$, or $C_x = \dfrac{C}{x + y} = \bar{C}$. Similarly, if $\bar{C}_y = 0$ then $C_y = \bar{C}$. This is reasonable because if the average cost is decreasing with increasing x, then the average cost is greater than the marginal cost C_x. Similarly, if the average cost is increasing with increasing x, then the average cost is less than the marginal cost C_x. Thus, if the average cost is stationary with increasing x, then the average cost equals the marginal cost C_x. (The situation is similar for the case of increasing y.)

53. The equation of the tangent plane at the point (a, b) is $z = f(a, b) + f_x(a, b)(x - a) + f_y(a, b)(y - b)$. If f has a relative extremum at (a, b), then $f_x(a, b) = 0 = f_y(a, b)$. Substituting these into the equation of the tangent plane gives $z = f(a, b)$, a constant. But the graph of $z =$ *constant* is a plane parallel to the xy-plane.

8.5

1. The objective function is $1 - x^2 - y^2 - z^2$. Substituting the constraint equation $z = 2y$ gives the objective function as

$$h(x, y) = 1 - x^2 - y^2 - (2y)^2$$
$$= 1 - x^2 - 5y^2$$
$$h_x = -2x,\ h_y = -10y$$

Critical points:

$$-2x = 0$$
$$-10y = 0$$

$(x, y) = (0, 0)$ is the only critical point.

$$h_{xx} = -2,\ h_{xy} = 0,\ h_{yy} = -10$$
$$H = h_{xx}h_{yy} - (h_{xy})^2 = 20 > 0$$

Since $h_{xx} < 0$, the critical point is a local maximum. That it is an absolute maximum can be seen by considering the graph of $h(x, y)$. The corresponding value of the objective function is

$$h(0, 0) = 1 - (0)^2 - 5(0)^2 = 1$$

When $x = 0$ and $y = 0$, $z = 2y = 0$ as well, so f has an absolute maximum of 1 at the point $(0, 0, 0)$.

3. The objective function is $1 - x^2 - x - y^2 + y - z^2 + z$. Substituting the constraint equation $y = 3x$ gives the objective function as

$$h(x, z) = 1 - x^2 - x - 9x^2 + 3x - z^2 + z$$
$$= 1 - 10x^2 + 2x - z^2 + z$$
$$h_x = -20x + 2,\ h_z = -2z + 1$$

Critical points:

$$-20x + 2 = 0$$
$$\Rightarrow x = 1/10$$
$$-2z + 1 = 0$$
$$\Rightarrow z = 1/2$$

$(x, z) = (1/10, 1/2)$ is the only critical point.

$$h_{xx} = -20,\ h_{xz} = 0,\ h_{zz} = -2$$
$$H = h_{xx}h_{zz} - (h_{xz})^2 = 40 > 0$$

Since $h_{xx} < 0$, the critical point is a local maximum. That it is an absolute maximum can be seen by considering the graph of $h(x, y) = 1 - 10x^2 + 2x - z^2 + z$. The corresponding value of the objective function is

$$h(1/10, 1/2)$$
$$= 1 - 10/100 + 2/10 - 1/4 + 1/2$$
$$= 1.35$$

When $x = 1/10$ and $z = 1/2$, $y = 3x = 3/10$, so f has an absolute maximum of 1.35 at the point $(1/10, 3/10, 1/2)$

5. The objective function is $xy + 4xz + 2yz$. Solving the constraint equation $xyz = 1$ for z gives

$$z = \frac{1}{xy}$$
$$S = xy + 4x\frac{1}{xy} + 2y\frac{1}{xy}$$
$$= xy + \frac{4}{y} + \frac{2}{x}$$
$$S_x = y - \frac{2}{x^2},\ S_y = x - \frac{4}{y^2}$$

Critical points:

$$y - \frac{2}{x^2} = 0$$
$$x - \frac{4}{y^2} = 0$$

Substituting the first equation in the second gives

$$x - \frac{4}{(2/x^2)^2} = 0$$
$$x - x^4 = 0$$
$$x(1 - x^3) = 0$$
$$x = 1$$

(We reject $x = 0$ because $x > 0$). Substituting $x = 1$ into the equation $y - \frac{2}{x^2} = 0$ gives

$$y = 2.$$

$(x, y) = (1, 2)$ is the only critical point.

$$S_{xx} = 4/x^3,\ S_{xy} = 1\ S_{yy} = 8/y^3$$
$$S_{xx}(1, 2) = 4/1^3 = 4$$
$$S_{xy}(1, 2) = 1$$
$$S_{yy}(1, 2) = 8/2^3 = 1$$
$$H = S_{xx}S_{yy} - (S_{xy})^2 = 4 - 1 = 3 > 0$$

Since $S_{xx}(1, 2) > 0$, the critical point is a local minimum. That it is an absolute minimum can be seen by considering the graph of $S(x, y) = xy + \frac{4}{y} + \frac{2}{x}$ for $x > 0$, $y > 0$. The corresponding value of the objective function is

$$S(1, 2) = (1)(2) + \frac{4}{2} + \frac{2}{1} = 6$$

When $x = 1$ and $y = 2$, $z = 1/xy = 1/2$, so S has an absolute minimum of 6 at (1, 2, 1/2).

7. The constraint is $x + 2y = 40$, or $x + 2y - 40 = 0$.

$f(x, y) = xy$

$g(x, y) = x + 2y - 4$

(1) $f_x = \lambda g_x \Rightarrow y = \lambda$

(2) $f_y = \lambda g_y \Rightarrow x = 2\lambda$

(3) Constraint $\Rightarrow x + 2y = 40$

Substitute Equation (1) into (2) to obtain

$x = 2y$

Substitute in Equation (3) to obtain

$2y + 2y = 40$

$4y = 40$

$y = 10$

The corresponding value of x is

$x = 2(10) = 20$

The corresponding value of the objective is

$f(20, 10) = (20)(10) = 200$

9. The constraint is $x^2 + y^2 = 8$, or $x^2 + y^2 - 8 = 0$.

$f(x, y) = 4xy$

$g(x, y) = x^2 + y^2 - 8$

(1) $f_x = \lambda g_x \Rightarrow 4y = 2\lambda x \Rightarrow 2y = \lambda x$

(2) $f_y = \lambda g_y \Rightarrow 4x = 2\lambda y \Rightarrow 2x = \lambda y$

(3) Constraint $\Rightarrow x^2 + y^2 = 8$

Divide Equation (1) by x to obtain

$$\lambda = \frac{2y}{x}$$

Substitute in Equation (2) to obtain

$$2x = \frac{2y}{x}y = \frac{2y^2}{x}$$

or $2x^2 = 2y^2$, giving

$y = \pm x$

Substituting into the constraint gives

$x^2 + x^2 = 8$

or $2x^2 = 8$, giving

$x^2 = 4$,

$x = \pm 2$, so $y = \pm 2$ as well.

Thus we have 4 critical points: $(-2, -2)$, $(-2, 2)$, $(2, -2)$ and $(2, 2)$. Substitute each of these into the objective function:

$f(-2, -2) = 4(-2)(-2) = 16$

$f(-2, 2) = 4(-2)(2) = -16$

$f(2, -2) = 4(2)(-2) = -16$

$f(2, 2) = 4(2)(2) = 16$

The first and last of these give a maximum of 16.

11. The constraint is $x + 2y = 10$, or $x + 2y - 10 = 0$.

$f(x, y) = x^2 + y^2$

$g(x, y) = x + 2y - 10$

(1) $f_x = \lambda g_x \Rightarrow 2x = \lambda$

(2) $f_y = \lambda g_y \Rightarrow 2y = 2\lambda \Rightarrow y = \lambda$

(3) Constraint $\Rightarrow x + 2y = 10$

Substitute Equation (2) in Equation (1) to obtain

$2x = y$

Substituting into the constraint gives

$x + 4x = 10$

$5x = 10$

$x = 2$

so $y = 2x = 4$. Thus we have one critical point: (2, 4). The corresponding value of the objective function is

$f(2, 10) = 2^2 + 4^2 = 20$

13. The constraint is $z = 2y$, or $z - 2y = 0$.

$f(x, y, z) = 1 - x^2 - y^2 - z^2$

$g(x, y, z) = z - 2y$

(1) $f_x = \lambda g_x \Rightarrow -2x = 0$

(2) $f_y = \lambda g_y \Rightarrow -2y = -2\lambda \Rightarrow y = \lambda$

(3) $f_z = \lambda g_z \Rightarrow -2z = \lambda$

(4) Constraint $\Rightarrow z = 2y$

The first equation tells us that $x = 0$. Substituting the third equation in the second gives

$y = -2z$, or

$y + 2z = 0$

Combining this with the constraint equation

$z - 2y = 0$

gives a system of 2 equations in 2 unknowns, whose solution is

$y = z = 0$

Thus the only critical point is $(x, y, z) = (0, 0, 0)$ and the corresponding value of the objective is

$f(0, 0, 0) = 1 - 0^2 - 0^2 - 0^2 = 1$

That this is an absolute maximum is seen from the fact that $f(x, y, z) = 1 - x^2 - y^2 - z^2$ can never be larger than 1.

15. The constraint is $3x = y$, or $3x - y = 0$.

$f(x, y, z) = 1 - x^2 - x - y^2 + y - z^2 + z$

$g(x, y, z) = 3x - y$

(1) $f_x = \lambda g_x \Rightarrow -2x - 1 = 3\lambda$

(2) $f_y = \lambda g_y \Rightarrow -2y + 1 = -\lambda$

(3) $f_z = \lambda g_z \Rightarrow -2z + 1 = 0$

(4) Constraint $\Rightarrow 3x = y$

Equation (3) tells us that $z = 1/2$. Substituting (2) in (1) gives

$-2x - 1 = 6y - 3$ or

$2x + 6y = 2$, that is

$x + 3y = 1$

Combining this with the constraint equation

$3x - y = 0$

gives a system of 2 equations in 2 unknowns, whose solution is

$x = 1/10,\ y = 3/10$

Thus the only critical point is $(x, y, z) = (1/10, 3/10, 1/2)$. and the corresponding value of the objective is

$f(1/10, 3/10, 1/2) = 1 - 1/100 - 1/10 - 9/100 + 3/10 - 1/4 + 1/2 = 1.35$

17. The constraint is $xyz = 1$, or $xyz - 1 = 0$.

$f(x, y, z) = xy + 4xz + 2yz$

$g(x, y, z) = xyz - 1$

(1) $f_x = \lambda g_x \Rightarrow y + 4z = \lambda yz$

(2) $f_y = \lambda g_y \Rightarrow x + 2z = \lambda xz$

(3) $f_z = \lambda g_z \Rightarrow 4x + 2y = \lambda xy$

(4) Constraint $\Rightarrow xyz = 1$

Solve Equation (1) for λ:

$$\lambda = \frac{1}{z} + \frac{4}{y}$$

Substituting in (2) gives

$$x + 2z = \left(\frac{1}{z} + \frac{4}{y}\right)xz$$

$$x + 2z = x + \frac{4xz}{y}$$

$$2z = \frac{4xz}{y}$$

$$1 = \frac{2x}{y}$$

$$y = 2x$$

Substituting the expression for λ in (3) gives

$$4x + 2y = \left(\frac{1}{z} + \frac{4}{y}\right)xy$$

$$4x + 2y = \frac{xy}{z} + 4x$$

$$2y = \frac{xy}{z}$$

$$2 = \frac{x}{z}$$

$$z = x/2$$

Substituting the expressions we obtained for y and z in the constraint equation gives:

$x(2x)(x/2) = 1$

$x^3 = 1$

$x = 1$

The corresponding values of y and z are

$y = 2x = 2$

$z = x/2 = 1/2$

Therefore, the only critical point is (1, 2, 1/2) and the corresponding value of the objective is $f(1, 2, 1/2) = (1)(2) + 4(1)(1/2) + 2(2)(1/2) = 6$. Therefore, the minimum value of the objective function is 6, and occurs at the point (1, 2, 1/2).

19. Let x be the length of the east and west sides, and let y be the length of the north and south sides. The problem translates to:

Maximize $A = xy$ subject to $8x + 4y = 80$

The constraint is $8x + 4y = 80$, or $8x + 4y - 80 = 0$.

$f(x, y) = xy$

$g(x, y) = 8x + 4y - 80$

(1) $f_x = \lambda g_x \Rightarrow y = 8\lambda$

(2) $f_y = \lambda g_y \Rightarrow x = 4\lambda$

(3) Constraint $\Rightarrow 8x + 4y = 80$

Solve Equation (1) for λ to obtain

$\lambda = y/8$

Substitute in Equation (2) to obtain

$x = 4y/8 = y/2$

or $2x - y = 0$

The constraint equation is

$8x + 4y = 80$

or $2x + y = 20$

So we have a system of two linear equations in two unknowns:

$2x - y = 0$

$2x + y = 20$

The solution is $(x, y) = (5, 10)$. The corresponding value of the objective function is

$A = xy = 5\times10 = 50$ sq. ft.

21. The problem translates to:

Maximize $R = pq$ subject to $q = 200{,}000 - 10{,}000p$

The constraint is $q = 200{,}000 - 10{,}000p$, or $q - 200{,}000 + 10{,}000p = 0$.

$f(p, q) = pq$

$g(p, q) = q - 200{,}000 + 10{,}000p$

(1) $f_p = \lambda g_p \Rightarrow q = 10{,}000\lambda$

(2) $f_q = \lambda g_q \Rightarrow p = \lambda$

(3) Constraint $\Rightarrow q = 200{,}000 - 10{,}000p$

Substitute Equation (2) in Equation (2) to obtain

$q = 10{,}000p$

The constraint equation is

$q = 200{,}000 - 10{,}000p$

Equating these two expressions for q gives

$10{,}000p = 200{,}000 - 10{,}000p$

$20{,}000p = 200{,}000$

$p = \$10$

We are not asked for any further information.

23. We want to maximize $f(x, y, z) = xyz$ subject to $x^2 + y^2 + z^2 - 1 = 0$. Using Lagrange multipliers, we need to solve the system $yz = 2\lambda x$, $xz = 2\lambda y$, $xy = 2\lambda z$, and $x^2 + y^2 + z^2 - 1 = 0$. We solve the first equation for $\lambda = \dfrac{yz}{2x}$ and substitute in the second to find $xz = \dfrac{y^2z}{x}$ or $x^2z = y^2z$, giving $x^2 = y^2$ (assuming that $z \neq 0$, but $z = 0$ makes $xyz = 0$ and we can easily find larger values). Substituting the expression for λ into the third equation gives $x^2 = z^2$ similarly. From the last equation we then get $3x^2 - 1 = 0$, or $x = \pm1/\sqrt{3}$. The corresponding values of y and z are also $\pm1/\sqrt{3}$, so we get eight points to check: $(\pm1/\sqrt{3}, \pm1/\sqrt{3}, \pm1/\sqrt{3})$ with all eight choices of signs. Checking, we see that the largest value of xyz occurs when all of the signs or exactly two of them are positive, that is, at the points $(1/\sqrt{3}, 1/\sqrt{3}, 1/\sqrt{3})$, $(-1/\sqrt{3}, -1/\sqrt{3}, 1/\sqrt{3})$, $(1/\sqrt{3}, -1/\sqrt{3}, -1/\sqrt{3})$, and $(-1/\sqrt{3}, 1/\sqrt{3}, -1/\sqrt{3})$.

25. Minimize $d(x, y, z) = x^2 + y^2 + z^2$ subject to $x^2 + y - 1 - z = 0$. Using Lagrange multipliers, we need to solve $2x = 2\lambda x$, $2y = \lambda$, $2z = -\lambda$, and $x^2 + y - 1 - z = 0$. From the first equation, either $x = 0$ or $\lambda = 1$. If $x = 0$, substitute $y = \lambda/2$ and $z = -\lambda/2$ in the last equation to get $\lambda/2 - 1 + \lambda/2 = 0$, $\lambda = 1$. This gives the point $(0, 1/2, -1/2)$. On the other hand, if $\lambda = 1$, then $y = 1/2$, $z = -1/2$, and $x^2 + 1/2 - 1 + 1/2 = 0$, so $x = 0$, giving the same point. Since the distance may get arbitrarily large, but can get no less than 0, there must be a minimum distance and it must be at the point we found, $(0, 1/2, -1/2)$.

27. Minimize $d(x, y, z) = (x + 1)^2 + (y - 1)^2 + (z - 3)^2$ subject to $-2x + 2y + z - 5 = 0$. Using Lagrange multipliers, we need to solve $2(x + 1) = -2\lambda$, $2(y - 1) = 2\lambda$, $2(z - 3) = \lambda$, and $-2x + 2y + z - 5 = 0$. Solve the first three equations for $x = -\lambda - 1$, $y = \lambda + 1$, and $z = \lambda/2 + 3$. Substitute these in the last equation: $-2(-\lambda - 1) + 2(\lambda + 1) + (\lambda/2 + 3) - 5 = 0$, $\frac{9}{2}\lambda + 2 = 0$, $\lambda = -4/9$. This gives $x = -5/9$, $y = 5/9$, and $z = 25/9$. Thus, the point on the plane closest to $(-1, 1, 3)$ is $(-5/9, 5/9, 25/9)$.

29. Let the length, width, and height be l, w, and h, respectively. We wish to minimize $C = 20\times2\times lw + 10(2lh + 2wh) = 40lw + 20lh + 20wh$ subject to $lwh = 2$. Solve the constraint for $l = \frac{2}{wh}$ and substitute to get $C(w, h) = \frac{80}{h} + \frac{40}{w} + 20wh$, $w > 0$ and $h > 0$. $C_w = -\frac{40}{w^2} + 20h = 0$ when $h = \frac{2}{w^2}$; $C_h = -\frac{80}{h^2} + 20w = 0$ when $w = \frac{4}{h^2} = w^4$, giving $w = 1$ (recall that $w > 0$). Thus, $h = 2$ and $l = 1$. Making either of h or w small or both large makes C large, so the point we found must be a minimum. Thus, the dimensions of the box of least cost are $l\times w\times h = 1\times1\times2$.

31. Let l = length, w = width, and h = height. We want to maximize the volume $V = lwh$. We are told that $l + 2(w + h) \le 108$; for the largest possible volume we will want $l + 2(w + h) = 108$. In the solution for Exercise 41 in Section 8.4 we showed one way to solve this problem. Here we use the alternative, which is Lagrange multipliers. So, we need to solve $wh = \lambda$, $lh = 2\lambda$, $lw = 2\lambda$, and $l + 2w + 2h - 108 = 0$. Substitute $\lambda = wh$ in the second equation to get $lh = 2wh$, so $l = 2w$ (we certainly must have $h > 0$ for a maximum volume). If we substitute $\lambda = wh$ in the third equation we get $lw = 2wh$, so $l = 2h$, hence $w = h = \frac{1}{2}l$. Substituting now in the last equation gives $3l = 108$, so $l = 36$, $w = 18$, and $h = 18$. Thus, the largest volume package has dimensions 18 in $\times$18 in $\times$ 36 in and volume = 11,664 cubic inches.

33. Let L be the cost of lightweight cardboard and let H be the cost of heavy-duty cardboard. Minimize $C = 2Hlw + 2Llh + 2Lwh$ subject to $lwh = 2$. Solve the constraint for $l = \frac{2}{wh}$ and substitute to get $C(w, h) = \frac{4H}{h} + \frac{4L}{w} + 2Lwh$. $C_w = -\frac{4L}{w^2} + 2Lh = 0$ when $h = \frac{2}{w^2}$; $C_h = -\frac{4H}{h^2} + 2Lw = 0$ when $w = \frac{2H}{Lh^2} = \frac{H}{2L}w^4$, so $w = \left(\frac{2L}{H}\right)^{1/3}$. Substituting in the expressions for l and h we find $h = 2^{1/3}\left(\frac{H}{L}\right)^{2/3}$ and $l = w = \left(\frac{2L}{H}\right)^{1/3}$. Thus, the box of least cost has dimensions $\left(\frac{2L}{H}\right)^{1/3} \times \left(\frac{2L}{H}\right)^{1/3} \times 2^{1/3}\left(\frac{H}{L}\right)^{2/3}$.

35. If (x, y, z) is one corner of the box (with $x > 0$, $y > 0$, and $z > 0$), the box will have dimensions $2x \times 2y \times z$. We need to maximize $V = 4xyz$ subject to $x^2 + y^2 + z - 1 = 0$. Using Lagrange multipliers we must solve $4yz = 2\lambda x$, $4xz = 2\lambda y$, $4xy = \lambda$, and $x^2 + y^2 + z - 1 = 0$. Solve the first equation for $\lambda = 2\frac{yz}{x}$ and substitute in the second and third equations: $4xz = 4\frac{y^2z}{x}$, $x^2 = y^2$; $4xy = 2\frac{yz}{x}$, $2x^2 = z$. Substituting $y^2 = x^2$ and $z = 2x^2$ in the last equation gives $x^2 + x^2 + 2x^2 = 1$, $x^2 = \frac{1}{4}$, $x = \frac{1}{2}$; $y = \frac{1}{2}$; $z = \frac{1}{2}$. The dimensions of the box are $1 \times 1 \times \frac{1}{2}$.

37. Method 1: Solve $g(x, y, z) = 0$ for one of the variables and substitute in $f(x, y, z)$. Then find the maximum value of the resulting function of two variables. Advantage (Answers may vary): We can use the second derivative test to check whether the resulting critical points are maxima, minima, saddle points, or none of these. Disadvantage (Answers may vary): We may not be able to solve $g(x, y, z) = 0$ for one of the variables.

Method 2: Use the method of Lagrange Multipliers. Advantage (Answers may vary): We do not need to solve the constraint equation for one of the variables. Disadvantage (Answers may vary): The method does not tell us whether the critical points obtained are maxima, minima, saddle points, or none of these.

39. If the only constraint is an equality constraint, and if it is impossible to eliminate one of the variables in the objective function by substitution (solving the constraint equation for a variable or some other method).

41. Answers may vary: Maximize $f(x, y) = 1 - x^2 - y^2$ subject to $x = y$.

43. Yes. There may be relative extrema at points on the boundary of the domain of the function. The partial derivatives of the function need not be 0 at such points.

45. If the solution were located in the interior of one of the line segments making up the boundary of the domain of f, then the derivative of a certain function would be 0. This function is obtained by substituting the linear equation $C(x, y) = 0$ in the linear objective function. But because the result would again be a linear function, it is either constant, or its derivative is a nonzero constant. In either event, extrema lie on the boundary of that line segment; that is, at one of the corners of the domain.

8.6

1. $\int_0^1\int_0^1 (x-2y)\,dx\,dy = \int_0^1 \left[\frac{1}{2}x^2 - 2xy\right]_{x=0}^1 dy = \int_0^1 \left(\frac{1}{2} - 2y\right) dy = \left[\frac{1}{2}y - y^2\right]_{y=0}^1 = -\frac{1}{2}$

3. $\int_0^1\int_0^2 (ye^x - x - y)\,dx\,dy = \int_0^1 \left[ye^x - \frac{1}{2}x^2 - xy\right]_{x=0}^2 dy = \int_0^1 (e^2y - 2 - 3y)\,dy =$
$\left[\frac{1}{2}e^2y^2 - 2y - \frac{3}{2}y^2\right]_{y=0}^1 = \frac{1}{2}e^2 - \frac{7}{2}$

5. $\int_0^2\int_0^3 e^{x+y}\,dx\,dy = \int_0^2 [e^{x+y}]_{x=0}^3\,dy = \int_0^2 (e^{3+y} - e^y)\,dy = [e^{3+y} - e^y]_{y=0}^2 = e^5 - e^2 - e^3 + 1 =$
$(e^3 - 1)(e^2 - 1)$. This may also be found by writing $\int_0^2\int_0^3 e^{x+y}\,dx\,dy = \int_0^2\int_0^3 e^xe^y\,dx\,dy = \left(\int_0^3 e^x\,dx\right)\left(\int_0^2 e^y\,dy\right)$
as in Exercise 57.

7. $\int_0^1\int_0^{2-y} x\,dx\,dy = \int_0^1 \left[\frac{1}{2}x^2\right]_{x=0}^{2-y} dy = \int_0^1 \frac{1}{2}(2-y)^2\,dy = \left[-\frac{1}{6}(2-y)^3\right]_{y=0}^1 = -\frac{1}{6} + \frac{8}{6} = \frac{7}{6}$

9. $\int_{-1}^1\int_{y-1}^{y+1} e^{x+y}\,dx\,dy = \int_{-1}^1 [e^{x+y}]_{x=y-1}^{y+1}\,dy = \int_{-1}^1 (e^{2y+1} - e^{2y-1})\,dy = \left[\frac{1}{2}e^{2y+1} - \frac{1}{2}e^{2y-1}\right]_{y=-1}^1 =$
$\frac{1}{2}(e^3 - e - e^{-1} + e^{-3})$

11. $\int_0^1\int_{-x^2}^{x^2} x\,dy\,dx = \int_0^1 [xy]_{y=-x^2}^{x^2}\,dx = \int_0^1 2x^3\,dx = \left[\frac{1}{2}x^4\right]_{x=0}^1 = \frac{1}{2}$

13. $\int_0^1\int_0^x e^{x^2}\,dy\,dx = \int_0^1 [ye^{x^2}]_{y=0}^x\,dx = \int_0^1 xe^{x^2}\,dx = \left[\frac{1}{2}e^{x^2}\right]_{x=0}^1 = \frac{1}{2}e - \frac{1}{2} = \frac{1}{2}(e-1)$

15. $\displaystyle\int_0^2\int_{1-x}^{8-x}(x+y)^{1/3}\,dy\,dx = \int_0^2\left[\frac{3}{4}(x+y)^{4/3}\right]_{y=1-x}^{8-x}dx = \int_0^2\left(\frac{3}{4}(16)-\frac{3}{4}\right)dx = \int_0^2\frac{45}{4}\,dx = \left[\frac{45}{4}x\right]_{x=0}^{2} = \frac{45}{2}$

17. $\displaystyle\int_{-1}^1\int_0^{1-x^2}2\,dy\,dx = \int_{-1}^1[2y]_{y=0}^{1-x^2}dx = \int_{-1}^1 2(1-x^2)\,dx = \left[2x-\frac{2}{3}x^3\right]_{x=-1}^{1} = 2-\frac{2}{3}+2-\frac{2}{3} = \frac{8}{3}$

19. $\displaystyle\int_{-1}^1\int_0^{1-y^2}(1+y)\,dx\,dy = \int_{-1}^1[x(1+y)]_{x=0}^{1-y^2}dy = \int_{-1}^1(1-y^2)(1+y)\,dy = \int_{-1}^1(1+y-y^2-y^3)\,dy =$

$\displaystyle\left[y+\frac{1}{2}y^2-\frac{1}{3}y^3-\frac{1}{4}y^4\right]_{y=-1}^{1} = 1+\frac{1}{2}-\frac{1}{3}-\frac{1}{4}+1-\frac{1}{2}-\frac{1}{3}+\frac{1}{4} = \frac{4}{3}$

21. $\displaystyle\int_{-1}^1\int_{-\sqrt{1-y^2}}^{\sqrt{1-y^2}}xy^2\,dx\,dy = \int_{-1}^1\left[\frac{1}{2}x^2y^2\right]_{x=-\sqrt{1-y^2}}^{\sqrt{1-y^2}}dy = \int_{-1}^1 0\,dy = 0$

23. $\displaystyle\int_{-1}^0\int_{-x-1}^{x+1}(x^2+y^2)\,dy\,dx + \int_0^1\int_{x-1}^{-x+1}(x^2+y^2)\,dy\,dx = \int_{-1}^0\left[x^2y+\frac{1}{3}y^3\right]_{y=-x-1}^{x+1}dx + \int_0^1\left[x^2y+\frac{1}{3}y^3\right]_{y=x-1}^{-x+1}dx =$

$\displaystyle\int_{-1}^0\left(2x^2(x+1)+\frac{2}{3}(x+1)^3\right)dx + \int_0^1\left(2x^2(-x+1)+\frac{2}{3}(-x+1)^3\right)dx = \int_{-1}^0\left(\frac{8}{3}x^3+4x^2+2x+\frac{2}{3}\right)dx +$

$\displaystyle\int_0^1\left(-\frac{8}{3}x^3+4x^2-2x+\frac{2}{3}\right)dx = \left[\frac{2}{3}x^4+\frac{4}{3}x^3+x^2+\frac{2}{3}x\right]_{x=-1}^{0} + \left[-\frac{2}{3}x^4+\frac{4}{3}x^3-x^2+\frac{2}{3}x\right]_{x=0}^{1} = -\frac{2}{3}+\frac{4}{3}$

$\displaystyle-1+\frac{2}{3}-\frac{2}{3}+\frac{4}{3}-1+\frac{2}{3} = \frac{2}{3}$

25. $\displaystyle\int_0^2\int_0^{2-x}y\,dy\,dx = \int_0^2\left[\frac{1}{2}y^2\right]_{y=0}^{2-x}dx = \int_0^2\frac{1}{2}(2-x)^2\,dx = \left[-\frac{1}{6}(2-x)^3\right]_{x=0}^{2} = \frac{8}{6} = \frac{4}{3}$. $A = \frac{1}{2}(2)(2) = 2$ by the geometric formula for the area of a triangle. So, the average value $= \dfrac{4/3}{2} = \dfrac{2}{3}$.

27. $\int_0^1 \int_{y-1}^{1-y} e^y \, dx\, dy = \int_0^1 [xe^y]_{x=y-1}^{1-y} \, dy = \int_0^1 2(1-y)e^y \, dy = [2(1-y)e^y + 2e^y]_{y=0}^1 = 2e - 4 = 2(e-2)$. $A = \frac{1}{2}(2)(1) = 1$ by the formula for the area of a triangle. So, the average value $= 2(e-2)$.

29. $\int_{-2}^{0} \int_{x/2+1}^{x+2} x^2 \, dy\, dx + \int_0^2 \int_{-x/2+1}^{-x+2} x^2 \, dy\, dx = \frac{4}{3}$ as calculated in Exercise 24. $A = 2 \cdot \frac{1}{2}(2)(1) = 2$, since A consists of the area of two triangles. So, the average value $= \dfrac{4/3}{2} = \dfrac{2}{3}$.

31. $\int_0^1 \int_0^{1-x} f(x, y) \, dy\, dx$

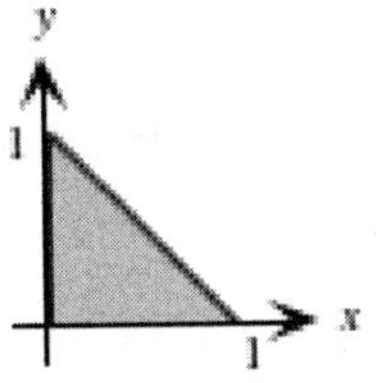

33. $\int_0^{\sqrt{2}} \int_{x^2-1}^{1} f(x, y) \, dy\, dx$

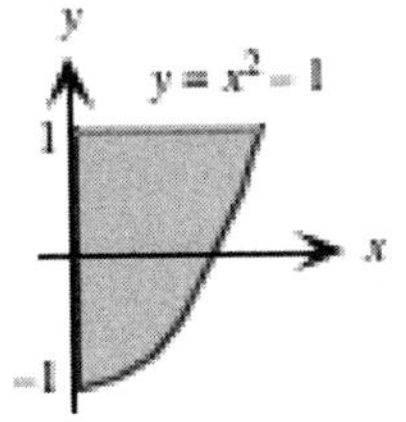

35. $\int_1^4 \int_1^{2/\sqrt{y}} f(x, y) \, dx\, dy$

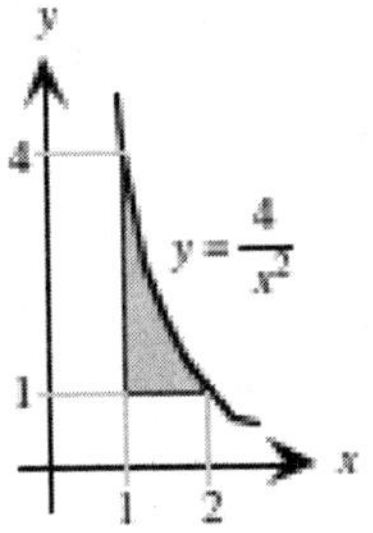

37. $V = \int_0^1 \int_0^2 (1 - x^2) \, dy\, dx = \int_0^1 [(1 - x^2)y]_{y=0}^2 \, dx = \int_0^1 2(1 - x^2) \, dx = \left[2x - \frac{2}{3}x^3\right]_{x=0}^1 = 2 - \frac{2}{3} = \frac{4}{3}$

39. The top face of the tetrahedron is the graph of $z = 1 - x - y$. So, the volume is $V = \int_0^1 \int_0^{1-x} (1 - x - y)\, dy\, dx = \int_0^1 \left[(1 - x)y - \frac{1}{2}y^2\right]_{y=0}^{1-x} dx = \int_0^1 \left[(1 - x)^2 - \frac{1}{2}(1 - x)^2\right] dx = \int_0^1 \frac{1}{2}(1 - x)^2\, dx = \left[-\frac{1}{6}(1 - x)^3\right]_{x=0}^{1} = \frac{1}{6}$.

41. $\int_{45}^{55} \int_8^{12} 10{,}000x^{0.3}y^{0.7}\, dy\, dx = 10{,}000 \int_{45}^{55} \left[\frac{1}{1.7}x^{0.3}y^{1.7}\right]_{y=8}^{12} dx = \frac{10{,}000(12^{1.7} - 8^{1.7})}{1.7} \int_{45}^{55} x^{0.3}\, dx = \frac{10{,}000(12^{1.7} - 8^{1.7})}{(1.7)(1.3)} [x^{1.3}]_{x=45}^{55} = \frac{10{,}000(12^{1.7} - 8^{1.7})(55^{1.3} - 45^{1.3})}{(1.7)(1.3)} \approx 6{,}471{,}000$. $A = (12 - 8)(55 - 45) = 40$. So, the average is approximately $\frac{6{,}471{,}000}{40} \approx 162{,}000$ gadgets.

43. $R(p, q) = pq$ with $40 \le p \le 50$ and $8000 - p^2 \le q \le 10{,}000 - p^2$. To find the average we first compute
$\int_{40}^{50} \int_{8000-p^2}^{10{,}000-p^2} pq\, dq\, dp = \frac{1}{2} \int_{40}^{50} [pq^2]_{q=8000-p^2}^{10{,}000-p^2} dp = \frac{1}{2} \int_{40}^{50} [p(10{,}000 - p^2)^2 - p(8000 - p^2)^2]\, dp = \frac{1}{12} [-(10{,}000 - p^2)^3 + (8000 - p^2)^3]_{p=40}^{50} = 6{,}255{,}000{,}000$. Next we compute $A = \int_{40}^{50} \int_{8000-p^2}^{10{,}000-p^2} dq\, dp = \int_{40}^{50} [q]_{q=8000-p^2}^{10{,}000-p^2} dp = \int_{40}^{50} [10{,}000 - p^2 - (8000 - p^2)]\, dp = \int_{40}^{50} 2000\, dp = [2000p]_{p=40}^{50} = 100{,}000 - 80{,}000 = 20{,}000$. The average revenue is thus $\frac{6{,}255{,}000{,}000}{20{,}000} = \$312{,}750$.

45. $R(p, q) = pq$ with $15{,}000/q \le p \le 20{,}000/q$ and $500 \le q \le 1000$. To find the average we first compute
$\int_{500}^{1000} \int_{15{,}000/q}^{20{,}000/q} pq\, dp\, dq = \frac{1}{2} \int_{500}^{1000} [p^2q]_{p=15{,}000/q}^{20{,}000/q} dq = \frac{1}{2} \int_{500}^{1000} \left(\frac{20{,}000^2}{q} - \frac{15{,}000^2}{q}\right) dq = \frac{175{,}000{,}000}{2} \int_{500}^{1000} \frac{1}{q}\, dq = \frac{175{,}000{,}000}{2} [\ln q]_{q=500}^{1000} = \frac{175{,}000{,}000}{2} (\ln 1000 - \ln 500) = \frac{175{,}000{,}000}{2} \ln 2$. Next we compute $A =$

$$\int_{500}^{1000}\int_{15{,}000/q}^{20{,}000/q} dp\,dq = \int_{500}^{1000}[p]_{p=15{,}000/q}^{20{,}000/q}\,dq =$$

$$5000\int_{500}^{1000}\frac{1}{q}\,dq = 5000[\ln q]_{q=500}^{1000} =$$

$5000(\ln 1000 - \ln 500) = 5000 \ln 2$. The average revenue is thus $\dfrac{175{,}000{,}000 \ln 2}{2 \cdot 5000 \ln 2} =$ \$17,500.

47. Total population $= \displaystyle\int_0^{20}\int_0^{30} e^{-0.1(x+y)}\,dy\,dx =$

$$-10\int_0^{20}[e^{-0.1(x+y)}]_{y=0}^{30}\,dx =$$

$$-10\int_0^{20}(e^{-0.1x-3} - e^{-0.1x})\,dx =$$

$$-10(e^{-3} - 1)\int_0^{20} e^{-0.1x}\,dx = 100(e^{-3} - 1)[e^{-0.1x}]_{x=0}^{20} = 100(e^{-3} - 1)(e^{-2} - 1) \approx 82.16$$ hundred people, or 8216 people.

49. $\displaystyle\int_0^1\int_0^1 (x^2 + 2y^2)\,dy\,dx = \int_0^1\left[x^2y + \frac{2}{3}y^3\right]_{y=0}^1 dx$

$$= \int_0^1\left(x^2 + \frac{2}{3}\right)dx = \left[\frac{1}{3}x^3 + \frac{2}{3}x\right]_{x=0}^1 = 1.\ A = 1,$$

the area of the square. Hence, the average temperature is 1/1 = 1 degree.

51. The area between the curves $y = r(x)$ and $y = s(x)$ and the vertical lines $x = a$ and $x = b$ is given by $\displaystyle\int_a^b\int_{r(x)}^{s(x)} dy\,dx$ assuming that $r(x) \le s(x)$ for $a \le x \le b$.

53. The first step in calculating an integral of the form $\displaystyle\int_a^b\int_{r(x)}^{s(x)} f(x, y)\,dy\,dx$ is to evaluate the integral $\displaystyle\int_{r(x)}^{s(x)} f(x, y)\,dy$, obtained by holding x constant and integrating with respect to y.

55. paintings per picasso per dali

57. The left-hand side is $\displaystyle\int_a^b\int_c^d f(x)g(y)\,dx\,dy =$ $\displaystyle\int_a^b\left(g(y)\int_c^d f(x)\,dx\right)dy$ (because $g(y)$ is treated as a constant in the inner integral) $= \displaystyle\left(\int_c^d f(x)\,dx\right)\int_a^b g(y)\,dy$ (because $\displaystyle\int_c^d f(x)\,dx$ is a constant and can therefore be taken outside the integral with respect to y). For example, $\displaystyle\int_0^1\int_1^2 ye^x\,dx\,dy = \frac{1}{2}(e^2 - e)$ if we compute it as an iterated integral or as $\displaystyle\int_1^2 e^x\,dx\int_0^1 y\,dy$.

Chapter 8 Review Exercises

1. $g(0, 0, 0) = 0$; $g(1, 0, 0) = 1$; $g(0, 1, 0) = 0$; $g(x, x, x) = x^3 + x^2$; $g(x, y + k, z) = x(y + k)(x + y + k - z) + x^2$

2. Decreases by 0.32 units; increases by 12.5 units

3. Reading left to right, starting at the top: 4, 0, 0, 3, 0, 1, 2, 0, 2

4. Answers may vary. Two examples are $f(x, y) = 3(x - y)/2$ and $f(x, y) = 3(x - y)^3/8$.

5. $f_x = 2x + y$, $f_y = x$, $f_{yy} = 0$

6. $f(x, y) = \dfrac{6}{xy} + \dfrac{xy}{6}$

$f_x = -\dfrac{6}{x^2y} + \dfrac{y}{6}$; $f_y = -\dfrac{6}{xy^2} + \dfrac{x}{6}$; $f_{yy} = \dfrac{12}{xy^3}$

7. $f(x, y) = 4x + 5y - 6xy$

$f_x(x, y) = 4 - 6y$

$f_{zz}(x, y) = 4$

$f_{xx}(1, 0) = f_{xx}(3, 2) = 4$

Therefore $f_{xx}(1, 0) - f_{xx}(3, 2) = 0$

8. $\dfrac{\partial f}{\partial x} = ye^{xy} + 6xe^{3x^2-y^2}$, $\dfrac{\partial^2 f}{\partial x \partial y} = (xy + 1)e^{xy} - 12xye^{3x^2-y^2}$

9. $\dfrac{\partial f}{\partial x} = \dfrac{-x^2 + y^2 + z^2}{(x^2 + y^2 + z^2)^2}$, $\dfrac{\partial f}{\partial y} = -\dfrac{2xy}{(x^2 + y^2 + z^2)^2}$, $\dfrac{\partial f}{\partial z} = -\dfrac{2xz}{(x^2 + y^2 + z^2)^2}$, $\left.\dfrac{\partial f}{\partial x}\right|_{(0,1,0)} = 1$

10. $f_{xx} + f_{yy} + f_{zz} = 2 + 2 + 2 = 6$

11. $f_x = 2(x - 1)$; $f_y = 2(2y - 3)$; $f_{xx} = 2$; $f_{yy} = 4$; $f_{xy} = 0$. $f_x = 0$ when $x = 1$; $f_y = 0$ when $y = 3/2$, so $(1, 3/2)$ is the only critical point. $H = 8$ and $f_{xx} > 0$, so f has an absolute minimum at $(1, 3/2)$.

12. $g_x = 2(x - 1)$; $g_y = -6y$; $g_{xx} = 2$; $g_{yy} = -6$; $g_{xy} = 0$. $g_x = 0$ when $x = 1$; $g_y = 0$ when $y = 0$, so $(1, 0)$ is the only critical point. $H = -12 < 0$, so g has a saddle point at $(1, 0)$.

13. $h_x = ye^{xy}$; $h_y = xe^{xy}$; $h_{xx} = y^2e^{xy}$; $h_{yy} = x^2e^{xy}$; $h_{xy} = (xy + 1)e^{xy}$. $h_x = 0$ when $y = 0$; $h_y = 0$ when $x = 0$, so $(0, 0)$ is the only critical point. $H(0, 0) = -1 < 0$, so h has a saddle point at $(0, 0)$.

14. $j_x = y + 2x$; $j_y = x$; $j_{xx} = 2$; $j_{yy} = 0$; $j_{xy} = 1$. $j_x = 0$ when $y = -2x$; $j_y = 0$ when $x = 0$, so $(0, 0)$ is the only critical point. $H = -1 < 0$, so j has a saddle point at $(0, 0)$.

15. Note that the domain of f contains every point except $(0, 0)$. $f_x = \dfrac{2x}{x^2 + y^2} - 2x$; $f_y = \dfrac{2y}{x^2 + y^2} - 2y$; $f_{xx} = \dfrac{-2x^2 + 2y^2}{(x^2 + y^2)^2} - 2$; $f_{yy} = \dfrac{2x^2 - 2y^2}{(x^2 + y^2)^2} - 2$; $f_{xy} = \dfrac{-4xy}{(x^2 + y^2)^2}$. $f_x = 0$ if either $x = 0$ or $x^2 + y^2 = 1$; $f_y = 0$ if either $y = 0$ or $x^2 + y^2 = 1$; since $(0, 0)$ is not in the domain, the critical points are all the points on the circle $x^2 + y^2 = 1$. At such a point we can substitute $y^2 = 1 - x^2$ to get $f_{xx} = (2 - 4x^2) - 2 = -4x^2$, $f_{yy} = 4x^2 - 4$; and $f_{xy} = \pm 4x(1 - x^2)^{1/2}$. Hence $H = -4x^2(4x^2 - 4) - 16x^2(1 - x^2) = 0$. To resolve what happens along the circle we graph the function:

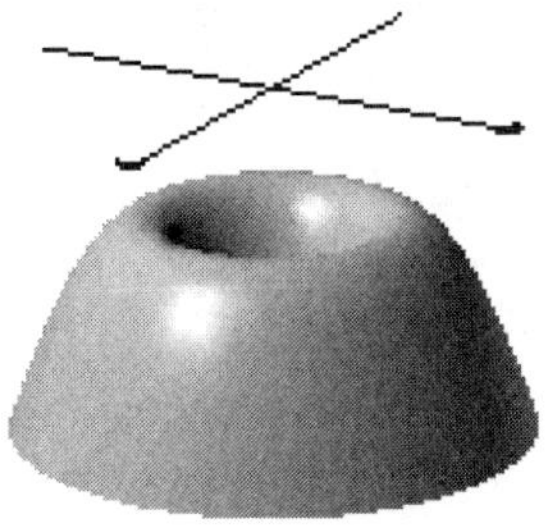

f has an absolute maximum at each point on the circle $x^2 + y^2 = 1$.

16. The objective function is $x^2 + y^2 + z^2 - 1$. Substituting the constraint equation $x = y + z$ gives the objective function as

$$h(y, z) = (y+z)^2 + y^2 + z^2 - 1$$
$$= 2y^2 + 2z^2 + 2yz - 1$$
$$h_y = 4y + 2z,\ h_y = 4z + 2y$$

Critical points:

$$4y + 2z = 0$$
$$4z + 2y = 0$$

$(y, z) = (0, 0)$ is the only critical point.

$$h_{yy} = 4,\ h_{yz} = 2,\ h_{zz} = 4$$
$$H = h_{yy}h_{zz} - (h_{yz})^2 = 14 > 0$$

Since $h_{yy} > 0$, the critical point is a local minimum. That it is an absolute minimum can be seen by considering the graph of $h(x, y) = 2y^2 + 2z^2 + 2yz - 1$. The corresponding value of the objective function is

$$h(0, 0) = 2(0)^2 + 2(0)^2 + 2(0)(0) - 1 = -1$$

When $y = 0$ and $z = 0$, $x = y + z = 0$ as well, so f has an absolute minimum of -1 at the point $(0, 0, 0)$

17. The objective function is $V = xyz$. Solving the constraint equation $x + y + z = 1$ for z gives

$$z = 1 - x - y$$
$$V = xy(1 - x - y)$$
$$= xy - x^2y - xy^2$$
$$V_x = y - 2xy - y^2,\ V_y = x - x^2 - 2xy$$

Critical points:

$$y - 2xy - y^2 = 0$$
$$x - x^2 - 2xy = 0$$

Dividing the first equation by y gives

$$1 - 2x - y = 0$$
$$2x + y = 1$$

Dividing the second equation by x gives

$$1 - x - 2y = 0$$
$$x + 2y = 1$$

The system of equations

$$2x + y = 1$$
$$x + 2y = 1$$

has solution $(x, y) = (1/3, 1/3)$. The corresponding value of z is

$$z = 1 - x - y = 1 - 2/3 = 1/3$$

$V_{xx} = -2y,\ V_{xy} = 1 - 2x - 2y,\ V_{yy} = -2x$

Evaluating these at $(1/3, 1/3)$ gives

$V_{xx} = -2/3,\ V_{xy} = -1/3,\ V_{yy} = -2/3$

$$H = V_{zz}V_{yy} - (V_{xy})^2 = 4/9 - 1/9 > 0$$
$$V_{xx} < 0$$

so V does have a relative maximum at $(1/3, 1/3, 1/3)$. The value of V at that point is $V = xyz = 1/27$. That it is the absolute maximum can be seen as interpreting the problems as maximizing the volume of a box whose dimensions add up to 1.

18. The objective function is $xy + x^2z^2 + 4yz$. Solving the constraint equation $xyz = 1$ for z gives

$$z = \frac{1}{xy}$$
$$S = xy + x^2\left(\frac{1}{xy}\right)^2 + 4y\left(\frac{1}{xy}\right)$$
$$= xy + \frac{1}{y^2} + \frac{4}{x}$$
$$S_x = y - \frac{4}{x^2},\ S_y = x - \frac{2}{y^3}$$

Critical points:

$$y - \frac{4}{x^2} = 0$$

$$x - \frac{2}{y^3} = 0$$

Substituting the first equation in the second gives

$$x - \frac{2}{(4/x^2)^3} = 0$$

$$x - x^6/32 = 0$$

$$x(1 - x^5/32) = 0$$

$$x^5 = 32, \text{ or } x = 2$$

(We reject $x = 0$ because $x > 0$). Substituting $x = 2$ into the equation $y - \frac{4}{x^2} = 0$ gives $y = 1$. $(x, y) = (2, 1)$ is the only critical point. That it is an absolute minimum can be seen by graphing $S(x, y) = xy + \frac{1}{y^2} + \frac{4}{x}$ for $x > 0$, $y > 0$. (The second derivative test can verify that it is a relative minimum.) The corresponding value of the objective function is

$$S(2, 1) = (2)(1) + 1 + 2 = 5$$

When $x = 2$ and $y = 1$, $z = 1/xy = 1/2$, so S has an absolute minimum of 5 at $(2, 1, 1/2)$.

19. We minimize the square of the distance of (x, y, z) to the origin: $d(x, y, z) = x^2 + y^2 + z^2$ subject to $z = \sqrt{x^2 + 2(y-3)^2}$, which we rewrite as $x^2 + 2(y - 3)^2 - z^2 = 0$ ($z \geq 0$). Solving the second equation for z^2 and substituting in the formula for d gives

$$d = x^2 + y^2 + x^2 + 2(y - 3)^2$$

$$= 2x^2 + 3y^2 - 12y + 18$$

$$d_x = 4x,\ d_y = 6y - 12$$

The critical points are

$$4x = 0 \text{ or } x = 0$$

$$6y - 12 = 0, \text{ or } y = 2$$

The corresponding value of z is

$$z = \sqrt{x^2 + 2(y-3)^2} = \sqrt{0^2 + 2(2-3)^2} = \sqrt{2}.$$

The critical point is therefore $(0, 2, \sqrt{2})$. Since there must be at least one point on the given surface closest to the origin, the critical point $(0, 2, \sqrt{2})$ must be that point.

20. The constraint is $y = e^{-x}$, or $y - e^{-x} = 0$.

$$f(x, y) = xy$$

$$g(x, y) = y - e^{-x}$$

$$(1)\ f_x = \lambda g_x \Rightarrow y = \lambda e^{-x}$$

$$(2)\ f_y = \lambda g_y \Rightarrow x = \lambda$$

$$(3)\ \text{Constraint} \Rightarrow y = e^{-x}$$

Substitute Equation (2) into (1) to obtain

$$y = xe^{-x}$$

Substitute in Equation (3) to obtain

$$xe^{-x} = e^{-x}$$

$$e^{-x}(x - 1) = 0$$

$$x = 1$$

The corresponding value of y is

$$y = xe^{-x} = (1)e^{-1} = 1/e$$

The corresponding value of the objective is

$$f(1, 1/e) = (1)(1/e) = 1/e$$

That the critical point must be an absolute minimum follows from the fact that the problem is equivalent to finding the largest rectangle in the first quadrant that is under the curve $y = e^{-x}$.

21. The constraint is $xy = 2$, or $xy - 2 = 0$.

$$f(x, y) = x^2 + y^2$$

$$g(x, y) = xy - 2$$

$$(1)\ f_x = \lambda g_x \Rightarrow 2x = \lambda y \Rightarrow \lambda = 2x/y$$

$$(2)\ f_y = \lambda g_y \Rightarrow 2y = \lambda x$$

$$(3)\ \text{Constraint} \Rightarrow xy = 2$$

Substitute Equation (1) in Equation (2) to obtain

$$2x^2/y = 2y \Rightarrow x^2 = y^2 \Rightarrow x = \pm y$$

Substituting into the constraint gives

$$x^2 = 2 \text{ (we must reject } y = -x \text{ here)}$$

$$x = \pm\sqrt{2}$$

so $y = 2/x = \pm\sqrt{2}$

Thus we have two critical points: $(\sqrt{2}, \sqrt{2})$ and $(-\sqrt{2}, -\sqrt{2})$. The corresponding value of the objective function is

$f(\pm\sqrt{2}, \pm\sqrt{2}) = 2 + 2 = 4$

22. The constraint is $x = y + z$, or $x - y - z = 0$.

$f(x, y, z) = x^2 + y^2 + z^2 - 1$

$g(x, y, z) = x - y - z$

(1) $f_x = \lambda g_x \Rightarrow 2x = \lambda$

(2) $f_y = \lambda g_y \Rightarrow 2y = -\lambda$

(3) $f_z = \lambda g_z \Rightarrow 2z = -\lambda$

(4) Constraint $\Rightarrow x = y + z$

Substituting (1) in (2) gives

$2y = -2x$ or

$y = -x$

Substituting (1) in (3) gives

$2z = -2x$, or

$z = -x$

Combining this information with the constraint equation

$x - y - z = 0$

gives

$x + x + x = 0$

$x = 0$

Thus the only critical point is $(x, y, z) = (0, 0, 0)$ and the corresponding value of the objective is

$f(0, 0, 0) = 0^2 + 0^2 + 0^2 - 1 = -1$

That this is an absolute minimum is seen from the fact that $f(x, y, z) = x^2 + y^2 + z^2 - 1$ can never be less than -1.

23. The constraint is $xyz = 1$, or $xyz - 1 = 0$.

$f(x, y, z) = xy + x^2z^2 + 4yz$

$g(x, y, z) = xyz - 1$

(1) $f_x = \lambda g_x \Rightarrow y + 2xz^2 = \lambda yz$

(2) $f_y = \lambda g_y \Rightarrow x + 4z = \lambda xz$

(3) $f_z = \lambda g_z \Rightarrow 2x^2z + 4y = \lambda xy$

(4) Constraint $\Rightarrow xyz = 1$

Solve Equation (1) for λ:

$\lambda = \frac{1}{z} + \frac{2xz}{y}$

Substituting in (2) gives

$x + 4z = (\frac{1}{z} + \frac{2xz}{y})xz$

$x + 4z = x + \frac{2x^2z^2}{y}$

$4z = \frac{2x^2z^2}{y}$

$2 = \frac{x^2z}{y}$

$y = x^2z/2$

Substituting the expression for λ in (3) gives

$2x^2z + 4y = (\frac{1}{z} + \frac{2xz}{y})xy$

$2x^2z + 4y = \frac{xy}{z} + 2x^2z$

$4y = \frac{xy}{z}$

$4 = \frac{x}{z}$

$z = x/4$

Substituting the expressions we obtained for y and z in the constraint equation gives:

$x[2x^2(x/4)/2](x/4) = 1$

$x^5 = 32$

$x = 2$

The corresponding values of y and z are

$z = x/4 = 2/4 = 1/2$

$y = x^2z/2 = 1$

Therefore, the only critical point is $(2, 1, 1/2)$ and the corresponding value of the objective is

$f(2, 1, 1/2)$

$= (2)(1) + (2)^2(1/2)^2 + 4(1)(1/2)$

$= 5$

Therefore, the minimum value of the objective function is 5, and occurs at the point $(2, 1, 1/2)$.

24. As in Exercise 19, we minimize the square of the distance of (x, y, z) to the origin: $d(x, y, z) = x^2 + y^2 + z^2$ subject to $z = \sqrt{x^2 + 2(y-3)^2}$, which we rewrite as $x^2 + 2(y - 3)^2 - z^2 = 0$ $(z \geq 0)$.

$d(x, y, z) = x^2 + y^2 + z^2$

$g(x, y, z) = x^2 + 2(y - 3)^2 - z^2$

(1) $d_x = \lambda g_x \Rightarrow 2x = 2\lambda x$

(2) $d_y = \lambda g_y \Rightarrow 2y = 4\lambda(y - 3)$

(3) $d_z = \lambda g_z \Rightarrow 2z = -2\lambda z$

(4) Constraint $\Rightarrow x^2 + 2(y - 3)^2 - z^2 = 0$

From (1) we see that either $x = 0$ or $\lambda = 1$. Let's try $\lambda = 1$ first: From (2) we get

$2y = 4(y - 3) = 4y - 12$, so

$y = 6$

and from (3) we get

$2z = -2z$, so

$z = 0$.

However, substituting in (4) we get

$0 + 2(6 - 3)^2 - 0 = 18 \neq 0$

so $\lambda = 1$ does not work. Therefore, $x = 0$. Now look at (3), which tells us that $z = 0$ or $\lambda = -1$. If we try $z = 0$, equation (4) then tells us that

$0 + 2(y - 3)^2 - z = 0$, so

$y = 3$.

However, substituting in (2) we get

$2y = 6 \neq 4\lambda(y - 3) = 0$,

so $z = 0$ does not work, and we must have $\lambda = -1$. Substituting in (2) we find

$2y = -4(y - 3) = -4y + 12$

$6y = 12$

$y = 2$

Substituting in (4) we get

$0 + 2(2 - 3)^2 - z^2 = 0$

$z^2 = 2$

$z = \sqrt{2}$

(Remember that $z \geq 0$.) Therefore, the point on the surface $z = \sqrt{x^2 + 2(y-3)^2}$ closest to the origin is $(0, 2, \sqrt{2})$.

25. $\int_0^1 \int_0^2 2xy\,dx\,dy = \int_0^1 [x^2y]_{x=0}^2\,dy = \int_0^1 4y\,dy = [2y^2]_{y=0}^1 = 2$

26. $\int_1^2 \int_0^1 xye^{x+y}\,dx\,dy = \int_1^2 [xye^{x+y} - ye^{x+y}]_{x=0}^1\,dy$

$= \int_1^2 ye^y\,dy = [ye^y - e^y]_{y=1}^2 = 2e^2 - e^2 - (e - e)$

$= e^2$

27. $\int_0^2 \int_0^{2x} \frac{1}{x^2+1}\,dy\,dx = \int_0^2 \left[\frac{y}{x^2+1}\right]_{y=0}^{2x} dy = \int_0^2 \frac{2x}{x^2+1}\,dy = [\ln(x^2+1)]_{x=0}^2 = \ln 5$

28. We've already computed $\int_1^2 \int_0^1 xye^{x+y}\,dx\,dy = e^2$. On the other hand, $A = 1$ because the domain is a square with sides of length 1. Hence the average value is $e^2/1 = e^2$

29. $\int_0^1 \int_y^{2-y} (x^2 - y^2)\,dx\,dy = \int_0^1 \left[\frac{1}{3}x^3 - xy^2\right]_{x=y}^{2-y} dy$

$= \int_0^1 \left[\frac{1}{3}(2-y)^3 - (2-y)y^2 - \frac{1}{3}y^3 + y^3\right] dy$

$= \int_0^1 \left(\frac{8}{3} - 4y + \frac{4}{3}y^3\right) dy = \left[\frac{8}{3}y - 2y^2 + \frac{1}{3}y^4\right]_{y=0}^{1}$

$= 1$

30. $\int_{-1}^{1} \int_0^{1-x^2} (1 - y)\, dy\, dx = \int_{-1}^{1} \left[y - \frac{1}{2}y^2\right]_{y=0}^{1-x^2} dx =$

$\int_{-1}^{1} \left[(1 - x^2) - \frac{1}{2}(1 - x^2)^2\right] dx = \frac{1}{2}\int_{-1}^{1} (1 - x^4)\, dx$

$= \frac{1}{2}\left[x - \frac{1}{5}x^5\right]_{x=-1}^{1} = \frac{1}{2}\left(1 - \frac{1}{5} + 1 - \frac{1}{5}\right) = \frac{4}{5}$

31. (a) $h(x, y) = 5000 - 0.8x - 0.6y$ hits per day (x = number of new customers at JungleBooks.com, y = number of new customers at FarmerBooks.com)
(b) Set $x = 100$ and $h = 4770$: $4770 = 5000 - 0.8(100) - 0.6y$, $y = 250$ new customers.

32. (a) $h(x, y, z) = 5000 - 0.8x - 0.6y + 0.0001z$ (z = number of new Internet shoppers)
(b) Set $x = y = 100$ and $h = 5000$: $5000 = 5000 - 0.8(100) - 0.6(100) + 0.0001z$, $z = 1.4$ million new Internet shoppers.

33. (a) $h(2000, 3000) = 2320$ hits per day
(b) $h_y = 0.08 + 0.00003x$ hits (daily) per dollar spent on television advertising per month; this increases with increasing x.
(c) Set $h_y = 1/5 = 0.2$: $0.2 = 0.08 + 0.00003x$, $x = \$4000$ per month

34. (A) because $h_x(x, 0) = 0.05$.

35. $P(10, 1000) \approx 15{,}800$ orders per day

36. Minimize $C(x, y) = 150x + y$ subject to $1000x^{0.9}y^{0.1} = 15{,}000$. Using Lagrange multipliers we need to solve $150 = 900\lambda x^{-0.1}y^{0.1}$, $1 = 100\lambda x^{0.9}y^{-0.9}$, and $1000x^{0.9}y^{0.1} - 15{,}000 = 0$. From the first equation we get $\lambda = \frac{x^{0.1}}{6y^{0.1}}$; substituting in the second equation gives $1 = \frac{100x}{6y}$, so $y = \frac{50}{3}x$. Substituting in the last equation gives $1000\left(\frac{50}{3}\right)^{0.1} x = 15{,}000$, so $x = 15\left(\frac{3}{50}\right)^{0.1} \approx 11$.

37. $\int_{1200}^{1500} \int_{1800}^{2000} (3x + 10y)\, dy\, dx =$

$\int_{1200}^{1500} [3xy + 5y^2]_{y=1800}^{2000}\, dx =$

$\int_{1200}^{1500} (600x + 3{,}800{,}000)\, dx =$

$[300x^2 + 3{,}800{,}000x]_{x=1200}^{1500} = 1{,}383{,}000{,}000.$

$A = (1500 - 1200)(2000 - 1800) = 60{,}000$, so the average profit is $1{,}383{,}000{,}000/60{,}000 = \$23{,}050$.

Chapter 9

9.1

1.

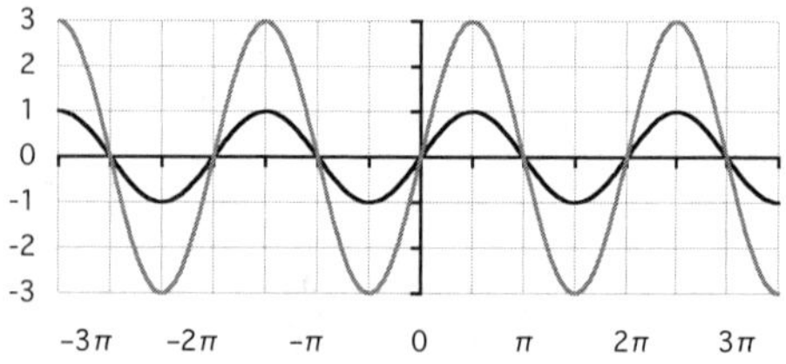

3.

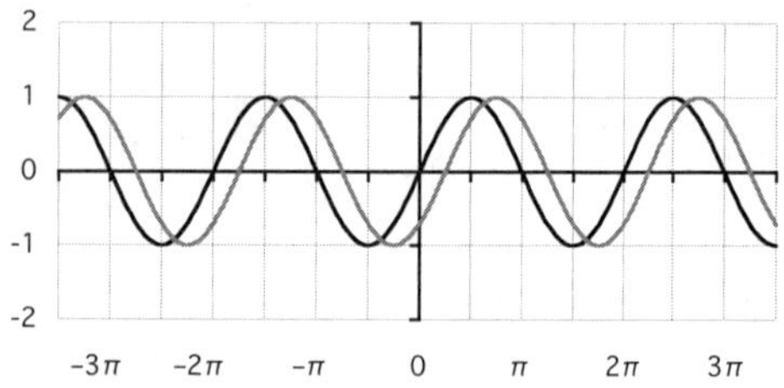

5.

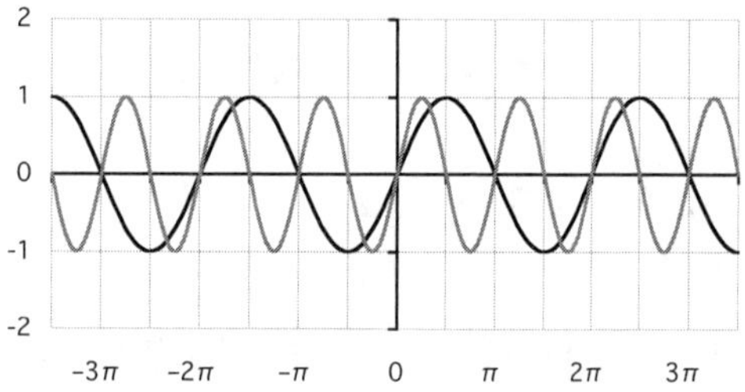

7.

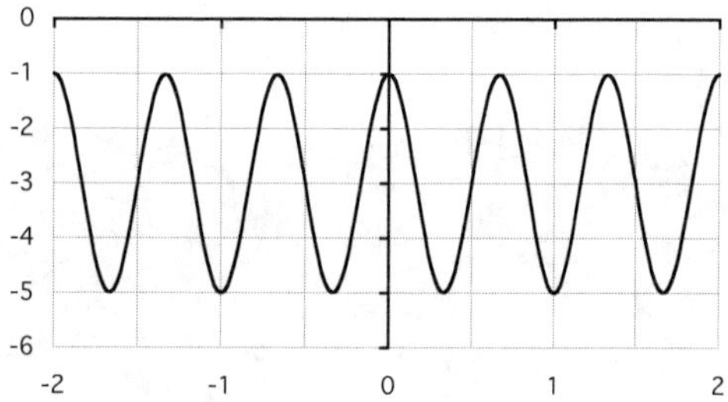

9.

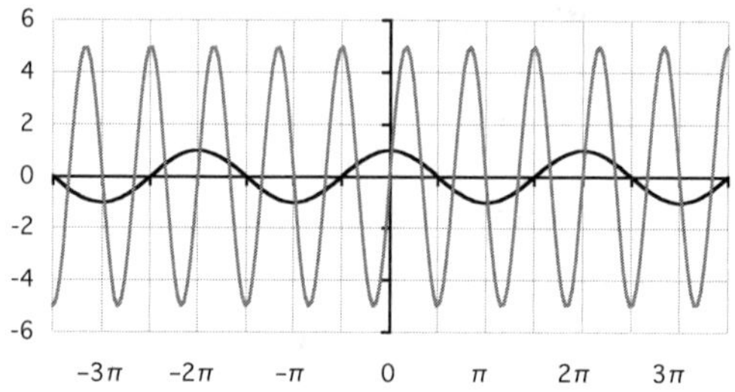

11.

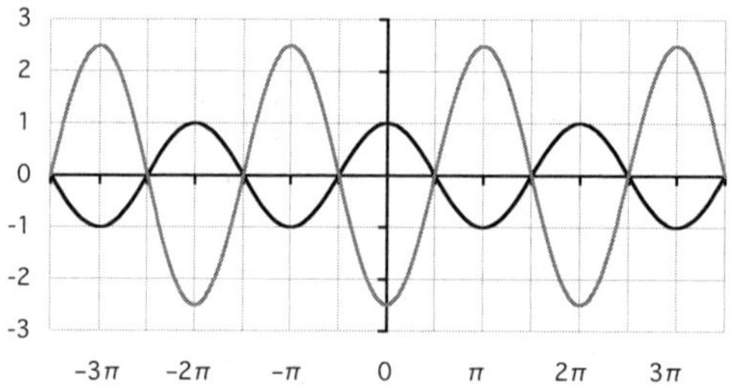

13. From the graph, $C = 1$, $A = 1$, $\alpha = 0$, and $P = 1$, so $\omega = 2\pi$. Thus, $f(x) = \sin(2\pi x) + 1$.

15. From the graph, $C = 0$, $A = 1.5$, $\alpha = 0.25$, and $P = 0.5$, so $\omega = 4\pi$. Thus, $f(x) = 1.5\sin[4\pi(x - 0.25)]$.

17. From the graph, $C = -50$, $A = 50$, $\alpha = 5$, and $P = 20$, so $\omega = \pi/10$. Thus, $f(x) = 50\sin[\pi(x - 5)/10] - 50$.

19. From the graph, $C = 0$, $A = 1$, $\alpha = 0$, and $P = 1$, so $\omega = 2\pi$. Thus, $f(x) = \cos(2\pi x)$.

21. From the graph, $C = 0$, $A = 1.5$, $\alpha = 0.375$, and $P = 0.5$, so $\omega = 4\pi$. Thus, $f(x) = 1.5\cos[4\pi(x - 0.375)]$. Alternatively, $\alpha = -0.125$, so $f(x) = 1.5\cos[4\pi(x + 0.125)]$.

23. From the graph, $C = 40$, $A = 40$, $\alpha = 10$, and $P = 20$, so $\omega = \pi/10$. Thus, $f(x) = 40\cos[\pi(x - 10)/10] + 40$.

25. $f(t) = 4.2\sin(\pi/2 - 2\pi t) + 3$

27. $g(x) = 4 - 1.3\sin[\pi/2 - 2.3(x - 4)]$

29. $\sin^2 x + \cos^2 x = 1$; $(\sin^2 x + \cos^2 x)/\cos^2 x = 1/\cos^2 x$; $\tan^2 x + 1 = \sec^2 x$.

31. $\sin(\pi/3) = \sin(\pi/6 + \pi/6) = \sin(\pi/6)\cos(\pi/6) + \cos(\pi/6)\sin(\pi/6) = (1/2)(\sqrt{3}/2) + (\sqrt{3}/2)(1/2) = \sqrt{3}/2$

33. $\sin(t + \pi/2) = \sin t\cos(\pi/2) + \cos t\sin(\pi/2) = \cos t$ because $\cos(\pi/2) = 0$ and $\sin(\pi/2) = 1$.

35. $\sin(\pi - x) = \sin\pi\cos x - \cos\pi\sin x = \sin x$ because $\sin\pi = 0$ and $\cos\pi = -1$.

37. $\tan(x + \pi) = \dfrac{\sin(x + \pi)}{\cos(x + \pi)} = \dfrac{\sin x\cos\pi + \cos x\sin\pi}{\cos x\cos\pi - \sin x\sin\pi} = \dfrac{-\sin x}{-\cos x} = \tan x$

39. (a) $P = 2\pi/0.602 \approx 10.4$ years.
(b) Maximum: $58.8 + 57.7 = 116.5 \approx 117$; minimum: $58.8 - 57.7 = 1.1 \approx 1$
(c) $1.43 + P/4 + P = 1.43 + 13.05 \approx 14.5$ years, or midway through 2011

41. (a)

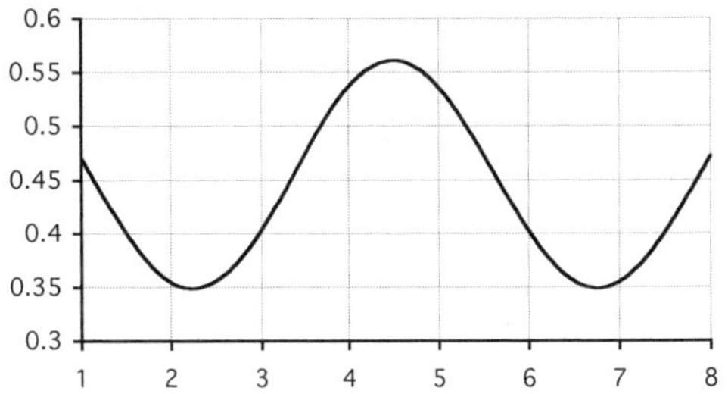

Maximum sales occurred when $t \approx 4.5$ (during the first quarter of 1996). Minimum sales occurred when $t \approx 2.2$ (during the third quarter of 1995) and $t \approx 6.8$ (during the third quarter of 1996). **(b)** Maximum quarterly revenues were \$0.561 billion; minimum quarterly revenues were \$0.349 billion. **(c)** The maximum and minimum values are $C \pm A$, so the maximum is $0.455 + 0.106 = 0.561$ and the minimum is $0.455 - 0.106 = 0.349$.

43. Amplitude = 0.106, vertical offset = 0.455, phase shift = $-1.61/1.39 \approx -1.16$, angular frequency = 1.39, period = 4.52. In 1995 and 1996, quarterly revenue from the sale of computers at Computer City fluctuated in cycles of 4.52 quarters about a baseline of \$0.455 billion. Every cycle, quarterly revenue peaked at \$0.561 billion (\$0.106 above the baseline) and dipped to a low of \$0.349 billion. Revenue peaked in the middle of the first quarter of 1996 (at $t = -1.16 + (5/4)\times 4.52 = 4.49$).

45. $C = 12.5$, $A = 7.5$, period = 52 weeks, so $\omega = \pi/26$, $\alpha = 52/4 = 13$. So, $P(t) = 7.5\sin[\pi(t - 13)/26] + 12.5$.

47. $C = 87.5$, $A = 7.5$, period = 12 months, so $\omega = \pi/6$, $\alpha = 6 + 12/4 = 9$. So, $s(t) = 7.5\sin[\pi(t - 9)/6] + 87.5$.

49. $C = 87.5$, $A = 7.5$, period = 12 months, so $\omega = \pi/6$, $\alpha = 0$. So, $s(t) = 7.5\cos(\pi t/6) + 87.5$.

51. $C = 10$, $A = 5$, period = 13.5 hours, so $\omega = 2\pi/13.5$, $\alpha = 5 - 13.5/4 = 1.625$. So, $d(t) = 5\sin[2\pi(t - 1.625)/13.5] + 10$.

53. **(a)** $C = 7.5$, $A = 2.5$, period = 1 year, so $\omega = 2\pi$, $\alpha = 0.75$. So, $u(t) = 2.5\sin[2\pi(t - 0.75)] + 7.5$. **(b)** $c(t) = 1.04^t\{2.5\sin[2\pi(t{-}0.75)] + 7.5\}$.

55. **(a)** $P \approx 8$, $C \approx 6$, $A \approx 2$, $\alpha \approx 8$ (Answers may vary)

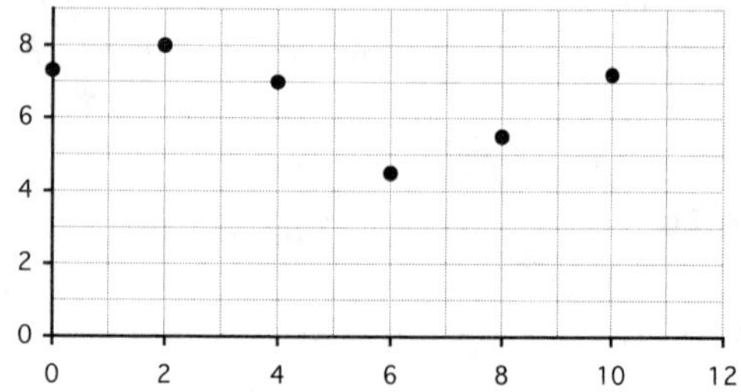

(b) $C(t) = 1.755\sin[0.636(t - 9.161)] + 6.437$

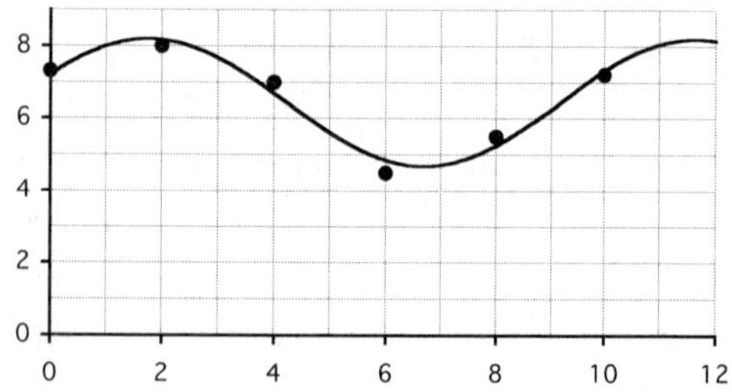

(c) 9.9 [$\approx 2\pi/0.636$], 4.7% [$\approx 6.437 - 1.755$], 8.2% [$\approx 6.437 + 1.755$]

57. **(a)**

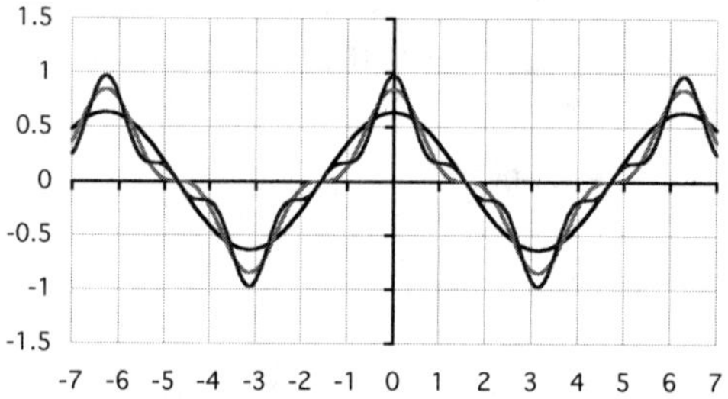

(b) $y_{11} = \dfrac{2}{\pi}\cos x + \dfrac{2}{3\pi}\cos 3x + \dfrac{2}{5\pi}\cos 5x + \dfrac{2}{7\pi}\cos 7x + \dfrac{2}{9\pi}\cos 9x + \dfrac{2}{11\pi}\cos 11x$

Graph of all four functions:

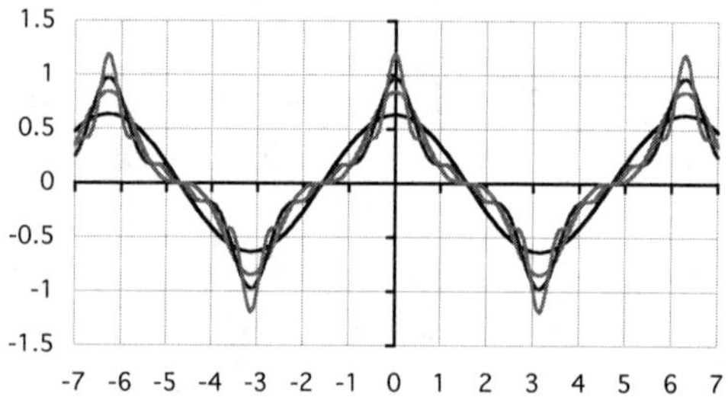

Graph of y_{11} alone:

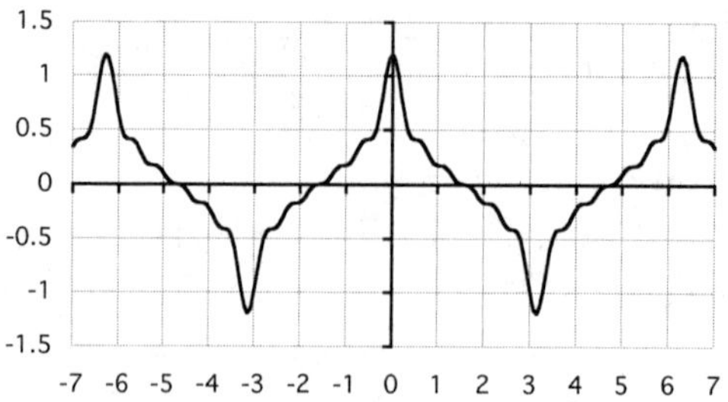

(c) Multiply the amplitudes by 3 and change ω to 1/2: $y_{11} = \dfrac{6}{\pi}\cos\dfrac{x}{2} + \dfrac{6}{3\pi}\cos\dfrac{3x}{2} + \dfrac{6}{5\pi}\cos\dfrac{5x}{2} + \dfrac{6}{7\pi}\cos\dfrac{7x}{2} + \dfrac{6}{9\pi}\cos\dfrac{9x}{2} + \dfrac{6}{11\pi}\cos\dfrac{11x}{2}$

59.

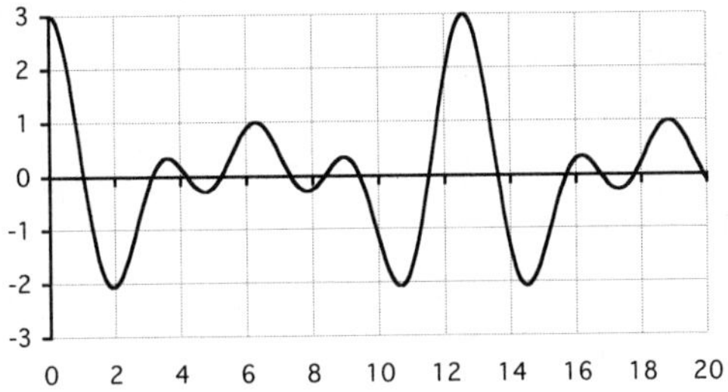

The period is approximately 12.6 units.

61. Lows: $B - A$; Highs: $B + A$

63. He is correct. The other trig functions can be obtained from the sine function by first using the formula $\cos x = \sin(x + \pi/2)$ to obtain cosine, and then using the formulas $\tan x = \dfrac{\sin x}{\cos x}$, $\cot x = \dfrac{\cos x}{\sin x}$, $\sec x = \dfrac{1}{\cos x}$, $\csc x = \dfrac{1}{\sin x}$ to obtain the rest.

65. The largest B can be is A. Otherwise, if B is larger than A, the low figure for sales would have the negative value of $A - B$.

9.2

1. $f'(x) = \cos x + \sin x$

3. $g'(x) = \cos x \tan x + \sin x \sec^2 x =$
$\sin x\,(1 + \sec^2 x)$

5. $h'(x) = -2 \csc x \cot x - \sec x \tan x + 3$

7. $r'(x) = \cos x - x \sin x + 2x$

9. $s'(x) = (2x - 1)\tan x + (x^2 - x + 1)\sec^2 x$

11. $t'(x) = -[\csc^2 x\,(1 + \sec x) +$
$\cot x \sec x \tan x]/(1 + \sec x)^2$

13. $k'(x) = -2 \cos x \sin x$

15. $j'(x) = 2 \sec^2 x \tan x$

17. $p'(x) = \pi \cos\left[\frac{\pi}{5}(x - 4)\right]$

19. $u'(x) = -(2x - 1)\sin(x^2 - x)$

21. $v'(x) = (2.2x^{1.2} + 1.2)\sec(x^{2.2} + 1.2x - 1) \cdot$
$\tan(x^{2.2} + 1.2x - 1)$

23. $w'(x) = \sec x \tan x \tan(x^2 - 1) +$
$2x \sec x \sec^2(x^2 - 1)$

25. $y'(x) = e^x[-\sin(e^x)] + e^x \cos x - e^x \sin x =$
$e^x[-\sin(e^x) + \cos x - \sin x]$

27. $z'(x) = \dfrac{\sec x \tan x + \sec^2 x}{\sec x + \tan x} =$
$\dfrac{\sec x\,(\tan x + \sec x)}{\sec x + \tan x} = \sec x$

29. $\dfrac{d}{dx} \sec x = \dfrac{d}{dx}\left[\dfrac{1}{\cos x}\right] = \dfrac{\sin x}{\cos^2 x} =$
$\dfrac{1}{\cos x}\dfrac{\sin x}{\cos x} = \sec x \tan x$

31. $\dfrac{d}{dx} \csc x = \dfrac{d}{dx}\left[\dfrac{1}{\sin x}\right] = -\dfrac{\cos x}{\sin^2 x} =$
$-\dfrac{1}{\sin x}\dfrac{\cos x}{\sin x} = -\csc x \cot x$

33. $\dfrac{d}{dx}[e^{-2x}\sin(3\pi x)] =$
$-2e^{-2x}\sin(3\pi x) + 3\pi e^{-2x}\cos(3\pi x) =$
$e^{-2x}[-2\sin(3\pi x) + 3\pi\cos(3\pi x)]$

35. $\dfrac{d}{dx}[\sin(3x)]^{0.5} = 0.5[\sin(3x)]^{-0.5}\, 3\cos(3x) =$
$1.5[\sin(3x)]^{-0.5}\cos(3x)$

37. $\dfrac{d}{dx}\sec\left(\dfrac{x^3}{x^2 - 1}\right) =$
$\dfrac{3x^2(x^2 - 1) - 2x^4}{(x^2 - 1)^2}\sec\left(\dfrac{x^3}{x^2 - 1}\right)\tan\left(\dfrac{x^3}{x^2 - 1}\right) =$
$\dfrac{x^4 - 3x^2}{(x^2 - 1)^2}\sec\left(\dfrac{x^3}{x^2 - 1}\right)\tan\left(\dfrac{x^3}{x^2 - 1}\right)$

39. $\dfrac{d}{dx}[\ln|x| \cot(2x - 1)] = \dfrac{\cot(2x - 1)}{x} -$
$2 \ln|x| \csc^2(2x - 1)$

41. **(a)** Not differentiable at 0:
$\lim\limits_{h\to 0^+} \dfrac{|\sin h| - |\sin 0|}{h} = \lim\limits_{h\to 0^+} \dfrac{\sin h}{h} = 1$ but
$\lim\limits_{h\to 0^-} \dfrac{|\sin h| - |\sin 0|}{h} = \lim\limits_{h\to 0^-} \dfrac{-\sin h}{h} = -1.$

(b) $\sin(1) \approx 0.84 > 0$ so $|\sin x| = \sin x$ for x near 1; hence $f'(1)$ exists, and $f'(1) \approx \dfrac{\sin 1.0001 - \sin 0.9999}{0.0002} \approx 0.5403.$

43. 0: Write $f(x) = (\sin^2 x)/x$.

x	$f(x)$
−0.1	−0.0997
−0.01	−0.01
−0.001	−0.001
−0.0001	−0.0001
0	
0.0001	0.0001
0.001	0.001
0.01	0.01
0.1	0.0997

By L'Hospital's rule: $\lim_{x\to 0} \frac{\sin^2 x}{x} = \lim_{x\to 0} \frac{2 \sin x \cos x}{1} = 0$.

45. 2: Write $f(x) = (\sin 2x)/x$.

x	$f(x)$
−0.1	1.98669331
−0.01	1.99986667
−0.001	1.99999867
−0.0001	1.99999999
0	
0.0001	1.99999999
0.001	1.99999867
0.01	1.99986667
0.1	1.98669331

By L'Hospital's rule: $\lim_{x\to 0} \frac{\sin 2x}{x} = \lim_{x\to 0} \frac{2 \cos 2x}{1} = 2$.

47. Does not exist: Write $f(x) = (\cos x - 1)/x^3$.

x	$f(x)$
−0.1	4.99583472
−0.01	49.9995833
−0.001	499.999958
−0.0001	4999.99997
0	
0.0001	−5000
0.001	−499.99996
0.01	−49.999583
0.1	−4.9958347

By L'Hospital's rule: $\lim_{x\to 0} \frac{\cos x - 1}{x^3} = \lim_{x\to 0} \frac{-\sin x}{3x^2} = \lim_{x\to 0} \frac{-\cos x}{6x}$ does not exist.

49. $1 = (\sec^2 y) \frac{dy}{dx}$, so $\frac{dy}{dx} = 1/\sec^2 y$

51. $1 + \frac{dy}{dx} + \left(y + x\frac{dy}{dx}\right) \cos(xy) = 0$, $[1 + x \cos(xy)] \frac{dy}{dx} = -[1 + y \cos(xy)]$, so $\frac{dy}{dx} = -[1 + y \cos(xy)]/[1 + x \cos(xy)]$

53. $c'(t) = 7\pi\cos[2\pi(t - 0.75)]$; $c'(0.75) \approx$ \$21.99 per *year* ≈ \$0.42 per week.

55. $N'(t) = 34.7354 \cos[0.602(t - 1.43)]$, $N'(6) \approx -32.12$. On January 1, 2003, the number of sunspots was decreasing at a rate of 32.12 sunspots per year.

57. $c'(t) = 1.035^t[\ln(1.035)(0.8 \sin(2\pi t) + 10.2) + 1.6\pi \cos(2\pi t)]$; $c'(1) \approx$ \$5.57 per year, or \$0.11 per week.

59. **(a)** $d(t) = 5\cos(2\pi t/13.5) + 10$ **(b)** $d'(t) = -(10\pi/13.5)\sin(2\pi t/13.5)$; $d'(7) \approx 0.270$. At noon, the tide was rising at a rate of 0.270 feet per hour.

61. **(a)** (C): From what we are told we must have $\alpha = -500$ and $P = 40$, so $\omega = 2\pi/40$. **(b)** $A'(t) = -(\pi/20)\sin[2\pi(t + 500)/40]$, $A'(-150) \approx 0.157$. The tilt was increasing at a rate of 0.157 degrees per thousand years.

63. -6; 6

65. Answers will vary. Examples: $f(x) = \sin x$; $f(x) = \cos x$

67. Answers will vary. Examples: $f(x) = e^{-x}$; $f(x) = -2e^{-x}$

69. The graph of $\cos x$ slopes down over the interval $(0, \pi)$, so that its derivative is negative over that interval. The function $-\sin x$, and not $\sin x$, has this property.

71. The derivative of $\sin x$ is $\cos x$. When $x = 0$, this is $\cos(0) = 1$. Thus, the tangent to the graph of $\sin x$ at the point (0, 0) has slope 1, which means it slopes upward at 45°.

9.3

1. $\int (\sin x - 2\cos x)\,dx = -\cos x - 2\sin x + C$

3. $\int (2\cos x - 4.3\sin x - 9.33)\,dx = 2\sin x + 4.3\cos x - 9.33x + C$

5. $\int \left(3.4\sec^2 x + \frac{\cos x}{1.3} - 3.2e^x\right) = 3.4\tan x + \frac{\sin x}{1.3} - 3.2e^x + C$

7. $\int 7.6\cos(3x - 4)\,dx = \frac{7.6}{3}\sin(3x - 4) + C$

9. $\int x\sin(3x^2 - 4)\,dx = -\frac{1}{6}\cos(3x^2 - 4) + C$: Substitute $u = 3x^2 - 4$.

11. $\int (4x + 2)\sin(x^2 + x)\,dx = -2\cos(x^2 + x) + C$: Substitute $u = x^2 + x$.

13. $\int (x + x^2)\sec^2(3x^2 + 2x^3)\,dx = \frac{1}{6}\tan(3x^2 + 2x^3) + C$: Substitute $u = 3x^2 + 2x^3$.

15. $\int x^2\tan(2x^3)\,dx = -\frac{1}{6}\ln|\cos(2x^3)| + C$: Substitute $u = 2x^3$.

17. $\int 6\sec(2x - 4)\,dx = 3\ln|\sec(2x - 4) + \tan(2x - 4)| + C$: Substitute $u = 2x - 4$.

19. $\int e^{2x}\cos(e^{2x} + 1)\,dx = \frac{1}{2}\sin(e^{2x} + 1) + C$: Substitute $u = e^{2x} + 1$.

21. $\int_{-\pi}^{0} \sin x\,dx = [-\cos x]_{-\pi}^{0} = -\cos(0) + \cos(-\pi) = -1 - 1 = -2$

23. $\int_{0}^{\pi/3} \tan x\,dx = -[\ln|\cos x|]_0^{\pi/3} = -\ln|\cos(\pi/3)| + \ln|\cos(0)| = -\ln(1/2) + \ln(1) = \ln(2)$

25. $\int_{1}^{\sqrt{\pi+1}} x\cos(x^2 - 1)\,dx = \int_0^{\pi} \frac{1}{2}\cos u\,du$ (substitute $u = x^2 - 1$) $= \frac{1}{2}[\sin u]_0^{\pi} = \frac{1}{2}[\sin(\pi) - \sin(0)] = 0$

27. $\int_{1/\pi}^{2/\pi} \frac{\sin(1/x)}{x^2}\,dx = -\int_{\pi}^{\pi/2} \sin u\,du$ (substitute $u = 1/x$) $= [\cos u]_{\pi}^{\pi/2} = \cos(\pi/2) - \cos(\pi) = 1$

29. $\int \cos(ax + b)\,dx = \int \frac{1}{a}\cos u\,du$ (substitute $u = ax + b$) $= \frac{1}{a}\sin u + C = \frac{1}{a}\sin(ax + b) + C$

31. $\int \cot x\,dx = \int \frac{\cos x}{\sin x}\,dx = \int \frac{1}{u}\,du$ (substitute $u = \sin x$) $= \ln|u| + C = \ln|\sin x| + C$

33. $\int \sin(4x)\,dx = -\frac{1}{4}\cos(4x) + C$

35. $\int \cos(-x+1)\,dx = -\sin(-x+1) + C$

37. $\int \sin(-1.1x - 1)\,dx$

$= \frac{1}{1.1}\cos(-1.1x - 1) + C$

39. $\int \cot(-4x)\,dx = -\frac{1}{4}\ln|\sin(-4x)| + C$

41. $\int_{-\pi/2}^{\pi/2} \sin x\,dx = 0$ because, by symmetry, there is as much area above the x axis as below.

43. $\int_0^{2\pi} (1 + \sin x)\,dx = 2\pi$ because, by symmetry, the average value of $1 + \sin x$ over $[0, 2\pi]$ is 1, hence the area under the curve is the same as the area under the line of height 1, which is 2π.

45.

	D		I
+	x	↘	$\sin x$
−	1	↘	$-\cos x$
$+\int$	0	→	$-\sin x$

$\int x\sin x\,dx = -x\cos x + \sin x + C$

47.

	D		I
+	x^2	↘	$\cos(2x)$
−	$2x$	↘	$\frac{1}{2}\sin(2x)$
+	2	↘	$-\frac{1}{4}\cos(2x)$
$-\int$	0	→	$-\frac{1}{8}\sin(2x)$

$\int x^2\cos(2x)\,dx = \left(\frac{x^2}{2} - \frac{1}{4}\right)\sin(2x) + \frac{x}{2}\cos(2x) + C$

49.

	D		I
+	$\sin x$	↘	e^{-x}
−	$\cos x$	↘	$-e^{-x}$
$+\int$	$-\sin x$	→	e^{-x}

$\int e^{-x}\sin x\,dx = -e^{-x}\sin x - e^{-x}\cos x - \int e^{-x}\sin x\,dx$, so $2\int e^{-x}\sin x\,dx = -e^{-x}\sin x - e^{-x}\cos x + C$; $\int e^{-x}\sin x\,dx = -\frac{1}{2}e^{-x}\sin x - \frac{1}{2}e^{-x}\cos x + C$

51.

	D		I
+	x^2	↘	$\sin x$
−	$2x$	↘	$-\cos x$
+	2	↘	$-\sin x$
$-\int$	0	→	$\cos x$

$$\int_0^{\pi} x^2 \sin x\, dx = [-x^2 \cos x + 2x \sin x + 2 \cos x]_0^{\pi}$$

$$= \pi^2 - 4$$

53. Average $= \dfrac{1}{\pi} \displaystyle\int_0^{\pi} \sin x\, dx = \dfrac{1}{\pi} [-\cos x]_0^{\pi} = \dfrac{2}{\pi}$

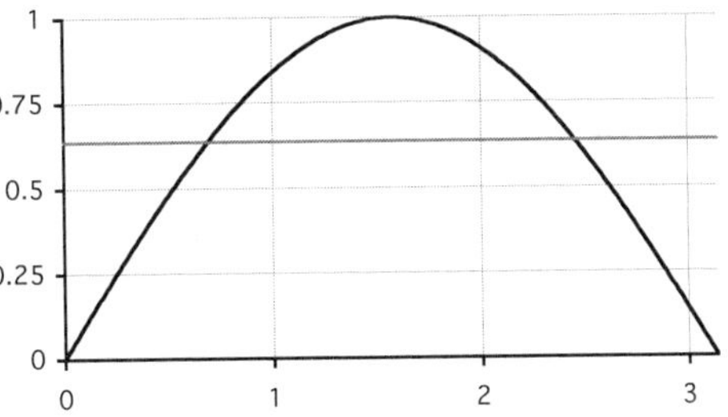

55. $\displaystyle\int_0^{+\infty} \sin x\, dx = \lim_{M \to +\infty} \int_0^{M} \sin x\, dx =$

$\displaystyle\lim_{M \to +\infty} [-\cos x]_0^{M} = \lim_{M \to +\infty} (1 - \cos M)$

diverges

57.

	D		I
+	$\cos x$	↘	e^{-x}
−	$-\sin x$	↘	$-e^{-x}$
$+\int$	$-\cos x$	→	e^{-x}

$\displaystyle\int e^{-x} \cos x\, dx = -e^{-x} \cos x + e^{-x} \sin x - \int e^{-x} \cos x\, dx$, so $\displaystyle 2\int e^{-x} \cos x\, dx = -e^{-x} \cos x + e^{-x} \sin x + C$; $\displaystyle\int e^{-x} \cos x\, dx = -\frac{1}{2} e^{-x} \cos x + \frac{1}{2} e^{-x} \sin x + C$; $\displaystyle\int_0^{+\infty} e^{-x} \cos x\, dx =$

$\displaystyle\lim_{M \to +\infty} \int_0^{M} e^{-x} \cos x\, dx = \lim_{M \to +\infty} [-\frac{1}{2} e^{-x} \cos x + \frac{1}{2} e^{-x} \sin x]_0^{M} = \lim_{M \to +\infty} (-\frac{1}{2} e^{-M} \cos M + \frac{1}{2} e^{-M} \sin M + \frac{1}{2}) = \frac{1}{2}$; converges

59. $\displaystyle C(t) = \int \left\{0.04 - 0.1 \sin\left[\frac{\pi}{26}(t - 25)\right]\right\} dt = 0.04t + \frac{2.6}{\pi} \cos\left[\frac{\pi}{26}(t - 25)\right] + K$; $C(12) = 1.50$ gives $K = 1.02$, so $\displaystyle C(t) = 0.04t + \frac{2.6}{\pi} \cos\left[\frac{\pi}{26}(t - 25)\right] + 1.02$

61. Position $\displaystyle = \int_0^{10} 3\pi \cos\left[\frac{\pi}{2}(t - 1)\right] dt = \left[6 \sin\left[\frac{\pi}{2}(t - 1)\right]\right]_0^{10} = 6 \sin \frac{9\pi}{2} - 6 \sin\left(-\frac{\pi}{2}\right) =$ 12 feet

63. Average =

$$\frac{1}{2}\int_5^7 \{57.7 \sin[0.602(t-1.43)] + 58.8\}\,dt =$$

$$\frac{1}{2}[-95.847 \cos[0.602(t-1.43)] + 58.8t]_5^7 \approx 79$$

sunspots

65. $P(t) = 7.5 \sin[\pi(t-13)/26] + 12.5$ (see Exercise 45 in Section 9.1). Average =

$$\frac{1}{13}\int_0^{13} \{7.5 \sin[\pi(t-13)/26] + 12.5\}\,dt =$$

$$\frac{1}{13}[-62.07 \cos[\pi(t-13)/26] + 12.5t]_0^{13} \approx 7.7\%$$

67. (a) Average voltage over [0, 1/6] is

$$6\int_0^{1/6} 165 \cos(120\pi t)\,dt = 2.63[\sin(120\pi t)]_0^{1/6} = 0.$$

In one second the voltage goes through 60 periods of the cosine wave, hence reaches its maximum 60 times; hence the electricity has a frequency of 60 cycles per second.

(b)

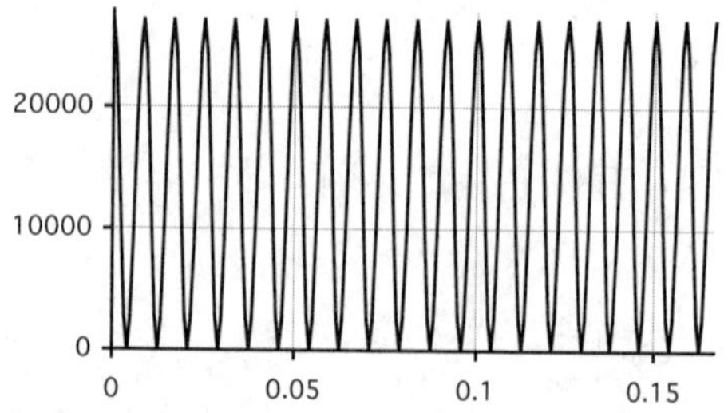

(c) We can use technology to estimate the average $\overline{S} \approx 13{,}612.5$, hence the RMS voltage is approximately $\sqrt{13{,}612.5} \approx 116.673$ volts. Or, we can notice that the graph in (b) appears to be a sinusoid with an average value of $\overline{S} = 165^2/2 = 13{,}612.5$, so that the RMS voltage is $165/\sqrt{2} \approx 116.673$ volts.

69. $TV = \int_0^1 [50{,}000 + 2000\pi \sin(2\pi t)]\,dt =$

$[50{,}000t - 1000 \cos(2\pi t)]_0^1 = \$50{,}000$

71. The integral over a whole number of periods is always zero, by symmetry.

73. 1, because the average value of $2 \cos x$ will be approximately 0 over a large interval.

75. Integrate twice to get $s = -\frac{K}{\omega^2}\sin(\omega t - \alpha) + Lt + M$ for constants L and M.

Chapter 9 Review Exercises

1. $C = 1$, $A = 2$, $\alpha = 0$, and $P = 2\pi$, so $\omega = 1$. Thus, $f(x) = 2 \sin x + 1$.

2. $C = 0$, $A = 10$, $\alpha = -\pi/4$, and $P = \pi$, so $\omega = 2$. Thus, $f(x) = 10 \sin[2(x + \pi/4)] = 10 \sin(2x + \pi/2)$

3. $C = 2$, $A = 2$, $\alpha = 1$, and $P = 2$, so $\omega = \pi$. Thus, $f(x) = 2 \sin[\pi(x - 1)] + 2$. We could also take $\alpha = -1$, getting $f(x) = 2 \sin[\pi(x + 1)] + 2$

4. $C = -2$, $A = 3$, $\alpha = 0.25$, and $P = 0.5$, so $\omega = 4\pi$. Thus, $f(x) = 3 \sin[4\pi(x - 0.25)] - 2$. We could also take $\alpha = 0.25$, getting $f(x) = 3 \sin[4\pi(x + 0.25)] - 2$.

In each of Exercises 5–8, substitute $\sin z = \cos(z - \pi/2)$ in the answer from the corresponding one of Exercises 1–4, and simplify as necessary.

5. $f(x) = 2 \cos(x - \pi/2) + 1$

6. $f(x) = 10 \cos(2x)$

7. $f(x) = 2 \cos[\pi(x - 3/2)] + 2 = 2 \cos[\pi(x + 1/2)] + 2$

8. $f(x) = 3 \cos[4\pi(x - 0.375)] - 2 = 3 \cos[4\pi(x + 0.125)] - 2$

9. $-2x \sin(x^2 - 1)$

10. $2x[\cos(x^2 + 1) \cos(x^2 - 1) - \sin(x^2 + 1) \sin(x^2 - 1)]$

11. $2e^x \sec^2(2e^x - 1)$

12. $\dfrac{(2x - 1) \sec\sqrt{x^2 - x} \tan\sqrt{x^2 - x}}{2\sqrt{x^2 - x}}$

13. $4x \sin(x^2) \cos(x^2)$

14. $4 \cos(2x) \cos[1 - \sin(2x)] \sin[1 - \sin(2x)]$

15. $\int 4 \cos(2x - 1)\, dx = 2 \sin(2x - 1) + C$

(substitute $u = 2x - 1$)

16. $\int (x - 1)\sin(x^2 - 2x + 1)\, dx = -\frac{1}{2}\cos(x^2 - 2x + 1) + C$ (substitute $u = x^2 - 2x + 1$)

17. $\int 4x \sec^2(2x^2 - 1)\, dx = \tan(2x^2 - 1) + C$

(substitute $u = 2x^2 - 1$)

18. $\displaystyle\int \frac{\cos\left(\frac{1}{x}\right)}{x^2 \sin\left(\frac{1}{x}\right)}\, dx = -\ln|\sin(1/x)| + C$

[substitute $u = \sin(1/x)$]

19. $\int x \tan(x^2 + 1)\, dx = -\frac{1}{2} \ln|\cos(x^2 + 1)| + C$

(substitute $u = x^2 + 1$)

20. $\displaystyle\int_0^{\pi} \cos(x + \pi/2)\, dx = [\sin(x + \pi/2)]_0^{\pi} = \sin(3\pi/2) - \sin(\pi/2) = -2$

21. $\int_{\ln(\pi/2)}^{\ln(\pi)} e^x \sin(e^x)\, dx = \int_{\pi/2}^{\pi} \sin u\, du$ (substitute $u = e^x$) $= [-\cos u]_{\pi/2}^{\pi} = -\cos \pi + \cos(\pi/2) = 1$

22. $\int_{\pi}^{2\pi} \tan(x/6)\, dx = [-6 \ln|\cos(x/6)|]_{\pi}^{2\pi} = -6 \ln|\cos(\pi/3)| + 6 \ln|\cos(\pi/6)| = -6 \ln(1/2) + 6 \ln(\sqrt{3}/2) = 3 \ln 3$

23.

	D		I
+	x^2	↘	$\sin x$
−	$2x$	↘	$-\cos x$
+	2	↘	$-\sin x$
$-\int$	0	→	$\cos x$

$\int x^2 \sin x\, dx =$

$-x^2 \cos x + 2x \sin x + 2 \cos x + C$

24.

	D		I
+	$\sin 2x$	↘	e^x
−	$2 \cos 2x$	↘	e^x
$+\int$	$-4 \sin 2x$		e^x

$\int e^x \sin 2x\, dx = e^x \sin 2x - 2e^x \cos 2x - 4\int e^x \sin 2x\, dx$, so $5\int e^x \sin 2x\, dx = e^x \sin 2x - 2e^x \cos 2x + C$, $\int e^x \sin 2x\, dx = \frac{1}{5} e^x \sin 2x - \frac{2}{5} e^x \cos 2x + C$

25. $C = 10{,}500$, $A = 1500$, $\alpha = 52/2 = 26$, and $P = 52$, so $\omega = 2\pi/52$. Thus, $s(t) = 1500 \bullet \sin[(2\pi/52)(t - 26)] + 10{,}500 \approx 1500 \bullet \sin(0.12083t - 3.14159) + 10{,}500$

26. $R'(t) = 15{,}000\{-0.12e^{-0.12t} \cos[(\pi/6)(t - 4)] - (\pi/6)e^{-0.12t} \sin[(\pi/6)(t - 4)]\}$; $R'(10) \approx \$542$ per month

27. $\int_0^{10} \{20{,}000 + 15{,}000e^{-0.12t}\cos[(\pi/6)(t-4)]\}\,dt = [20{,}000t]_0^{10} + 15{,}000\int_0^{10} e^{-0.12t}\cos[(\pi/6)(t-4)]\,dt =$

$200{,}000 + 15{,}000\int_0^{10} e^{-0.12t}\cos[(\pi/6)(t-4)]\,dt$. Evaluate the latter integral using integration by parts:

	D	*I*
+	$\cos[(\pi/6)(t-4)]$	$e^{-0.12t}$
–	$-(\pi/6)\sin[(\pi/6)(t-4)]$	$-\frac{1}{0.12}e^{-0.12t}$
$+\int$	$-(\pi/6)^2\cos[(\pi/6)(t-4)]$	$\frac{1}{0.12^2}e^{-0.12t}$

$\int e^{-0.12t}\cos[(\pi/6)(t-4)]\,dt = -\frac{1}{0.12}e^{-0.12t}\cos[(\pi/6)(t-4)] + \frac{\pi/6}{0.12^2}e^{-0.12t}\sin[(\pi/6)(t-4)] -$

$\left(\frac{\pi/6}{0.12}\right)^2\int e^{-0.12t}\cos[(\pi/6)(t-4)]\,dt$; $\int_0^{10} e^{-0.12t}\cos[(\pi/6)(t-4)]\,dt =$

$\left[1+\left(\frac{\pi/6}{0.12}\right)^2\right]^{-1}\left[-\frac{1}{0.12}e^{-0.12t}\cos[(\pi/6)(t-4)] + \frac{\pi/6}{0.12^2}e^{-0.12t}\sin[(\pi/6)(t-4)]\right]_0^{10} \approx 1.489.$ Hence,

$\int_0^{10} \{20{,}000 + 15{,}000e^{-0.12t}\cos[(\pi/6)(t-4)]\}\,dt \approx 200{,}000 + 15{,}000(1.489) \approx \$222{,}300$

28. Total consumption $= \int_0^t \{150 + 50\sin[(\pi/2)(x-1)]\}\,dx = [150x - (100/\pi)\cos[(\pi/2)(x-1)]]_0^t =$

$150t - \frac{100}{\pi}\cos\left[\frac{\pi}{2}(t-1)\right]$ grams